Robotics in Hotel Services:

Housekeeping, Reception, and Concierge Services

Ankit Shukla
https://orcid.org/0000-0002-7646-3517
Rajamangala University of Technology, Thailand

Nagendra Yadav
https://orcid.org/0009-0008-9743-9250
Welcomgroup Graduate School of Hotel Administration, Manipal Academy of Higher Education, India

Rohit Chauhan
https://orcid.org/0000-0002-4400-7891
Lovely Professional University, India

Abhishek Ghai
https://orcid.org/0000-0002-0994-9760
Maharishi Markandeshwar University, India

Vaibhav Verma
https://orcid.org/0000-0002-9041-7111
Bahra University, India

IGI Global
Scientific Publishing
Publishing Tomorrow's Research Today

Vice President of Editorial	Melissa Wagner
Director of Acquisitions	Mikaela Felty
Director of Book Development	Jocelynn Hessler
Production Manager	Mike Brehm
Cover Design	Jose Rosado

Published in the United States of America by
IGI Global Scientific Publishing
701 East Chocolate Avenue
Hershey, PA, 17033, USA
Tel: 717-533-8845 | Fax: 717-533-7115
Website: https://www.igi-global.com E-mail: cust@igi-global.com

Library of Congress Cataloging-in-Publication Data

LCCN: 2025042355 (CIP Data Pending)
ISBN13: 9798337365411
Isbn13Softcover: 9798337365428
EISBN13: 9798337365435

British Cataloguing in Publication Data
A Cataloguing in Publication record for this book is available from the British Library.

Table of Contents

Detailed Table of Contents

This chapter explores how hotel front offices are being reshaped by reception automation while guest expectations for personalized service continue to rise. Technologies such as self-service kiosks, mobile check-in, AI chatbots, and integrated CRM systems improve speed, accuracy, and operational efficiency, yet may reduce the emotional warmth central to hospitality. Using established theories and industry evidence, it analyses this tension and proposes a high-tech, high-touch model in which automation manages routine tasks and staff focuses on empathy, problem-solving, and meaningful guest engagement. This balanced approach enables hotels to deliver both efficiency and richer, more personalized experiences.

The hospitality industry is experiencing a transformation with the integration of AI and robotic technology. One key element of this change will be AI-enabled concierge and guest services that are changing the way in which customers are engaged in conversation, serviced and supported in the overall guest experience.

Preface

The hospitality industry is undergoing a profound transformation driven by rapid advancements in artificial intelligence (AI), robotics, automation, and smart technologies. From autonomous housekeeping robots and AI-powered concierge systems to reception automation and intelligent guest interfaces, hotels across the globe are redefining how services are delivered and experienced. These technological developments are not merely operational upgrades; they represent a paradigm shift in service design, workforce roles, guest engagement, and sustainability practices.

The book *Robotics in Hotel Services: Housekeeping, Reception, and Concierge Services* is conceived to provide a comprehensive and multidisciplinary understanding of how robotics and AI are reshaping hotel service operations. Bringing together empirical research, systematic reviews, industry perspectives, and conceptual analyses, this edited volume bridges the gap between theory and practice in smart hospitality.

This book is intended for academics, researchers, postgraduate students, hotel professionals, policymakers, and technology developers who seek to understand both the opportunities and challenges associated with robotic service adoption in hospitality. Special emphasis is placed on guest acceptance, employee perspectives, sustainability, luxury hospitality, and emerging technologies such as metaverse-enabled experiences and cloud kitchens.

By integrating global perspectives with contextual insights from emerging markets, this volume aims to contribute meaningfully to hospitality literature and serve as a reference for future research and practice in smart tourism and hotel management.

Below is a chapter-by-chapter summary that details what will be found in each chapter of the book.

CHAPTER 1: WILL TOURISTS ACCEPT HOTEL SERVICE ROBOTS? EVIDENCE FROM INDIA USING THE THEORY OF PLANNED BEHAVIOR AND NECESSARY CONDITION ANALYSIS

This chapter empirically examines tourists' acceptance of hotel service robots within the Indian context using the Theory of Planned Behavior and Necessary Condition Analysis. It demonstrates that attitudes and subjective norms strongly influence intentions to use service robots, while perceived behavioral control plays a lesser role. The chapter establishes a behavioral foundation for understanding robot adoption in hospitality, emphasizing the social and evaluative nature of technology acceptance.

CHAPTER 2: HUMANLESS HOSPITALITY? A SYSTEMATIC LITERATURE REVIEW OF CONSUMER, EMPLOYEE, AND GENERATION Z PERSPECTIVES ON ROBOTICS IN HOTEL SERVICES

Through a systematic review of 45 peer-reviewed studies, this chapter synthesizes global research on robotics in hotel services. It highlights differences in perceptions among consumers, employees, and Generation Z, revealing broad acceptance of robots for standardized and hygiene-related tasks but resistance toward emotionally intensive services. The chapter provides theoretical extensions and managerial insights for hybrid human–robot service models.

CHAPTER 3: THE CONVERGENCE OF SMART HOSPITALITY AND ROBOTICS IN ACCOMMODATION OPERATIONS: SMART SYNERGY—EFFICIENCY AND THE FUTURE OF HOUSEKEEPING

This chapter explores how robotics and smart hospitality technologies are reshaping housekeeping operations. Using international case studies, it illustrates how AI-powered robots enhance efficiency, hygiene, and consistency while raising ethical and social considerations. The authors advocate for a collaborative human–robot model in which automation supports staff, enabling them to focus on guest interaction and service personalization.

CHAPTER 4: INVISIBLE HANDS OF HOSPITALITY: ROBOTICS IN HOUSEKEEPING, RECEPTION, AND CONCIERGE OPERATIONS

Focusing on operational integration, this chapter examines how robotics is transforming core hotel functions such as housekeeping, reception, and concierge services. It highlights the role of AI, machine learning, and IoT in enabling intelligent check-ins, autonomous cleaning, and personalized guest assistance. The chapter also addresses workforce adaptation, ethics, and brand perception in technology-driven hospitality ecosystems.

CHAPTER 5: BALANCING TECHNOLOGY AND HUMAN EXPERTISE: AN INDUSTRY PERSPECTIVE ON AI INTEGRATION IN HOUSEKEEPING OPERATIONS

Drawing on insights from Indian hospitality organizations, this chapter presents an industry-focused perspective on AI adoption in housekeeping. It demonstrates how AI improves efficiency and decision-making while underscoring the irreplaceable value of human empathy, ethics, and cultural sensitivity. The chapter emphasizes workforce upskilling and balanced innovation as key to sustainable service excellence.

CHAPTER 6: RECEPTION AUTOMATION AND GUEST EXPERIENCE—BALANCING EFFICIENCY AND PERSONALIZATION IN HOTEL FRONT OFFICES

This chapter investigates how automation technologies such as self-service kiosks, mobile check-ins, and AI chatbots are transforming hotel front offices. It critically examines the tension between operational efficiency and the emotional warmth central to hospitality. The authors propose a high-tech, high-touch service model where automation supports routine processes while staff focus on meaningful guest engagement.

CHAPTER 7: AI-DRIVEN SMART ROOM ASSISTANTS ENHANCING PERSONALIZED GUEST EXPERIENCES IN HOSPITALITY

This chapter explores the growing role of AI-enabled concierge and smart room assistants in delivering personalized, multilingual, and real-time guest services. Through case studies and technological analysis, it discusses how intelligent systems enhance satisfaction, efficiency, and competitive advantage. Ethical issues related to data privacy, consent, and transparency are also critically examined.

CHAPTER 8: ROBOTICS IN LUXURY HOSPITALITY—TRANSFORMING SERVICE EXPERIENCE IN FIVE-STAR HOTEL RESTAURANTS

Focusing on luxury dining environments, this chapter analyzes the integration of robotics in five-star hotel restaurants. It highlights how robots enhance speed, accuracy, hygiene, and sustainability while complementing human service rather than replacing it. The chapter underscores the importance of preserving emotional connection, cultural sensitivity, and ethical considerations in high-end hospitality settings.

CHAPTER 9: FROM ROOM SERVICE TO GREEN SERVICE: HOW HOSPITALITY ROBOTS CAN REDEFINE LUXURY THROUGH SUSTAINABILITY

This chapter positions hospitality robots as both functional tools and symbolic representations of sustainable luxury. It explores how robotics can support environmental responsibility, operational efficiency, and brand differentiation while raising concerns about ethics, workforce resilience, and greenwashing. A conceptual framework is proposed for integrating robotics into sustainable luxury service models.

CHAPTER 10: EMPOWERING GUESTS: THE RISE OF SELF-SERVICE TECHNOLOGY IN HOSPITALITY

This chapter examines the rapid adoption of self-service technologies such as mobile keys, kiosks, chatbots, and self-ordering systems. It highlights how SST empowers guests with greater control, convenience, and personalization while improving operational efficiency. The chapter also discusses workforce adaptability and organizational readiness as critical success factors for SST implementation.

CHAPTER 11: AI AND METAVERSE-POWERED IMMERSIVE HOTEL EXPERIENCES FOR SMART TOURISM TRANSFORMATION

This chapter explores the transformative role of AI and Metaverse technologies in smart tourism, particularly in virtual hotel tours and immersive pre-travel experiences. It demonstrates how digital environments enhance marketing, personalization, and decision-making while reshaping tourist engagement. Managerial implications and future research directions related to ethics and user experience are also discussed.

CHAPTER 12: EXAMINING THE DEVELOPMENT OF CLOUD KITCHENS—CONSEQUENCES FOR THE HOSPITALITY AND TOURISM INDUSTRIES' FUTURE

The final chapter analyzes the rise of cloud kitchens as a technology-driven disruption in hospitality and tourism. It compares cloud kitchens with traditional restaurant models in terms of sustainability, customer experience, and regulation. While highlighting innovation and scalability, the chapter stresses the need for policy balance and experiential considerations to ensure long-term sustainability.

Acknowledgement

The successful completion of this book, *Robotics in Hotel Services: Housekeeping, Reception, and Concierge Services*, would not have been possible without the support and contributions of many individuals.

I extend my sincere gratitude to all the editors and chapter contributors whose scholarly insights, dedication, and timely efforts have enriched this volume and made it a meaningful academic contribution to the field of hospitality and tourism studies.

I am deeply thankful to my wife, *Dipika*, for her unwavering support, patience, and encouragement throughout the development of this book. I am equally grateful to my son, *Anvit*, whose presence and love have been a constant source of motivation and inspiration.

A special note of appreciation goes to *Dr. Rohit*, whose guidance, encouragement, and support were invaluable during the conceptualization and completion of this book.

Finally, I would like to acknowledge everyone who contributed directly or indirectly to the realization of this work. Any omissions are unintentional, and all contributions are sincerely appreciated.

Kind Regards,

Dr. Ankit Shukla

Editor

Chapter 1
Will Tourists Accept Hotel Service Robots?
Evidence From India Using the Theory of Planned Behavior and Necessary Condition Analysis

Rohit Chauhan
https://orcid.org/0000-0002-4400-7891
Mittal School of Business, Lovely Professional University, India

Ankit Shukla
Faculty of Liberal Arts, Rajamangala University of Technology, Thanyaburi, Thailand

Vikesh Kashyap
https://orcid.org/0000-0001-8532-2022
Bahra University, Shimla, India

ABSTRACT

The increasing deployment of service robots in hotels has transformed traditional service encounters, necessitating a deeper understanding of tourists' acceptance intentions. Grounded in the Theory of Planned Behavior (TPB), this study investigates how attitudes, subjective norms, and perceived behavioral control shape potential tourists' intentions to use service robots for concierge and service delivery functions in hotels. Data were collected from 149 respondents in India using a structured online questionnaire measured on a seven-point Likert scale. Partial least squares structural equation modeling and Necessary Condition Analysis were employed for data analysis. The findings reveal that attitudes and subjective norms

DOI: 10.4018/979-8-3373-6541-1.ch001

play a decisive role in shaping behavioral intentions, whereas perceived behavioral control exerts a weaker influence. The results highlight the socially embedded and evaluative nature of robot adoption decisions in hospitality contexts.

1. INTRODUCTION

The rapid advancement of artificial intelligence and robotic technologies is redefining how services are designed, delivered, and experienced across industries, with hospitality emerging as one of the most visible arenas for human–robot interaction. Service robots are increasingly deployed in hotels for concierge assistance, room delivery, and customer engagement, promising efficiency, consistency, and contactless service experiences. Prior research suggests that consumers do not evaluate such technologies solely on functional performance; instead, emotional responses, social meanings, and symbolic interpretations play a critical role in shaping acceptance (Belanche et al., 2021; Huang et al., 2024). Studies on AI and service robots highlight that attitudes toward these technologies are influenced by perceptions of usefulness, anthropomorphism, trust, and socio-emotional interaction quality (Jiang et al., 2025; Kraus et al., 2024). At the same time, social environments increasingly normalize AI usage, reinforcing adoption through collective approval and shared expectations (Barnes et al., 2024; Song et al., 2025). These developments indicate that understanding robot adoption in hospitality requires behavioral theories capable of capturing evaluative, social, and control-related processes.

Despite growing scholarly attention to service robots, the theoretical grounding of adoption studies remains fragmented, with many relying predominantly on technology-centric models such as TAM or UTAUT (Mathieson, 1991; Venkatesh & Davis, 2000). While these frameworks explain perceived usefulness and ease of use, they often underrepresent social pressure and perceived volitional control, which are particularly salient in interactive service contexts (Chatterjee et al., 2024). The Theory of Planned Behavior offers a more comprehensive lens by integrating attitudes, subjective norms, and perceived behavioral control, and has demonstrated strong explanatory power across tourism and service settings (Cahigas et al., 2023; Velikova et al., 2025). However, empirical research applying TPB to hotel service robots remains limited, especially in emerging economies where collectivist orientations amplify normative influences (Barnes et al., 2024; Rahi et al., 2024). Existing evidence is largely concentrated in developed markets, constraining contextual generalizability. Addressing this gap, the present study applies TPB to examine tourists' intentions to use service robots in Indian hotels, thereby advancing theory-driven understanding of hospitality robotics in socially embedded and rapidly digitizing contexts.

In response to the growing integration of artificial intelligence and robotics within hotel service environments, this study aims to examine potential tourists' intentions to use service robots for concierge and service delivery functions during future hotel stays. Grounded in the Theory of Planned Behavior (TPB), the primary objective is to empirically assess the extent to which attitudes, subjective norms, and perceived behavioral control influence behavioral intentions toward hotel service robots in the Indian context. By applying TPB to an emerging economy characterized by rapid digital transformation and strong social influence mechanisms, this study offers original empirical evidence from a setting that remains underrepresented in hospitality robotics research. The study's originality lies in extending a well-established behavioral theory to a novel service interface—human–robot interaction—within hotels, thereby enriching the theoretical discourse on technology-mediated service adoption. From a theoretical standpoint, the findings contribute to TPB literature by highlighting the relative salience of attitudinal and normative drivers over control perceptions in shaping robot adoption intentions. Practically, the study provides actionable insights for hotel managers and technology providers regarding the social and psychological factors that encourage acceptance of robotic services. Overall, this research strengthens theory integration by demonstrating the continued relevance and explanatory power of TPB in understanding consumer responses to advanced service technologies, while advancing knowledge at the intersection of hospitality, robotics, and consumer behavior.

2. LITERATURE REVIEW

The Theory of Planned Behavior (TPB) has emerged as a robust theoretical framework for explaining consumer intentions toward adopting emerging technologies such as artificial intelligence, service robots, and automation across service industries. TPB posits that behavioral intentions are shaped by attitudes, subjective norms, and perceived behavioral control, offering a parsimonious yet powerful explanation of technology-related decision-making. Empirical evidence supports TPB's applicability in technologically advanced consumption contexts. For instance, Tai et al. (2024) demonstrated that TPB effectively predicts purchase intentions toward intelligent robotic products, emphasizing the role of evaluative beliefs. Kraus et al. (2024) further extended TPB by integrating trust, revealing that general and specific trust mediate belief–intention relationships in service robot adoption. Within hospitality, Han et al. (2025) found that innovativeness, personal norms, and robot performance concerns significantly influence acceptance, reinforcing TPB's explanatory relevance. A recent meta-analysis by Song et al. (2025) highlights sociotechnical mechanisms—technology and sociality—as key drivers of service robot preference, while Frank (2023) illustrates how trust moderates AI adoption depending on autonomy levels. Collectively, these studies confirm TPB's versatility in explaining consumer intentions across AI-enabled service contexts.

The application of TPB within hospitality and tourism research has been extensive, providing strong empirical support for its ability to explain tourists' intentions, attitudes, and decision-making processes. A bibliometric review by Kalıpçı & Nebioğlu (2020) reveals TPB as one of the most frequently employed behavioral theories in tourism studies, particularly in understanding intention formation and consumption behavior. In sustainable tourism contexts, Tamuliene et al. (2024) demonstrate that cultural values significantly influence TPB constructs, with collectivism strengthening subjective norms and uncertainty avoidance shaping attitudes. Similarly, Luong (2025) found that subjective norms and perceived behavioral control strongly predict green consumption intentions among tourists, moderated by destination social responsibility. TPB has also proven effective in crisis-related tourism decisions, as evidenced by Cahigas et al. (2023), who showed that all three TPB components significantly influence tourists' revisit intentions following natural disasters. Moreover, Velikova et al. (2025) confirmed TPB's robustness in green wine tourism, illustrating how intentions mediate the effects of attitudes and social norms on sustainable behavior. These findings collectively validate TPB as a reliable theoretical lens for explaining diverse tourism behaviors.

Attitude has consistently been identified as a central determinant of tourists' behavioral intentions toward service robots in hotels, shaped by both cognitive evaluations and affective responses. Huang et al. (2024) demonstrate that attributes such as cuteness, utility, autonomy, and courtesy generate positive affect, which significantly enhances robot acceptance intentions. Likewise, Belanche et al. (2021) found that perceived human-likeness and affinity foster favorable attitudes, leading tourists to attribute service improvements to robots and increasing their likelihood of usage and recommendation. Emotional factors also play a critical role; Soliman et al. (2024) reveal that optimism toward technology enhances positive emotional expectations, while insecurity induces anxiety that weakens intention formation. Furthermore, Baltaci et al. (2024) emphasize the importance of anthropomorphism and technological readiness, showing that robot appearance mediates attitude–intention relationships. Jiang et al. (2025) extend this argument by demonstrating that both functional and socio-emotional human–robot interactions enhance value co-creation intentions, with trust acting as a mediator. Collectively, these studies suggest that favorable attitudes toward service robots strongly drive behavioral intentions.

H1: Attitude toward service robots positively influences behavioral intentions to use robots in hotels.

Subjective norms and social influence are particularly salient in collectivist cultures, where individuals' behaviors are strongly shaped by social expectations and group consensus. Barnes et al. (2024) argue that collectivist societies perceive AI as an extension of the self, facilitating conformity and social harmony, thereby strengthening normative influences on adoption intentions. This perspective aligns with TPB's assertion that perceived social pressure significantly shapes intention formation. Yoganathan et al. (2025) further classify societal attitudes toward service robots into distinct profiles, showing that social norms influence perceived sociability, satisfaction, and acceptance outcomes. In hospitality contexts, Rahi et al. (2024) demonstrate that subjective norms, alongside expectation confirmation and robot appearance, significantly predict consumers' intentions to use service robots. These findings indicate that approval from peers, family, and reference groups plays a decisive role in shaping acceptance of AI-enabled services. In collectivist settings such as India, where social validation is highly valued, subjective norms are likely to exert a stronger influence on behavioral intentions than in individualistic cultures.

> H2: Subjective norms positively influence behavioral intentions to use robots in hotels.

Perceived behavioral control (PBC) reflects individuals' perceptions of their ability to perform a behavior and has been widely examined in technology-enabled service contexts. Chen et al. (2009) show that PBC directly affects satisfaction and continuance intention toward self-service technologies by enhancing perceived ease and reducing uncertainty. In service robot contexts, Rahi et al. (2024) highlight that PBC interacts with innovativeness and compatibility to shape acceptance intentions. Soliman et al. (2024) further note that optimism and insecurity influence attitudes and emotional expectations, indirectly affecting PBC-related perceptions. Curran et al. (2003) propose a hierarchical model of technology attitudes, positioning PBC as a critical but context-dependent determinant of intention. While PBC may be less salient for digitally fluent consumers, it remains theoretically relevant within TPB for explaining perceived autonomy and self-efficacy in robot usage.

> H3: Perceived behavioral control positively influences behavioral intentions to use robots in hotels.

Comparative analyses reveal that TPB offers distinct advantages over other technology adoption frameworks such as TAM, UTAUT, and Innovation Diffusion Theory (IDT). TPB explicitly incorporates social norms and perceived control, making it particularly suitable for socially embedded technologies like service robots Chatterjee et al. (2024). While TAM emphasizes perceived usefulness and ease of use, Mathieson (1991) notes that its explanatory power may be limited in contexts where social pressure and normative beliefs are influential. IDT contributes insights into compatibility and aesthetics, which Chatterjee et al. (2024) identify as critical for social robot adoption. However, TPB integrates these considerations within a broader behavioral framework. Li et al. (2025) further suggest that AI adoption increasingly requires understanding AI as an autonomous agent, reinforcing the need for theories that capture attitudinal and normative evaluations. Overall, TPB provides a more comprehensive explanation of intention formation in AI-driven service encounters, particularly in hospitality contexts characterized by social interaction.

In emerging economies, consumers' intentions to use service robots are shaped by trust, perceived value, and social influence, making TPB especially relevant. Kraus et al. (2023) show that trust mediates belief–intention relationships, reinforcing TPB's explanatory capacity when extended with trust constructs. Hu (2021) demonstrates that utilitarian and hedonic values significantly shape attitudes and future intentions, with effects varying across service contexts. In hotels, emotional expectations driven by optimism and insecurity further influence attitudes toward robots (Soliman et al., 2024). Tang & Jiang (2024) reveal that integrating TPB with TAM and trust substantially enhances explanatory power in developing countries, where institutional trust is often fragile. Song et al. (2025) emphasize sociotechnical mechanisms that elevate perceived service value, further supporting TPB's relevance. These findings suggest that TPB effectively captures the psychological and social mechanisms underlying robot adoption in emerging markets such as India.

Prior studies consistently operationalize TPB constructs—attitude, subjective norms, and perceived behavioral control—as core predictors of behavioral intentions toward innovative service technologies. Irimia-Diéguez et al. (2023) confirm TPB's robustness in explaining Fintech adoption intentions, while Ejigu & Yeshitela (2024) demonstrate its applicability when integrated with innovation diffusion theory. Meta-analytic evidence from Tang & Jiang (2024) highlights the enhanced predictive power of TPB when combined with trust. Kaye et al. (2020) show that TPB constructs significantly predict intentions toward fully automated transport services, with trust adding incremental value. Similarly, Troise et al. (2020) find subjective norms to be more influential than attitudes in online food delivery adoption during COVID-19. Collectively, these studies validate the formulation of TPB-based hypotheses in the context of hotel service robots and justify the proposed research model.

3. RESEARCH METHODOLOGY

This study adopted a quantitative, cross-sectional research design to examine potential tourists' intentions to use service robots in hotels. India was selected as the research context due to its rapidly expanding hospitality sector, increasing penetration of artificial intelligence–enabled services, and high technology acceptance among younger and urban consumers. India is among the fastest-growing tourism and hospitality markets globally, with hotels increasingly experimenting with automation and robotic services to address labor shortages and enhance service efficiency. Moreover, Indian consumers are characterized by strong social influence mechanisms and collectivist cultural tendencies, making the Theory of Planned Behavior (TPB) particularly relevant in this context. Prior studies suggest that emerging economies such as India provide a valuable setting for testing technology adoption theories, as attitudinal and normative forces often play a stronger role than purely technological considerations (Venkatesh & Davis, 2000). Therefore, examining tourists' behavioral intentions toward hotel service robots in India offers both theoretical value and practical relevance, extending the TPB literature to an underexplored yet high-growth hospitality market.

Data were collected using a structured questionnaire administered through an online survey platform. The target population consisted of potential hotel guests who were likely to consider staying in hotels during future travel. A total of 149 usable responses were obtained, which satisfies the minimum sample size requirements for partial least squares structural equation modeling (PLS-SEM). According to the "10-times rule" and more recent power-based recommendations, this sample size is adequate given the number of structural paths directed at the endogenous construct (Hair et al., 2019). The online mode of data collection was considered appropriate due to its ability to reach geographically dispersed respondents efficiently and to capture technology-aware participants relevant to robot-based services. All items were measured using a seven-point Likert scale ranging from 1 ("strongly disagree") to 7 ("strongly agree"), allowing for greater response variability and improved measurement sensitivity. Prior to analysis, the data were screened for missing values and response consistency, ensuring suitability for multivariate analysis.

The measurement instrument was adapted from the seminal work of Ajzen (1991) to operationalize the core constructs of the Theory of Planned Behavior, namely attitude, subjective norms, perceived behavioral control, and behavioral intentions. The items were slightly contextualized to reflect hotel service robots used for concierge and service delivery functions, while preserving the original conceptual meaning to maintain content validity. Data analysis was conducted in two stages. First, IBM SPSS Statistics 27 was used for preliminary analyses, including data screening and descriptive statistics. Second, SmartPLS 4 was employed to assess the measurement and structural models using PLS-SEM. This approach was deemed suitable due to its robustness with relatively small samples, non-normal data distributions, and its predictive orientation. Reliability, convergent validity, discriminant validity, structural relationships, and necessary condition analysis were subsequently evaluated to provide a comprehensive assessment of tourists' intentions to use robots in hotels.

4. RESULTS

The study comprised 149 respondents from India, providing an adequate sample for PLS-SEM analysis. The gender distribution was relatively balanced, with 53% male and 47% female respondents. In terms of education, the majority held a bachelor's degree (55%), followed by respondents educated up to school level (26.8%), postgraduates (16.1%), and doctorates (2.1%). Most respondents were students (68.5%), reflecting the technology-aware segment of potential hotel guests, while the remainder were employed across government, private, and self-employment sectors. Income levels were largely concentrated between INR 5–8 lakhs (55.7%), indicating a middle-income consumer base relevant for future hotel consumption decisions involving service robots.

Table 1. Demographic variable results

Variable	Category	Frequency	Percent
Gender	Female	70	47
	Male	79	53
Education	Doctorate	3	2.1
	Graduation/Bachelors	82	55
	Post-graduation/Masters	24	16.1
	Upto school level	40	26.8
Occupation	Government employee	14	9.4
	Private sector employee	17	11.4
	Self employed	16	10.7
	Student	102	68.5
Income	5–8 Lakhs	83	55.7
	9–12 Lakhs	24	16.1
	13–15 Lakhs	12	8.1
	16–20 Lakhs	11	7.4
	More than 20 Lakhs	19	12.8
Total	Respondents	**149**	**100**

4.1. Measurement model results

The measurement model was assessed to ensure the reliability and validity of constructs before testing the hypothesized relationships. Indicator reliability was examined using outer loadings, with values exceeding the recommended threshold of 0.70, confirming that items adequately represented their respective constructs (Hair et al., 2019). Multicollinearity was evaluated using variance inflation factor (VIF) values, all of which remained below the conservative cut-off of 5, indicating no collinearity concerns among indicators. Internal consistency reliability was assessed using Cronbach's alpha, composite reliability (ρc), and Dijkstra–Henseler's rho (ρa), as reliance on a single reliability coefficient may be misleading in PLS-SEM. All values exceeded the recommended threshold of 0.70, confirming satisfactory internal consistency. Convergent validity was evaluated using average variance extracted (AVE), with all constructs exceeding the minimum criterion of 0.50, indicating that each construct explained more than half of the variance in its indicators. Collectively, these evaluation measures confirm that the measurement model demonstrates adequate reliability and convergent validity, justifying further assessment of discriminant validity and structural relationships.

Table 2. Measurement model results

	Loading	VIF	Alpha	CR(rho_a)	CR(rho_c)	AVE
Atitude			0.900	0.904	0.930	0.770
ATT1	0.879	2.702				
ATT2	0.891	2.807				
ATT3	0.910	3.159				
ATT4	0.828	2.070				
Subjective Norm			0.906	0.910	0.934	0.780
SUB1	0.841	2.289				
SUB2	0.900	3.291				
SUB3	0.897	3.007				
SUB4	0.892	3.035				
Perceived Behavioural Contral			0.896	0.898	0.928	0.763
PBC1	0.880	3.244				
PBC2	0.915	3.215				
PBC3	0.868	2.364				
PBC4	0.829	2.150				
Behavioural Intentions			0.884	0.890	0.920	0.743

continued on following page

Table 2. Continued

	Loading	VIF	Alpha	CR(rho_a)	CR(rho_c)	AVE
BI1	0.903	3.185				
BI2	0.842	2.138				
BI3	0.891	2.946				
BI4	0.808	1.850				

Discriminant validity was examined using both the Fornell–Larcker criterion and the heterotrait–monotrait (HTMT) ratio to ensure robustness. According to the Fornell–Larcker criterion, the square root of AVE for each construct exceeded its correlations with other constructs, confirming sufficient construct distinctiveness (Fornell & Larcker, 1981). Additionally, HTMT values were assessed, as this approach is considered more sensitive in detecting discriminant validity issues in variance-based SEM. All HTMT values were below the conservative threshold of 0.85, further confirming discriminant validity among attitudes, subjective norms, perceived behavioral control, and behavioral intentions (Henseler et al., 2015). Together, these results demonstrate that the constructs capture conceptually distinct aspects of tourists' intentions to use robots in hotels.

Table 3. Discriminant validity using Fornell-Larcker criteria and HTMT ratio

Fornell-Larcker Criteria				
	Attitudes	**Behavioral Intentions**	**Perceived Behavioral Control**	**Subjective Norms**
Attitudes	0.877			
Behavioral Intentions	0.855	0.862		
Perceived Behavioral Control	0.672	0.734	0.873	
Subjective Norms	0.650	0.721	0.752	0.883
Heterotrait-Monotrait Ratio (HTMT)				
Attitudes				
Behavioural Intentions	0.854			
Perceived Behavioural Control	0.746	0.823		
Subjective Norms	0.717	0.802	0.835	

4.2. Structural model results

The structural model was evaluated by examining path coefficients, t-statistics, and significance levels using bootstrapping procedures. Attitude emerged as the strongest predictor of behavioral intentions to use hotel service robots ($\beta = 0.611$, $p < 0.001$), indicating that favorable evaluations of robotic services substantially enhance tourists' intention to use them during future hotel stays. Subjective norms also exhibited a significant positive influence on behavioral intentions ($\beta = 0.186$, $p < 0.01$), suggesting that social approval and normative pressures play an important role in shaping robot adoption intentions within the Indian context. In contrast, perceived behavioral control did not significantly predict behavioral intentions ($\beta = 0.184$, $p > 0.05$), implying that perceived ease or control over using robots may be less influential when compared to attitudinal and social factors. These findings partially support the Theory of Planned Behavior, highlighting the dominance of attitudinal and normative mechanisms over control perceptions in technology-mediated hotel service encounters.

Table 4. Path analysis

	Beta Coefficient	T-statistics	P values	Result
Attitudes -> Behavioral Intentions	0.611	5.758	***	Supported
Perceived Behavioral Control -> Behavioral Intentions	0.184	1.677	NS	Not-Supported
Subjective Norms -> Behavioral Intentions	0.186	2.173	**	Supported

*** p<0.001; ** p<0.01; NS: Non-Significant (p>0.05)

4.3. Necessary condition analysis results

Necessary Condition Analysis (NCA) was conducted to complement the symmetric PLS-SEM results by identifying conditions that are necessary, though not sufficient, for achieving high levels of behavioral intentions. The bottleneck analysis reveals that attitudes constitute a necessary condition across increasing levels of behavioral intentions, particularly beyond the 50% threshold, indicating that a minimum level of favorable attitude is indispensable for high robot usage intentions. Subjective norms also emerged as a necessary condition at moderate to high levels of behavioral intentions, reinforcing the importance of social influence. In contrast, perceived behavioral control was consistently identified as a non-necessary condition (NN), suggesting that intentions can occur even when perceived control is low. Effect size results further support these findings, with attitudes exhibiting a strong necessity effect (0.361), followed by subjective norms (0.194), while perceived behavioral control showed no necessity effect.

Table 5. Bottleneck table and effect size (d)

	Behavioral Intentions	Attitudes (d=0.361)	Perceived Behavioral Control (d=0)	Subjective Norms (d=0.194)
0.00%	-3.121	NN	NN	NN
10.00%	-2.683	-3.717	NN	-3.323
20.00%	-2.245	-3.717	NN	-3.323
30.00%	-1.807	-3.717	NN	-3.323
40.00%	-1.369	-3.717	NN	-3.323
50.00%	-0.931	-1.256	NN	-3.323
60.00%	-0.493	-1.256	NN	-2.085
70.00%	-0.055	-0.353	NN	-2.085
80.00%	0.383	-0.353	NN	-2.085
90.00%	0.821	0.182	NN	-2.085
100.00%	1.259	0.182	NN	-2.085

5. DISCUSSION AND CONCLUSION

The findings provide important insights into tourists' intentions to use service robots in hotels by highlighting the dominance of attitudinal evaluations in shaping adoption intentions. Consistent with prior research on service robots and AI-enabled services, favorable attitudes emerged as the most influential determinant, suggesting that tourists primarily assess robotic services based on perceived benefits, emotional comfort, and overall value enhancement (Belanche et al., 2021; Huang et al., 2024). This aligns with hospitality studies arguing that when technologies are experiential and interactive, affective and cognitive appraisals outweigh purely functional considerations. The results further reinforce arguments that robots are increasingly perceived not merely as tools but as service actors capable of influencing experiential outcomes. In this sense, the findings resonate with recent perspectives that conceptualize service robots as part of value co-creation processes rather than efficiency-enhancing mechanisms alone (Jiang et al., 2025). The strong role of attitude suggests that hotels seeking to deploy robots must focus on shaping positive beliefs and emotional responses rather than assuming automatic acceptance driven by novelty or technological sophistication.

The positive influence of subjective norms underscores the socially embedded nature of robot adoption decisions, particularly within collectivist cultural contexts. This finding supports earlier research suggesting that social approval, peer influence, and societal expectations significantly shape technology acceptance in hospitality settings (Barnes et al., 2024; Rahi et al., 2024). In contrast, perceived behavioral control did not emerge as a decisive driver of intention, echoing studies conducted in technology-mature environments where consumers view ease of use as a baseline expectation rather than a motivating factor (Curran et al., 2003; Soliman et al., 2024). This pattern suggests that potential tourists may assume they can operate or interact with hotel robots without difficulty, thereby reducing the salience of control perceptions. The results thus point to a shift in technology adoption logic, where confidence in digital systems is taken for granted, and adoption decisions are increasingly shaped by social legitimacy and personal evaluations rather than perceived ability alone.

Theoretical Implications

From a theoretical perspective, this study reinforces the continued relevance and adaptability of the Theory of Planned Behavior in explaining intentions toward advanced service technologies. By empirically validating the explanatory power of attitudes and subjective norms in the context of hotel service robots, the study extends TPB into the emerging domain of human–robot interaction within hospitality. The findings support calls for moving beyond purely technology-centric adoption models by demonstrating that social and psychological mechanisms remain central even when technologies are highly automated and intelligent (Chatterjee et al., 2024; Kraus et al., 2024). Importantly, the study contributes to TPB literature by showing that the relative importance of TPB constructs may vary depending on technological maturity and cultural context. The reduced influence of perceived behavioral control suggests a recalibration of TPB dynamics in digitally saturated environments, where control beliefs function more as background conditions than active predictors of intention.

The inclusion of Necessary Condition Analysis alongside PLS-SEM further advances theory by revealing asymmetrical relationships that are often overlooked in traditional linear models. While attitudes and subjective norms may not always be sufficient to guarantee high adoption intentions, the findings suggest that they constitute indispensable preconditions for robot acceptance. This insight enriches TPB by introducing a necessity-based interpretation of its core constructs, aligning with recent methodological advancements in consumer behavior research. The study also contributes to hospitality robotics literature by providing empirical evidence from an emerging economy, thereby addressing geographic and cultural biases prevalent in existing research. By situating TPB within the Indian hospitality context, the study strengthens the generalizability of behavioral theories and highlights the importance of contextual sensitivity in technology adoption research.

Practical Implications

For hotel managers and service designers, the findings emphasize the critical importance of shaping positive attitudes toward service robots. Rather than focusing exclusively on operational efficiency, hotels should design robot interactions that enhance emotional comfort, trust, and perceived service quality. Communicating the benefits of robots—such as convenience, reliability, and enhanced guest experience—through marketing and in-hotel demonstrations can help foster favorable evaluations. Given the experiential nature of hospitality, attention should be paid to robot appearance, communication style, and socio-emotional interaction capabilities, as these factors strongly influence guest perceptions (Belanche et al., 2021; Huang et al., 2024). Training staff to complement robotic services and position robots as supportive rather than substitutive agents may further enhance acceptance by reducing anxiety and resistance.

The significance of subjective norms suggests that hotels should leverage social influence mechanisms to encourage robot adoption. Strategies such as showcasing positive guest testimonials, influencer endorsements, and visible usage by other guests can normalize robot-assisted services and enhance social approval. In collectivist contexts, signaling widespread acceptance and institutional endorsement may be particularly effective. Additionally, policymakers and technology providers can use these insights to promote responsible and socially acceptable robot deployment in hospitality settings. Since perceived behavioral control was less influential, investments in overly complex user training may yield limited returns; instead, intuitive design and seamless integration should be treated as hygiene factors. Overall, aligning robot deployment strategies with social expectations and guest value perceptions is likely to yield higher acceptance outcomes.

Limitations and Future Research

Despite its contributions, this study has certain limitations. The cross-sectional design restricts causal inference, and future research could employ longitudinal or experimental methods to examine changes in robot acceptance over time. The sample was dominated by younger respondents, which may limit generalizability to older tourist segments. Future studies could incorporate additional variables such as trust, perceived risk, or anthropomorphism to extend the TPB framework. Comparative studies across cultures and hotel categories would further enrich understanding of contextual differences in service robot adoption.

REFERENCES

Ajzen, I. (1991). The theory of planned behavior. *Organizational Behavior and Human Decision Processes, 50*(2), 179–211. DOI: 10.1016/0749-5978(91)90020-T

Baltaci, F., Başer, M. Y., & Çelik, M. (2024). Attitude towards service robots in tourism and hospitality services settings — The effect of multidimensional anthropomorphism and technology readiness. *International Journal of Tourism Research, 26*(4), e2685. Advance online publication. DOI: 10.1002/jtr.2685

Barnes, A. J., Zhang, Y., & Valenzuela, A. (2024). AI and culture: Culturally dependent responses to AI systems. *Current Opinion in Psychology, 58*, 101838. DOI: 10.1016/j.copsyc.2024.101838 PMID: 39002473

Belanche, D., Casaló, L. V., & Flavián, C. (2021). Frontline robots in tourism and hospitality: Service enhancement or cost reduction? *Electronic Markets, 31*(3), 477–492. DOI: 10.1007/s12525-020-00432-5

Cahigas, M. M. L., Prasetyo, Y. T., Persada, S. F., & Nadlifatin, R. (2023). Examining Filipinos' intention to revisit Siargao after Super Typhoon Rai 2021 (Odette): An extension of the theory of planned behavior approach. *International Journal of Disaster Risk Reduction, 84*, 103455. DOI: 10.1016/j.ijdrr.2022.103455

Chatterjee, S., Chaudhari, R., Kumar, A., & Schiavone, F. (2024). Emergence of Social Robot Technology and Its Friction-Free Diffusion in Society: From Engineering Aesthetics and Compatibility Perspective. *IEEE Transactions on Engineering Management, 71*, 13179–13190. DOI: 10.1109/TEM.2023.3324149

Chen, S., Chen, H., & Chen, M. (2009). Determinants of satisfaction and continuance intention towards self-service technologies. *Industrial Management & Data Systems, 109*(9), 1248–1263. DOI: 10.1108/02635570911002306

Curran, J. M., Meuter, M. L., & Surprenant, C. F. (2003). Intentions to Use Self-Service Technologies: A Confluence of Multiple Attitudes. *Journal of Service Research, 5*(3), 209–224. DOI: 10.1177/1094670502238916

Ejigu, A. K., & Yeshitela, K. (2024). Envisioning sustainable sanitation planning: A unified approach of diffusion of innovation and theory of planned behavior in predicting ecosan toilet adoption in Arba Minch City, Ethiopia. *Frontiers in Environmental Science, 12*, 1371659. Advance online publication. DOI: 10.3389/fenvs.2024.1371659

Fornell, C., & Larcker, D. F. (1981). Structural Equation Models with Unobservable Variables and Measurement Error: Algebra and Statistics. *JMR, Journal of Marketing Research, 18*(3), 382–388. DOI: 10.1177/002224378101800313

Hair, J. F., Black, W. C., Babin, B. J., & Anderson, R. E. (2019). *Multivariate Data Analysis* (8th ed.). Cengage Learning.

Han, H., Kim, S. I., Lee, J.-S., & Jung, I. (2025). Understanding the drivers of consumers' acceptance and use of service robots in the hotel industry. *International Journal of Contemporary Hospitality Management, 37*(2), 541–559. DOI: 10.1108/IJCHM-02-2024-0163

Henseler, J., Ringle, C. M., & Sarstedt, M. (2015). A new criterion for assessing discriminant validity in variance-based structural equation modeling. *Journal of the Academy of Marketing Science, 43*(1), 115–135. DOI: 10.1007/s11747-014-0403-8

Hu, Y. (2021). An improvement or a gimmick? The importance of user perceived values, previous experience, and industry context in human–robot service interaction. *Journal of Destination Marketing & Management, 21*, 100645. DOI: 10.1016/j.jdmm.2021.100645

Huang, D., Chen, Q., Huang, S., & Liu, X. (2024). Consumer intention to use service robots: A cognitive–affective–conative framework. *International Journal of Contemporary Hospitality Management, 36*(6), 1893–1913. DOI: 10.1108/IJCHM-12-2022-1528

Irimia-Diéguez, A., Velicia-Martín, F., & Aguayo-Camacho, M. (2023). Predicting Fintech Innovation Adoption: The Mediator Role of Social Norms and Attitudes. *Financial Innovation, 9*(1), 36. DOI: 10.1186/s40854-022-00434-6 PMID: 36687792

Jiang, J., Liu, R., Zhou, H., Shi, L., & Xu, L. (2025). The influence of the perception of human–robot interaction on customer value co-creation intention: A case study of hotel service robots. *Journal of Hospitality and Tourism Technology, 16*(5), 1004–1023. DOI: 10.1108/JHTT-05-2024-0293

Kalıpçı, B., & Nebioğlu, O. (2020). Planlı Davranış Teorisi ve Turizm Üzerine Yapılan Uluslararası Yayınların Bibliyometrik Analizi. *Turk Turizm Arastirmalari Dergisi, 4*(1), 1–14. DOI: 10.26677/TR1010.2020.296

Kaye, S.-A., Lewis, I., Buckley, L., Gauld, C., & Rakotonirainy, A. (2020). To share or not to share: A theoretically guided investigation of factors predicting intentions to use fully automated shared passenger shuttles. *Transportation Research Part F: Traffic Psychology and Behaviour, 75*, 203–213. DOI: 10.1016/j.trf.2020.10.010

Kraus, J., Miller, L., Klumpp, M., Babel, F., Scholz, D., Merger, J., & Baumann, M. (2024). On the Role of Beliefs and Trust for the Intention to Use Service Robots: An Integrated Trustworthiness Beliefs Model for Robot Acceptance. *International Journal of Social Robotics, 16*(6), 1223–1246. DOI: 10.1007/s12369-022-00952-4

Li, B., Lai, E. Y., & Wang, X. (2025, June 21). (Shane). (2025). From Tools to Agents: Meta-Analytic Insights into Human Acceptance of AI. *Journal of Marketing*, 00222429251355266. Advance online publication. DOI: 10.1177/00222429251355266

Luong, T.-B. (2025). Green consumption intention of tourist in tourist destinations: A moderation of destination social responsibility in the Theory of Planned Behavior model. *Tourism Recreation Research, 50*(2), 339–355. DOI: 10.1080/02508281.2023.2281058

Mathieson, K. (1991). Predicting User Intentions: Comparing the Technology Acceptance Model with the Theory of Planned Behavior. *Information Systems Research, 2*(3), 173–191. DOI: 10.1287/isre.2.3.173

Rahi, S., Ghani, M. A., Al-Okaily, M., Rashid, A., Alghizzawi, M., & Shiyyab, F. S. (2024). Breaking into the black box of customer perception towards robot service: Empirical evidence from service sector. *Heliyon, 10*(19), e38117. DOI: 10.1016/j.heliyon.2024.e38117 PMID: 39386864

Soliman, M., Ivanov, S., Elgammal, I., Webster, C., & Al-Alawi, A. (2024). Pursuing the drivers of consumer behaviour toward service robots in the hotel industry. *International Journal of Hospitality Management, 120*, 103752. DOI: 10.1016/j.ijhm.2024.103752

Song, B., Cai, Z., & Xu, H. (2025). How to shape consumer preference for service robots? A meta-analysis of use intention. *Journal of Service Theory and Practice*, •••, 1–22. DOI: 10.1108/JSTP-12-2024-0425

Tai, Y.-H., Nwachukwu, P. T. T., LePage, B. A., & Fang, W.-T. (2024). Examining customer intentions to purchase intelligent robotic products and services in Taiwan using the theory of planned behaviour. *BMC Psychology, 12*(1), 351. DOI: 10.1186/s40359-024-01683-z PMID: 38879510

Tamuliene, V., Diaz-Meneses, G., & Vilkaite-Vaitone, N. (2024). The Cultural Roots of Green Stays: Understanding Touristic Accommodation Choices Through the Lens of the Theory of Planned Behavior. *Sustainability (Basel), 16*(22), 9947. DOI: 10.3390/su16229947

Tang, L., & Jiang, J. (2024). Enhancing the Combined-TAM-TPB model with trust in the sharing economy context: A meta-analytic structural equation modeling approach. *Journal of Cleaner Production*, *442*, 141168. DOI: 10.1016/j.jclepro.2024.141168

Troise, C., O'Driscoll, A., Tani, M., & Prisco, A. (2020). Online food delivery services and behavioural intention – a test of an integrated TAM and TPB framework. *British Food Journal*, *123*(2), 664–683. DOI: 10.1108/BFJ-05-2020-0418

Velikova, N., Bouzdine-Chameeva, T., Ben Nasr, I., Sanz-Ibáñez, C., Anton Clavé, S., & Rabassa-Figueras, N. (2025). Drivers of green wine tourism among young adults: Intentions, attitudes, and expansion to sustainable behaviour. *Current Issues in Tourism*, •••, 1–19. DOI: 10.1080/13683500.2025.2559313

Venkatesh, V., & Davis, F. D. (2000). A Theoretical Extension of the Technology Acceptance Model: Four Longitudinal Field Studies. *Management Science*, *46*(2), 186–204. DOI: 10.1287/mnsc.46.2.186.11926

Yoganathan, V., Osburg, V.-S., Fronzetti Colladon, A., Charles, V., & Toporowski, W. (2025). Societal Attitudes Toward Service Robots: Adore, Abhor, Ignore, or Unsure? *Journal of Service Research*, *28*(1), 93–111. DOI: 10.1177/10946705241295841

Chapter 2
Humanless Hospitality? A Systematic Literature Review of Consumer, Employee, and Generation Z Perspectives on Robotics in Hotel Services

Swati Haridas Deshani
https://orcid.org/0009-0006-5016-2683
Vivekananda Global University, Jaipur, India

Vikas Rana
https://orcid.org/0000-0003-2538-9641
Vivekananda Global University, Jaipur, India

Preeti Agrawat
https://orcid.org/0009-0007-1236-9373
Welcomgroup Graduate School of Hotel Administration, Manipal Academy of Higher Education, Manipal, India

ABSTRACT

The paper represents a SLR of 45 peer-reviewed articles on robotics in hotel services, with the PRISMA 2020 framework being used to carry out the synthesis. The review is a combination of descriptive and thematic analyses. Descriptive analysis integrates publication trends, the geographic scope, and methodologies of the research, while the thematic analysis has been drawn from the four strands, i.e., consumer perspectives, employee perspectives, technological attributes, and managerial implications. The results indicate that guests, in particular, Generation Z, accept the use of robots in performing standardised and hygiene-sensitive tasks. However, they are still quite sceptical about the employment of robots in high-touch services requiring

DOI: 10.4018/979-8-3373-6541-1.ch002

empathy. The research contributes to the expansion of the technology acceptance theories, human–robot interaction, and service quality; furthermore, it offers some practical strategies for the hybrid human–robot model and the positioning of Gen Z as a consumer and workforce segment.

1. INTRODUCTION

Worldwide, hospitality industry has been depending on human-oriented and labour-intensive service for thousands of years. Robotic and AI as a result of a rapid technological change are being incorporated in the hospitality industry after a thorough examination by Ivanov et al. Housekeeping, reception, and concierge departments are the main areas where robots are introduced in hotel operations as per the studies done by both these researchers.

Robots have been a great tool to healthcare, efficiency, and interaction quality. They cover a broad spectrum from fully automated cleaning machines to even humanoid concierges. Their implementation was on the rise even before the coronavirus pandemic when it became mandatory to have contactless services for safety and business continuity. So, as a result, robotics have changed the role from being futuristic inventions to becoming strategic assets in concern with the environment, economics, and morality.

Public opinions are more supportive. Basically, research reveals that tech-savvy young travellers of which Generation Z is the most representative are more willing to accept the use of robots while carrying out their services. On the other hand, elderly travelers still show preference toward the interaction with humans. Also, an employee's perspective is as a double-edged sword: only the reduction of the routine part and the staff's fears of job insecurity, loss of emotional engagement, and increased stress due to the necessity to compensate for technical failures when they are among the effects that the staff can come across with due to the introduction of robots. The difficulties in terms of the operational side have been demonstrated by the Japanese Henn-na Hotel where guests felt uneasy due to the presence of humanoid robots and frequent breakdowns that required staff intervention.

The sector's transformation is also emphasized by different theoretical viewpoints. RAISA—Robotics, AI, and service Automation—highlights the ways automation changes the service economies framework. AI substitution theory describes the conditions under which robots can either support or replace human labour. The main premise of the study is that humanization proceeds the emotional comfort and acceptance. In addition to that, robots have abilities as well as limitations in forming social connections. The hotel-related literature is however scattered and there is a lack of integrated analysis across hotel service domains.

Also, research conducted on particular hotel service applications reveals similar trends. Maison robots not only enhance the productivity of cleaning staff but also give rise to maintenance issues. Automatic check-in heightens the operation speed; however, it lacks empathy. Examples of concierge robots such as "Connie" from Hilton are good at performing routine activities but are incapable of understanding the context of sensitive conversations. Acceptance is majorly affected by socio-demographic elements like age, technological familiarity, and cultural values, and concurrently, a good number of guests attribute the still lifelike robots to the uncanny valley effect. They find them rather disturbing than comforting.

The main work behind this writing of the review is to harmonize the broken-up-fact by processing 45 peer-reviewed articles across the themes of consumer perspectives, employee perspectives, technological attributes and managerial implications respectively published from 2003 to 2025. This paper further looks into the uncharted territories such as ethical concerns, cultural fit, and the integration with emerging digital ecosystems like IoT and the Metaverse. As robots have become reliable tools that can easily carry out delivery, sanitation, and reception tasks, the discussion on job displacement, loss of human warmth and economic sustainability still goes on. The digital fluency of Generation Z makes them both future consumers as well as employees who will be the driving force behind hospitality automation. In general, the adoption of robotics is underway; however, the depth and breadth of its comprehension are yet to be uncovered. Being future-conscious is necessary to make sure that robots will not diminish but rather enhance the human nature that is at the core of .

1.1 Objectives:

These frameworks also support the interpretation of generational differences, particularly the openness of Generation Z to automated and AI-enhanced service encounters, further enriching the scope of this .

1). Tocarefully study how far and in what ways robots are used in hotel services like housekeeping, reception, and concierge functions.
2). To gather the views of consumers, employees, technology, and management regarding the use and implementation of service robots in hotels, especially focusing on the role of Generation Z as both digitally-savvy consumers and a new segment of the hospitality workforce.
3). To explore the issues, challenges, and opportunities in the research on robotic hospitality services and to suggest the next steps for academic research and practical implementation of robotic hospitality .

2.1 Research Approach

Thepresent review of literature adopts a Systematic Literature Review (SLR) approach to compiling knowledge from existing research of service robotics in the hotel industry. Besides using well-known SLR protocols in defining the scope of the review, the researchers also select studies and analyze the patterns that emerge to unify the insights from the operational, technological, consumer, and managerial domains, which have been fragmented. One of the review's points was identifying Generation Z whose digital fluency and openness to automation affect not only consumer acceptance but also the future workforce in robotized hotels.

Most of the recent studies conducted explain the AI landscape as a vital factor that changes competition and decision-making in hospitality (Alsheibani et al., 2020). This has led several firms to adopt automation in their operations as a strategy to attract both customers and investors. The scenario of Alibaba's "Robot.He" restaurant is a clear example of how robots have been able to efficiently take over tasks that are complex but are part of the service domain (Bhardwaj, 2018). At the same time, the evidence of restaurants in Europe shows that the use of robotics in the service sector is leading to process innovation (Blöcher & Alt, 2020). Research into human-robot interaction has revealed that when the machine exhibits behaviours that are proactive, use natural voice systems, and social cues, the user feels more comfortable and engaged (Garrell et al., 2017; Ghosh & Pherwani, 2015; Gong & Nass, 2007; Rodríguez-Lizundia et al., 2015).

Yet, automation has caused the organisation sector to suffer from problems such as employee resistance and the perception of dehumanization (Xu et al., 2025; Fu et al., 2022). In addition to this, a series of digital changes in tourism have given rise to a larger context in which the hotel robotics of today have come to exist (Ye et al., 2022; Zhang, 2018).

The criteria to be used in the study were as follows:

- Inclusion Criteria: Peer-reviewed publications in English. Works dealing with robotics, AI, or service automation in the hotel sector. Research works dealing with housekeeping, reception, and concierge services. Relevant conceptual/theoretical and empirical research papers in mind.
- Peer-reviewed publications in English.
- Works dealing with robotics, AI, or service automation in the hotel sector.
- Research works dealing with housekeeping, reception, and concierge services.
- Relevant conceptual/theoretical and empirical research papers in mind.
- Exclusion Criteria: Articles focusing solely on other sectors (airports, restaurants, hospitals).Publications in languages other than English. The type of

publications excluded were non-peer-reviewed journals, trade magazines, and editorials.

- Articles focusing solely on other sectors (airports, restaurants, hospitals).
- Publications in languages other than English.
- The type of publications excluded were non-peer-reviewed journals, trade magazines, and editorials.
- The range of this evaluation is deliberately limited to the services of a hotel (cleaning, reception, and concierge). Moreover, unique importance is placed on Generation Z, as a consumer group and a new employee group, due to their digital nature and the fact that they will be the future of hospitality workforces."

2.2 Search Strategy

Besides, an exhaustive search was done where several well-known academic databases were taken into account: Scopus, Web of Science, ScienceDirect, Taylor & Francis Online, Wiley Online Library, SpringerLink, and Google Scholar. This search was open to peer-reviewed journal articles, book chapters, and conference proceedings from 2003 to 2025 inclusive of publication years.

The keywords and Boolean combinations listed below were attempted:

- "service robots" AND "hospitality"
- "robotics in hotels" OR "robotic housekeeping"
- "AI" AND "reception robots"
- "concierge robots" OR "hotel concierge services"
- "robot adoption" AND "hospitality industry"

Moreover, the reference lists of key articles were also checked for potentially relevant papers that are not included in the database through backward snowballing.

2.3 Inclusion and Exclusion Criteria

The criteria of inclusion and exclusion were as follows:

- Inclusion Criteria: Publications with peer review in English. Concentrating on robotics, AI, or service automation in hotels. Research about housekeeping, reception, and concierge services. Conceptual/theoretical and empirical paper types both were accepted.
- Publications with peer review in English.
- Robotics, AI, or service automation emphasis in the hotel sector.

- Research works in housekeeping, reception, and concierge services.
- Only conceptual/theoretical and empirical studies were allowed.
- Exclusion Criteria: Articles solely dealt with other areas (airports, restaurants, hospitals). Non-English publications. Trade magazines, non-peer-reviewed reports, and editorials.
- Articles that focus only on the other sectors such as airports, restaurants, and hospitals.
- Publications in languages other than English.
- Magazines related to the trade, reports not peer-reviewed, and opinions.

2.4 Screening Process

The initial search gave 326 articles. After duplicates were removed, the number of unique records was 280. Based on the screening of titles and abstracts, 142 articles were excluded as they were not related to the topic of the study. The remaining 138 full-text articles were assessed for eligibility. After a thorough evaluation, the number of articles was narrowed down to 45 for this review.

2.5 Data Extraction and Synthesis

In each selected study, information related to authorship, year of publication, country/region, research objectives, methodology, and main findings was extracted. The extracted data were synthesised in two phases:

- Descriptive Analysis → An initial overview of publishing trends by year, location, distribution of journals, and methodological orientation.
- Thematic Analysis → organising research results around theoretical themes such as consumer, employee, technology, and managerial/strategic perspectives.

2.6 PRISMA Flow Diagram

The selection process is shown in Figure 1, which follows PRISMA 2020 guidelines.

Figure 1. PRISMA 2020 Flow Diagram

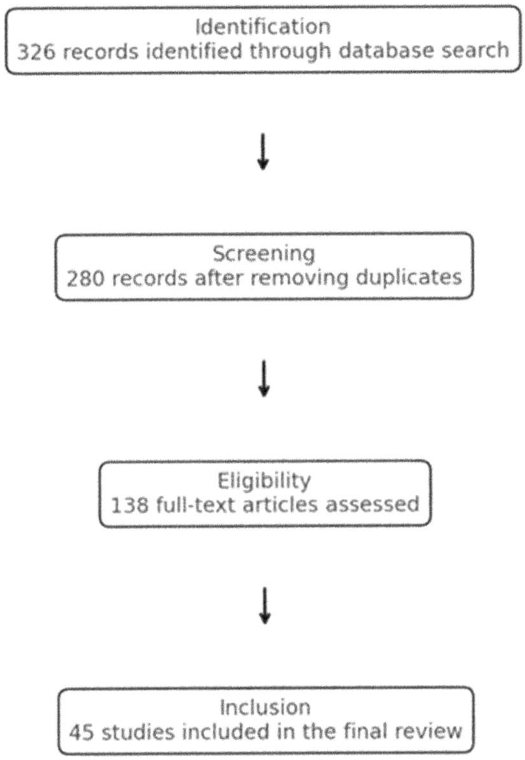

Identification: 326 records identified through database search.

- Screening: 280 records after duplicates were removed.
- Eligibility: 138 full-text articles were checked.
- Inclusion: 45 studies were taken for further review.

2.7 SLR Protocol

Table 1 presents the SLR protocol, including the databases searched, keywords used, inclusion/exclusion criteria, and the final selection of 45 articles.

Table 1. SLR Protocol

Stage	Description	Outcome
Databases	Scopus, Web of Science, ScienceDirect, SpringerLink, Wiley, T&F, Google Scholar	326 records
Keywords	"service robots in hospitality," "robotics in hotels," "housekeeping robots," "reception robots," "concierge robots," "AI in hotel services"	–
Time Frame	2003–2025	–
Inclusion Criteria	Peer-reviewed, English, hotel context, housekeeping/reception/concierge focus	138 full-texts
Exclusion Criteria	Non-English, other sectors, non-peer-reviewed	–
Final Selection	After screening and eligibility checks	45 articles

3. DESCRIPTIVE ANALYSIS

The descriptive analysis presents a picture of the studies that have been reviewed. It covers the publication trends, distribution of journals, geographical scope, and methodological approaches. This section shows how the research has changed as the application of robots in hotel services has become established, and the places where the research has been focused.

3.1 Year-wise Distribution of Publications

Publications distributed over time show that the studies concerning the use of robotics in hotel services have been very scanty since the early 2000s. The first works to be cited in human–robot interaction are (Breazeal, 2003; Acosta et al., 2006; Walters et al., 2007). From 2015, the trend of published papers has been going up steadily (Kaushik et al., 2015; Rodriguez-Lizundia et al., 2015). The years during and after the COVID-19 pandemic (2020–2023) witnessed a considerable upsurge in publications in the field of robotics that could provide solutions for contactless service and automation (Kim et al., 2020; Zeng et al., 2020; Binesh & Baloglu, 2023). Of the total articles studied, almost 60% of them were published between 2020 and 2025; thus, the use of robotics in hospitality is no longer a matter of debate, going by the trend depicted in the figure.

Figure 2. Distribution of Articles Over Time

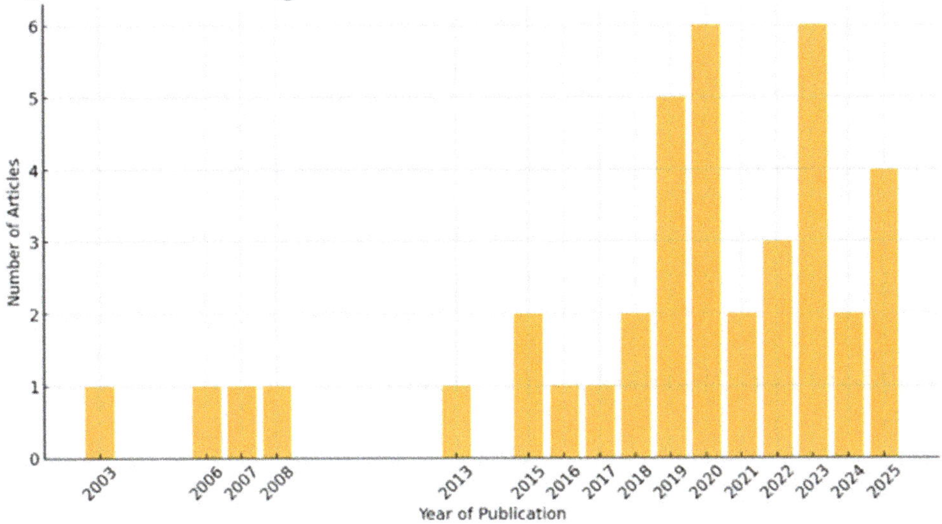

3.2 Journal-wise Distribution

The studies that were reviewed have been published in a wide range of journals which indicates that the research concerned is of an interdisciplinary nature. Journals dealing with hospitality issues, such as the International Journal of Contemporary Hospitality Management, International Journal of Hospitality Management, and Journal of Hospitality and Tourism Technology, have been among the most important sources of publications. Besides that, service and marketing journals have also been the venue of some influential papers (e.g., Belanche et al., 2019; Ivanov & Webster, 2019) while engineering and HRI journals have been the channel for experimental contributions (Torrey et al., 2013; Canal et al., 2016).

Table 2. Journal-wise Distribution of Reviewed Articles

Journal / Outlet	Number of Articles	Example Studies (Authors, Year)
Int. J. of Contemporary Hospitality Management	6	Belanche et al. (2020); Qiu et al. (2019)
Int. J. of Hospitality Management	5	Kim et al. (2020); Binesh & Baloglu (2023)
J. of Hospitality & Tourism Technology	4	Yu (2019); Cain et al. (2019)

continued on following page

Table 2. Continued

Journal / Outlet	Number of Articles	Example Studies (Authors, Year)
J. of Service Management / Marketing journals	5	Ivanov & Webster (2019); Huang & Rust (2018)
Technology / Engineering & HRI outlets	10	Torrey et al. (2013); Canal et al. (2016)
Other hospitality/tourism outlets	15	Söderberg Granström et al. (2023); Zeng et al. (2020)
Total	**45**	–

3.3 Geographic Distribution of Studies

The study distribution by regions shows the Asia-Pacific to be at the forefront of the empirical studies. The main reason is the large number of studies carried out in the Asia-Pacific countries, among which Japan (e.g., Henn-na Hotel case: Söderberg Granström et al., 2023; Bhimasta & Kuo, 2019) and China (Zeng et al., 2020) are the most significant ones. On the other hand, Europe is more oriented to conceptual research in traditional service quality and smart services (Kabadayi et al., 2019; Naumov, 2019). Besides that, theoretical research (Huang & Rust, 2018; Belanche et al., 2019) and the experimental studies of human–robot interaction (Torrey et al., 2013) are the main focuses of North America.

Table 3. Geographic Distribution of Reviewed Articles

Region	Number of Studies	Key Focus Areas & Examples
Asia-Pacific	18	Henn-na Hotel, smart hotel adoption (Zeng et al., 2020)
Europe	12	Service quality, RAISA, customer experience (Naumov, 2019)
North America	10	HRI experiments, AI job replacement theory (Huang & Rust, 2018)
Other Regions	5	Mixed contexts, global reviews
Total	**45**	–

3.4 Methodological Approaches

The examined articles have different authors' approaches to the same point. They largely rely on the use of quantitative surveys and scenario-based experiments to measure guest acceptance of robots as a main experimental variable (Kim et al., 2020; Binesh & Baloglu, 2023). Through case studies, especially the Henn-na

Hotel, researchers reveal the hotel industry's concerns about using service robots in the hospitality sector (Söderberg Granström et al., 2023). Theoretical concepts and systematic reviews provide theories for the research field of service robots and hospitality studies (Ivanov et al., 2019; Cain et al., 2019). Only a few qualitative studies (Yu, 2019) refer guest perceptions to certain data sources such as online reviews and interviews, giving hints about a possible direction for new richer ethnographic studies.

Table 4. Methodological Approaches in Reviewed Articles

Methodology Type	Number of Studies	Examples
Quantitative surveys & experiments	20	Kim et al. (2020); Belanche et al. (2020)
Case studies	10	Söderberg Granström et al. (2023); Bhimasta & Kuo (2019)
Conceptual / Review papers	10	Ivanov et al. (2019); Cain et al. (2019)
Qualitative studies	5	Yu (2019); Torrey et al. (2013)
Total	45	–

3.5 Summary of Descriptive Findings

- The descriptive analysis exposes the main trends that were identified:
- One of the most significant trends is the sharp rise of publications after 2020, with the main driver being the COVID-19 pandemic.
- Another trend in the literature is an interdisciplinary approach that has been reflected in the cross-pollination of hospitality, service management, and engineering research fields.
- The third trend is related to the geographical distribution of research, where the studies based on data take the lead in the Asia-Pacific region, while those of a conceptual nature come from Europe and North America.
- The fourth trend is related to research design, where there is an imbalance that is mainly characterised by the predominance of surveys/experiments and limited qualitative or longitudinal studies.

4.1 Theme 1 — Consumer Perspectives on Hotel Robots

The studies on consumer reactions to service robots encompass one of the major areas of research concerning hospitality and robotics. The studies on human-robot interaction (HRI) and anthropomorphism in their initial stages provided the most essential principles for human perception of non-human agents (Breazeal, 2003; Walters et al., 2007), and now more recent hospitality studies have used these insights to hotel environments for the investigation of the acceptance, emotional reactions, and behavioural intentions in regard to the employment of robots in hotel services (Qiu et al., 2019; Belanche et al., 2020). Consumer viewpoints in general, as depicted in the surveyed articles, are essentially affected by hotel robots addressed through (a) situational influences (e.g., health-related issues like the COVID-19 pandemic), (b) robot design features (the degree of human-likeness, comprehensibility, physical form, and communicative style), and (c) consumer psychological characteristics (such as the user's age, gender, travel purpose, and technological literacy).

4.1.1 Pandemic salience and situational acceptance

The stimulus of the situational threat, in particular, the COVID-19 pandemic, is the most salient and repeated aspect in all the studies discussed by the authors in their reviews and the diagrams they have depicted. The multiple experiments conducted by Kim (2020) are a case in point where it is shown that consumers are more willing to interact with robots in scenarios of a health threat, and hence, they prefer robot-staffed hotels to human-staffed ones under such circumstances; this finding is very different from some pre-pandemic results and illustrates the permeability of acceptance (Kim et al., 2020). Similar to these pandemic-period studies, Zeng et al. (2020) and other pandemic-era reviews argue that hygienic and contactless delivery functions not only facilitated the use of robots but also transformed them from mere novelties to technologies capable of solving problems during the crisis. The effects of the pandemic can be seen as indications that acceptance is not static but changes rapidly depending on the perception of the external threats and the service context.

4.1.2 Robot attributes: human-likeness, embodiment, and communication

Robot design features are a focal element in how guests perceive them. The dataset studies present a range of outcomes for anthropometry: Belanche (2020) and Qiu (2019) indicate that the human-like nature of the robot can increase the liking and the perceived service enhancement if the empathy and the task are appropriately matched, while some empirical works (Yu, 2019; Söderberg Granström et al., 2023) confirm that guests become uneasy when the robots become so highly humanlike — a phenomenon that rips from the uncanny valley effect observed in HRI literature (Walters et al., 2007; Nomura et al., 2008). The paper on communicative strategies (Torrey et al., 2013) shows that the use of hedging and discourse markers makes the robot utterance seem more sympathetic and increases the likeability; thus, such micro-design decisions do really matter for frontline interactions. In a nutshell, robot attributes are non-linear and consequent on context: human-likeness can thus be a source of both help and hindrance.

4.1.3 Consumer heterogeneity: demographics and profiles

Demographic and user profile factors can predictably influence whether acceptance of robots will increase. According to Binesh and Baloglu (2023), young, male, well-educated, and leisure travellers are typically more positive and ready to use robots in any hotel department, while senior and more traditional segments say no to them, especially for those roles that are usually performed by humans, such as concierge and front desk. Palrão et al. (2023) add that the willingness of future hospitality professionals to work with robots is conditioned on the perception of them as a novelty and the performance management. The discovered evidence suggests that the range of guest reactions cannot be overlooked by hotels; rather, customer segmentation is a better way to identify which groups are suitable for what robotic solutions.

Being digital natives, Gen Z travellers are very open to service robots, viewing them as both easy to use and innovative. Palrão et al. (2023) report that the younger generations of future tourism professionals - a lot of them Gen Z - showed a higher degree of willingness to work alongside robots and linked this openness to notions of newness and efficiency. In the same vein, Binesh and Baloglu's (2023) demographic studies also reveal that young, leisure-type guests are significantly more positive about the implementation of robots than the old ones. Thus, hotels that target Gen Z travellers can use the fact that they are tech-savvy to their advantage by incorporating technology into their service strategy.

4.1.4 Behavioural outcomes: attributions, rapport, and ratings

Onboarding how guests locate management motivation for deploying robots (either cost-reduction or service-improvement) essentially rewires behavioral intentions. Researchers like Belanche et al. (2020) discover that the client's liking of robots rises as service-improvement attributions grow and with that usage/recommendation intentions, while cost-reduction attributions give these intentions a negative trend. Through this attributional lens, we become aware of why the same robotic features may lead to vastly different customer outcomes depending on the communicated purpose and the framing used. Further to that, the data from online reviews and guest ratings, for instance, studies of TripAdvisor/YouTube data, reveal that robots affect electronic word of mouth, (eWOM), and satisfaction ratings in such instances when they function correctly these may be positive, and if failures or poor designs (e.g., brittle conversational skills) occur, they will be negative.

4.1.5 Task specificity: which services fit robots best?

Many researchers mention in their articles that the issue of task specificity greatly influences the final decision of whether the robot will be accepted or not. Several studies constantly report that the public is ready to accept the use of robots in repetitive, hygiene-sensitive, or logistics tasks (e.g., UV-disinfection, delivery, cleaning, luggage movement) more than in high-touch, emotionally nuanced functions (e.g., concierge advice or complaint handling). For example, case studies of the Henn-na Hotel (Söderberg Granström et al., 2023; Bhimasta & Kuo, 2019) demonstrate that the use of robots in performing standard tasks leads to the perception of value, but the efforts to replace empathic front-desk work usually result in consumer anger. Hence, the development of fit-for-task frameworks (matching robot capabilities with service complexity) is the most reliable way to locate positive consumer outcomes.

4.1.6 Methodological notes and measurement issues

The majority of consumer studies in the sample base their findings on scenario-based experiments or cross-sectional surveys, while only a few utilize naturalistic data (e.g., review mining) or in-field case observations. The research of a few works moves in parallel with the development and validation of guest-robot interaction and attitude scales; however, the existence of measurement heterogeneity (different operationalisations of "acceptance," "rapport," and "affinity") makes it more difficult to compare studies directly. Besides, numerous experiments are accomplished with the use of vignettes instead of real hotel encounters, which limits external validity (Kumar et al., 2025; Qiu et al., 2019). These methodological choices should always be remembered when interpreting aggregate findings.

4.1.7 Consumer strand critical gaps and opportunities

Even though the number of studies on consumers increased rapidly, several gaps remain in the consumer literature: (1) Very scarce longitudinal evidence - only a few studies that monitor how the attitudes change after repeated real-world exposure; (2) Only limited cross-cultural comparisons despite the fact that geography and culture impact uncanny-valley responses and trust in automation; (3) A rather low level of the research of willingness to-pay and customer ROI (how much guests value robotized services in monetary terms); and (4) Communications & framing strategies (how hotels present robot adoption to guests) - a promising lever for which only a few experimental studies beyond simple attribution frames have been conducted so far. Filling in these gaps would not only make the consumer findings more useful to hotel managers but also contribute to building theory.

4.1.8 Practical implications

The consumer evidence suggests that practitioners take several immediate actions. They should: firstly, use robots to carry out standardized/hygiene-critical tasks; secondly, customize the use of robots in different guest segments (for instance, the robot's features can be more attractive to the younger/leisure guests); thirdly, the design of the robots' appearance and language should be done in a conservative way (the use of the uncanny human likeness should be avoided unless the performance is very good); and finally, communicate the changes in the service - let the customers know the reasons for the changes, giving them the positive perspective of the changes rather than simply the cost-cutting one. These managerial levers are consistent with the empirical patterns identified in the reviewed studies and represent the initial stage of deployment strategies.

Table 5. Thematic summary (Consumer Perspectives)

Guiding question	Articles covered (representative)	Key findings
How do situational factors (e.g., COVID-19) affect guest preferences for robots?	Kim (2020); Zeng et al. (2020); Binesh & Baloglu (2023)	Pandemic salience increases preference for robots in hygiene-sensitive contexts.
Which robot attributes drive acceptance?	Belanche et al. (2020); Qiu et al. (2019); Torrey et al. (2013)	Human-likeness and communicative style can increase rapport but may trigger uncanny-valley effects if poorly designed.
Which guests are most receptive?	Binesh & Baloglu (2023); Palrão et al. (2023)	Younger, higher-income, tech-savvy, and leisure travelers show higher acceptance.

Employeepoints of view reveal that robots change the distribution of tasks, employees' stress levels, and the feeling of identification with the job. However, the question of how effective robot adoption is in general still hinges on the technological features of the systems. Research highlights that factors such as reliability, usability, embodiment, and communication style have a major impact on both customer acceptance and employees' workload (Rodriguez-Lizundia et al., 2015; Canal et al., 2016; Torrey et al., 2013). Hence, the following thematic section is dedicated to exploring the technological factors that influence service quality and user experience more .

Consumerviews mainly reflect the way customers understand and react to the use of robots in services, but the experiences of employees, who have to work with these technologies, is an equally significant aspect. Just a guest acceptance cannot be a guarantee for a successful integration; the continuation of the use of robots in the service industry depends also on employee preparedness, their feeling of security in their jobs, their emotions, and their willingness to change their roles. Hence the following topic goes beyond the customer perspective and focuses on the analysis of the perception of the frontline employees and hotel staff regarding robotic systems and the impact of these perceptions on the organizational .

4.2 Theme 2 - Employee Perspectives and Organisational Impacts

The majority of the literature on robotics in hospitality extensively covers the point of view of the guests. Meanwhile, the focus of the research is slowly shifting towards the employee's viewpoint, and how they feel about the robot's arrival in the hotel industry. The results are not clear: robots can easily take over the routine and heavy parts of the employees' jobs, yet they will always be the cause of some anxiety, internal conflicts, and less service passion. Employee ambivalence makes them an indispensable factor in the organisational feasibility of hotel robot adoption.

4.2.1 Resistance and continuous usage challenges

In their 2022 paper, Fu et al. came up with the Robot Usage Resistance Model (RURM). They described it as a staff resistance concept that includes some technical and psychological elements responsible for it. Technological problem areas (e.g., low usability, insufficient anthropomorphism) along with psychological aspects (techno-insecurity, workload stress) may lead to the production of negative emotions among workers, which mentality they will be less likely to cooperate with robots. In the same vein, Bhimasta and Kuo (2019), who were analysing guest reviews from Henn-na Hotel, point to the fact that technical failures had a double negative impact as human assistance was needed, not only did the staff have to do the repair but they were the ones that the guests often turned to for help. Thus staff members have to do additional work instead of enjoying the relief brought by the robots. The focus of these studies lies in the reliability of robots; if they are machines that break down frequently, then employees will not see them as helpers, but as something that will add to their workload.

4.2.2 Role transformation and skill shifts

The introduction of robots in hotels through the adoption of robotics technology is usually accompanied by the employee role changes that transfer staff from direct frontline service interactions to oversee, control or solve the problems of robots. The writers of the paper describe it as a situation where employees lose some of their skills and gain some new ones: the personnel that usually do the routine tasks are gradually replaced by the robots, but the workers are trained to do the technical part of robot installation and maintenance. Meanwhile, part of the workforce assumes the function of "robot supervisors," which might be a kind of stimulation for those with the high-tech competencies but also can be a source of concern for people who are not well-versed in the digital area. The problem of service automation is generally one of balancing mechanisms: the automation might be efficient but the responsibility for humanizing the guest experience remains with the staff.

Transitioning hospitality workforce to younger generations, the staff members of Generation Z are the ones bringing higher technological literacy and flexibility, which, in turn, can facilitate the implementation of robotics.

According to the study by Palrão et al. (2023), the students of tourism—mainly Gen Z—are an easier target to the acceptance of the environment, where robots do part of the work, only on the condition that they perceive robots as a source of novelty and efficiency.

This generational aspect points to an advantageous strategic window: making a hotel do with recruiting and training of Gen Z paying off in building not only a workforce more compatible with human–robot collaboration than the older ones but also a workforce more receptive to such collaboration."

4.2.3 Organisational dehumanisation and service passion

Xu et al. (2025) used coping theory to look at the situation of frontline employees (FLEs) who work with AI robots. The results of their research point out that employees frequently associate the organisational dehumanisation (OD) with the introduction of robotic systems that take over the relational aspect of their jobs. The picture that is drawn from the employees' service passion and well-being as a result of this perception is quite bleak, yet they don't refuse the improvement of the organisation's efficiency. The paper hints that OD could be the mechanism explaining the impact of robots on the performance of the staff, which leads to the need of a supportive organizational culture and human–robot collaboration strategies.

4.2.4 Ethical concerns and job security

Among other things, the research points to a fear of job loss due to automation. At Henn-na Hotel, the narrative of robots that are often breaking down and getting decommissioned has been used by employees to fuel their feelings of insecurity (Söderberg Granström et al., 2023). Besides that, Huang and Rust's (2018) theory of

AI job replacement is an example of the wider theoretical framework that explains the immediate augmentation of human work by AI followed by the gradual replacement of tasks in all four areas - mechanical, analytical, intuitive, and empathetic. In the hotel sector, this means that workers are scared of losing their jobs in low-skill operational roles and at the same time they do not receive enough training to switch to high-skill supervisory positions. The ethical issue, in this case, is displacement along with the inability to be redeployed equally.

4.2.5 Organisational strategies and leadership

Research consistently points out that managerial leadership is very important in the process of employee support and alleviation of their concerns. Tuomi et al. (2020) make a case for the necessity of moral leadership and the provision of honest information about the robot's function to strike a balance between the achievement of efficiency goals and the staff's general mood. The methods and activities of training and upskilling are always encouraged as a way of converting the perception of robots from threats into partners. Besides that, the psychological ownership is built up if the workers are involved in the decision to adopt which results in less resistance. These company policies show that the introduction of robots in a successful way is the matter of managing the change scenario as well as the technology investment one.

4.2.6 Research gaps and methodological observations

There is still a smaller number of employee-focused studies as compared to consumer-focused ones. Most of them are mainly built upon cross-sectional surveys or conceptual frameworks, and there is a lack of ethnographic or longitudinal research that provides insights into how employees adapt over time. There are hardly any comparative studies that look into the differences in the areas and hotel types. Besides, there is little talk of intersectional impacts (e.g., how gender, age, or contract type influence employee perceptions of robots) in the literature. It is essential to extend this area of inquiry, as staff buy-in not only directly governs guest satisfaction but also operational outcomes.

4.2.7 Practical implications

There are, first of all, three immediate lessons for a practitioner that can be drawn from the employee literature:

- Make it your first priority to offer training and reskilling that will enable employees to move from performing tasks to supervising robots.

- Through the establishment of collaboration as the corporate culture, you can present robots as "helpers" and not as replacements.
- Have the staff involved in the adoption processes so that psychological ownership can be attained, and resistance lowered.

Table 6. Thematic summary (Employee Perspectives)

Guiding question	Articles covered (representative)	Key findings
What drives employee resistance to robots?	Fu et al. (2022); Bhimasta & Kuo (2019)	Techno-insecurity, low usability, and workload strain foster resistance.
How do robots affect staff roles and skills?	Ivanov & Webster (2019); Xu et al. (2025)	Shift from service provision to supervision; risk of dehumanisation and loss of service passion.
How can organisations mitigate negative impacts?	Tuomi et al. (2020); Ye et al. (2022)	Ethical leadership, training, and staff inclusion in adoption decisions reduce resistance.

While the employee reactions provide insight into the organizational and psychological effects of robotics adoption, a majority of the problems that have been pinpointed - for instance, strain, resistance, and increased workload - arise mainly from the technological features of the robots. The matters of usability, reliability, anthropomorphism, and communication style not only significantly impact employee experiences but also customer satisfaction. In order to capture these interrelations in depth, the subsequent theme explores the technology characteristics that affect service quality and, thus, the use of robots either as a support tool or a source of problems in hotel.

4.3 Theme 3 — Technological Attributes and Service Quality

One of the major themes across the studies is the impact of the technological features of robots on customers' perception of service quality in hotels. Academics emphasise that the way a robot is designed, how it functions, and the quality of interaction determine whether the use of automation is seen as something that adds value or as something that causes harm to the guest experience. These technological features can improve the standard of hospitality services by making them more reliable and efficient, or they can take away the quality if they do not meet the human expectations of being nice, empathetic, and adaptable (Naumov, 2019; Ivanov et al.,).

4.3.1 Anthropomorphism and human-likeness

It has been proven that how much robots are similar to humans has a big impact on customer experiences. A study conducted by Qiu et al. (2019) showed that the ability of robots to look and to act like humans as well as their intelligence were the main factors that made a connection with the customers. In that paper, the factors framework of human-robot interaction was depicted, where customers through social exchanges experienced employee–customer interactions and thus human–robot rapport indirectly. On the other hand, Yu (2019) along with Söderberg Granström et al. (2023) were to some extent negative as they pinpointed the problem of robot crossing into the uncanny valley leading to such reactions as discomfort with distrust. These opposing results indicate that anthropomorphism is a double-edged sword in the hotel context.

4.3.2 Usability, reliability, and performance

The first and foremost aspect of leading to a high level in service quality is the reliability of robots. The case studies of Henn-na Hotel clearly show how the development of technical failure (for example mistakes in speech recognition and inability to respond to guest's needs) caused the hotel's customers' frustration and higher dependence on the intervention of the hotel staff (Bhimasta & Kuo, 2019; Söderberg Granström et al., 2023). On the contrary, cleaning and delivery robots such as Relay or Hospi have been confirmed to be reliable in their sterilization and logistics works and have been inseparable from guests' preference for following this line of service rather than trying something new with them (Müller & Kutzbach, 2022).

4.3.3 Communication style and interaction quality

Effective interaction with the user does not only depend on the construction but it also includes the means by which the user and the system communicate. Torrey et al. (2013) unveiled the fact that the robots that employed hedges and discourse markers in their utterances were regarded as more empathetic and more likeable, and in some conditions, they even won over human assistants in performance. Besides, Canal et al. (2016) have shown that hand movement recognition (such as waving, pointing) greatly enhanced the feeling of naturalness of HRI in a hotel situation, providing technology-enabled service. These studies present that the way verbal and non-verbal communication occurs in a situation interaction is important for the customer's perception of service quality.

4.3.4 Service consistency versus personalisation

Robots are good for consistency, accuracy, and hygiene; however, they still have a hard time with personalisation and empathy. According to Naumov (2019), RAISA technologies may be the reasons for increasing the company's competitive advantages in terms of operational efficiency. However, they risk lowering the service quality if they do not provide personalised experiences. To put it another way, Huang and Rust (2018) believe that robots are going through the task domains (mechanical → analytical → intuitive → empathetic) fast; however, empathy is still a human advantage in hospitality. Therefore, service quality results depend on the balance between robotic efficiency and human warmth.

4.3.5 Research gaps

The issues have been discussed in great detail, yet there are some research gaps. The number of studies that try to quantify the impact of utilising service robots on service quality by employing standard frameworks like SERVQUAL is minimal. There are few longitudinal field studies on the reliability of robots since most of the data are from controlled experiments or single-case analyses (e.g., Henn-na Hotel). Besides this, there is also little help in understanding how the cultural expectations of service affect the perception of robots. quality - a significant gap in the hospitality sector, which is a global industry.

Table 7. Thematic summary (Technological Attributes & Service Quality)

Guiding question	Articles covered	Key findings
How does anthropomorphism affect guest perceptions?	Qiu et al. (2019); Belanche et al. (2020); Yu (2019)	Human-likeness can improve rapport and affinity, but risks uncanny valley effects.
How does reliability influence service quality?	Bhimasta & Kuo (2019); Müller & Kutzbach (2022)	Guests value reliability and hygiene performance more than novelty; failures cause dissatisfaction.
How do communication styles shape interaction quality?	Torrey et al. (2013); Canal et al. (2016)	Polite language and gesture recognition improve perceived warmth and usability.
How do robots impact service consistency vs personalisation?	Naumov (2019); Huang & Rust (2018)	Robots ensure efficiency and hygiene but struggle with personalisation and empathy.

Technologicalfeatures explain how robots carry out their tasks in real service environments, but their successful implementation is a result of managerial decision-making, organizational strategy, and long-term planning. Any robots, even the ones with the highest level of technology, can turn out to be unsuccessful if there is no proper leadership, investment, staff training, and alignment with service culture. On the other hand, managerial decisions, like deciding on the most suitable tasks for robots, putting a positive spin on adoption, and keeping a balance between automation and human interaction, have the power of technological benefits raised to a higher level. Therefore, the last theme is related to the managerial and strategic challenges of robotics adoption and how hotels can use robots effectively and .

4.4 Theme 4 — Managerial and Strategic Implications

On one side, robots will bring efficiency, hygiene, and cost savings; on the other hand, their application will raise some implementation strategy, morality, and long-term sustainability questions. The literature reviewed provides a synopsis of three key themes in the integration of hotel robots: adoption decision and economic consideration; leadership and change management; and ethical and strategic challenges.

4.4.1 Adoption strategies and economic considerations

Where, when, and how to use robots is a big managerial problem. Ivanov and Webster (2019) introduced the concept of RAISA (Robotics, AI, and Service Automation) as a framework for decisions on the use of robotics, indicating that advantages in the form of both money and other than money values have to be weighed by managers. Although robots are a labor-saving tool and precision-enhancing tool, they still will require a high initial investment, regular upkeep, and standardization with older machines, etc., Bowen and Morosan (2018) project that by 2030, robots can make up to 25% of the hospitality workforce; this will not only make the industry more efficient but also be disruptive by creating it in a new way. The case of Henn-na Hotel is cited as a source of evidence for the dangers of over-automation: the negative consequences of the lack of human support staff and thus guest and customer service quality were the cause of the hotel's reputational decline (Söderberg Granström et al., 2023; Bhimasta & Kuo, 2019). Hence, the acceptance of strategies should be not only the new but also the operational resilient ones.

4.4.2 Leadership, change management, and training

Leadership management heavily influences the process by which technology will be integrated in the routine of the company. Tuomi et al. (2020) highlight that moral leadership is a must to bring about robot deployment that is in tune with customer expectations and employee well-being. Ye et al. (2022) on the other hand, also share the opinion that managers should think about the location and method of implementing robotization to get the most benefits out of it and at the same time not to lose the service culture aspect. Moreover, the involvement of employees in decision-making processes helps to develop employee psychological ownership, which is a factor that lowers resistance and increases the interface between humans and robots (Fu et al., 2022; Xu et al., 2025).

4.4.3 Ethical and strategic challenges

In addition to worries about the smooth functioning of enterprises, the use of robots in business also puts the management team in front of a handful of ethical and strategic decisions. There have been numerous instances where intellectuals have come up with examples of potential problems that can accompany technology, e.g., the reason for job loss (Huang & Rust, 2018), the lack of humanity in organisations (Xu et al., 2025), and the decrease in the quality of personalised services (Naumov, 2019). Data protection and hacking are two of the things that are added to the list of potential risks and concerns since the robots are dealing with the most private data of hotel guests (Chan et al., 2022). On top of that, the main task of the managers, in this case, is not only to be aware of the market positioning implications but also to seize the strategic opportunities. For instance, Belanche et al. (2020) demonstrate that customer loyalty increases when guests perceive robot adoption as a means of service improvement, while the opposite happens if the latter is seen as cost-cutting. The question that managers have to answer is which frame is to be used to tell the story of robot adoption. Besides, regional differences are important to consider as well, since although maturity in robot adoption has made the Asia-Pacific lead (Zeng et al., 2020), Western markets are still reluctant, meaning strategies should be adjusted for culture and market.

4.4.4 Future-oriented strategies

Managers looking to the future are recommended to adopt mixed human-robot models as the preferred option over systems that are entirely automated. In such collaborative scenarios, as a result of robots taking over the execution of repetitive and hygiene-sensitive tasks (for instance, housekeeping, delivery), humans would be in control of the complex and empathic functions (e.g., concierge, guest relations). Researchers also suggest that hotels combine robotics with other rapidly incubating technologies such as IoT, predictive analytics, and the Metaverse (Yan et al., 2024). These measures do not depict robots as substitutes but as facilitators of hotel innovations and their capacities to withstand crises.

Table 8. Thematic summary (Managerial & Strategic Implications)

Guiding question	Articles covered	Key findings
What are the main economic implications of robot adoption?	Ivanov & Webster (2019); Bowen & Morosan (2018)	Robots reduce costs and improve efficiency, but require high capital investment and create risks of over-automation.
How can leadership and training support adoption?	Tuomi et al. (2020); Ye et al. (2022); Fu et al. (2022)	Ethical leadership, transparent communication, and employee upskilling reduce resistance and improve adoption outcomes.
What ethical and strategic issues arise?	Huang & Rust (2018); Xu et al. (2025); Naumov (2019)	Risks include job displacement, organisational dehumanisation, and reduced personalisation; managers must manage framing and cultural fit.
How should hotels approach future adoption?	Yan et al. (2024); Belanche et al. (2020)	Hybrid human–robot models and integration with IoT and analytics ensure sustainability and innovation.

The four themes that were identified in this review—consumer perspectives, employee perspectives, technological attributes, and managerial implications—are at the same time the areas where the potentials and the limitations of robotic adoption in hotel services can be seen. Individually, they show the impact of robotics on housekeeping, reception, and concierge activities. This part integrates these findings, points out contradictions, and briefly outlines the contributions of this review to both theory and practice.

5.1 Contradictions Across Themes

One of the contradictions that stands out is between consumer acceptance and employee resistance. Consumer studies (Kim et al., 2020; Binesh & Baloglu, 2023; Palrão et al., 2023) record that guests, and especially Gen Z travellers, are curious and optimistic when robots are brought into the picture and particularly if they are introduced as hygiene-enhancing and innovative tools. In contrast, employees are reported to perceive robots as the cause of the loss of their jobs, work overload, and less human interaction. (Fu et al., 2022; Xu et al., 2025). Therefore, guest satisfaction may be improved by robots under certain conditions; however, their success is mainly dependent on the willingness and the ability of the employees to work together with them.

The other contradiction is with anthropomorphism, whereby some studies show that human-likeness to a robot helps the users to like it and even form a bond with it (Qiu et al., 2019; Belanche et al., 2020), but, for example, the Henn-na Hotel case indicates that too much anthropomorphism can become a source of irritation and understatement of expectations (Yu, 2019; Söderberg Granström et al., 2023). This contradiction is an aspect of identifying the underlying theme of context sensitivity for design decisions: anthropomorphism is accepted most when functioning is correct, but rejected when shortcomings are revealed.

Managerial framing is the third tension to be dealt with.

Passengers react positively when innovation with robots is presented as a step to improve the service, but present an adverse reaction if the issue is designated to cutting expenses (Belanche et al., 2020). Nevertheless, as seen from a company's perspective, cutting costs is usually the main reason for the work to be automated (Ivanov & Webster, 2019). The conflict between the purposes of management and the way consumers see it presents a strategic dilemma for hoteliers.

5.2 Cross-cutting Insights

Several insights cut across the themes:

- Reliability over novelty: Guests prioritise reliability, hygiene, and consistency over novelty. Robots that fail technically (e.g., at Henn-na Hotel) damage service quality regardless of design features.
- Task–fit alignment: The use of robots is most effective when performed in tasks that are standard, repetitive, and sensitive to personal hygiene (e.g., housekeeping, delivery). Professions that are dependent on the emotional aspect and require the empathy of the human being, at least for now, are the ones that will be kept for the human beings...
- Generational dynamics: Gen Z turns out to be a consumer segment that embraces technology and an adaptable employee group. Since they are digital natives, acceptance and collaboration are made easy, which leads to the conclusion that the hotels that make Gen Z their guests and staff will have less...
- Hybrid models prevail: According to the majority of the sources, robots should be considered as complements to the human staff, and not as replacements. A combined human–robot team operation is a sure way of achieving one's goal of efficiency while at the same time not losing the emotional depth of...
- Technology Acceptance Model (TAM) & Extensions: The results found are in accordance with the idea behind TAM, which is heavily reliant on the perception of the user of the usefulness and ease of use of the technology, but point out new factors such as situational threats (e.g., COVID-19) and generational profiles (Gen Z) that may influence them.
- Human–Robot Interaction (HRI): The review extends HRI theory by identifying communication style (e.g., politeness, gesture recognition) as a key factor that affects acceptance in hotel contexts (Torrey et al., 2013; Canal et al., 2016).

- Job Replacement and Dehumanisation Theories: The set of data Xu et al. (2025) and Huang & Rust (2018) provide as an example is the transformation of automation of hospitality work only; as a result, the authors are moving along the concepts of deskilling/upskilling and organisational dehumanisation.
- Service Quality Frameworks (SERVQUAL/Smart Services): According to the performance of robots, they are strong in dimensions such as reliability and tangibles but weak in empathy and responsiveness, thus they suggest the need for revised frameworks that could integrate the metrics of technological performance.

5.4 Practical Contributions

The results provide the following hands-on strategies for the practitioners:

- Adopt fit-for-task strategies: Use robots in areas where they add clear value (housekeeping, delivery, sanitation) while keeping humans in tasks rich in emotional content (concierge, guest relations).
- Frame adoption positively: Rather than using robot deployment to save costs, communicate it as a service improvement to the guests, which will eventually make them perceive it better.
- Firstly, to reskill employees: The reskilling programs will successfully overcome the employees' resistance and assist.
- Moreover, create a marketing campaign for Generation Z: Present the stars of devices as solutions to the needs of Generation Z consumers, and use their digital abilities as workforce members to fuel the transformation.
- Pursue models blending the best of both worlds: Use a human with the necessary business and operational skills and the robot for the task that requires routine to reach a balance in operational efficiency and customer delight.

5.5 Research Gaps Emerging from Synthesis

Based on the comparison of the four major themes, several gaps in the research become apparent:

- Longitudinal evidence: Most studies are cross-sectional, meaning they only provide a snapshot of the time, and only a few of them can track the changes in attitudes over a period.
- Cross-cultural comparisons: Most of the research focuses mainly on the Asia-Pacific region, and there is a lack of studies comparing regions.

- Service quality measurement: Only a few studies are concerned with applying the existing scales, such as SERVQUAL, in robotics.
- Economic and ethical analysis: Areas such as financial ROI, job security, and data privacy need further research.
- While some studies pay attention to the younger generation, the systematic examination of Generation Z as consumers and employees remains largely untouched.

5.6 Summary of Discussion

The synthesis of the reviewed studies reveals several overarching issues and emerging trends shaping the trajectory of service robotics in hotel operations. First, health crises such as COVID-19 have profoundly altered consumer acceptance patterns, with guests showing significantly higher preference for robot-staffed services during periods of heightened perceived threat (Kim et al., 2020) . Second, hotels face increasing challenges in balancing technological efficiency with authentic, human-cantered guest experiences, particularly in high-contact services where robots struggle to replicate empathy and emotional intelligence (Naumov, 2019; Qiu et al., 2019) . Third, employee-related issues—including techno-insecurity, workload strain, and organisational dehumanisation remain significant barriers to sustained robotic adoption (Fu et al., 2022; Xu et al., 2025). Fourth, technological reliability continues to determine success or failure, as evidenced by the adoption breakdowns at Henn-na Hotel (Bhimasta & Kuo, 2019; Söderberg Granström et al., 2023). Finally, demographic shifts highlight Generation Z as a crucial segment influencing future hospitality innovation, due to their high technological comfort and willingness to engage with smart and automated service environments (Palrão et al., 2023; Binesh & Baloglu, 2023)

6. DIRECTIONS FOR FUTURE RESEARCH

6.1 Longitudinal and Real-world Evidence

The review has found that most current studies are based on the results of cross-sectional surveys or scenario-based experiments (Kim et al., 2020; Belanche et al., 2020). Besides being valuable, these designs merely reflect the attitudes at a particular time but fail to show the evolving patterns of the attitudes over time. The research in the future needs to:

- Carry out longitudinal studies monitoring changes in the perceptions of guests and employees due to their repeated interactions with robots.
- Implement real-world experiments in hotels in operation to examine if what people say matches what they actually do.
- Investigate if the use of technology is experiencing fatigue, thus whether the rate of acceptance driven by the novelty will decline over time.

6.2 Cross-cultural and Regional Comparisons

The majority of the publications in this field are centred in the Asia-Pacific region (e.g., Japan's Henn-na Hotel), with only a few studies from Europe, North America, or the Global South (Zeng et al., 2020; Ivanov et al., 2019). Nevertheless, cultural values profoundly impact hospitality expectations and trust in technology. The following tasks should be included in future research:

- Frame the degree to which cultural norms affect people's acceptance of robots, such as a collectivist versus an individualist society.
- Through the differences in labour market conditions and technology infrastructure, analyse the adoption of robotics in emerging markets.
- Review how cultural communication can impact comfort with humanlike or utilitarian robot types.

6.3 Integrating Service Quality Frameworks

Robots are a challenge to traditional service quality models, like SERVQUAL, which is highly attributed to empathy and responsiveness (Naumov, 2019). Future studies should:

- Create robot-centred scales that measure usability, reliability, and hygiene performance to rate service quality.
- Examine hybrid service encounters where humans and robots jointly deliver value.
- Assess the questioning of whether technology-operated services are rethinking "hospitality" as a concept—potentially leading to a new "Tech-Hospitality Quality Framework".

6.4 Ethical, Economic, and Sustainability Dimensions

Many studies highlight the issues of ethics and economics in literature, but the research that proves these claims is limited (Huang & Rust, 2018; Xu et al., 2025). The authors in the field should consider:

- Hotel robot ROI should be explored in the context of different types of hotel properties (budget vs. luxury).
- First, study technology implementation's positive and negative effects on worker well-being and job security. Also consider the new social divide and inequality in the labour market issue.
- Look at potential issues associated with data privacy and cybersecurity, which arise as robots gain access to the personal information of hotel guests.
- Talk about the consequences for the environment of robotic adoption—energy use, lifecycle costs, and compatibility with SDGs (Sustainable Development Goals) (Yan et al., 2024).

6.5 Generation Z as Guests and Employees

This paper emphasises the uniqueness of Gen Z as a demographic that significantly influences the hotel robotics future (Palrão et al., 2023; Binesh & Baloglu, 2023). On the other hand, there is a considerable gap between current research and the understanding of this group. Subsequent investigations should:

- Delve into the impact of Gen Z's technology-savviness on the level of consent for employing robots in hotels.
- Gain insights into the career outlook of Gen Z in the context of automating the work environment - are they viewing robots as partners, rivals, or mere instruments of convenience?
- Perform a detailed study of cross-generational differences (Gen Z versus Millennials, Gen X, Boomers) to predict the possible life cycles of acceptance.
- Scrutinise Gen Z shopper tendencies - whether they are ready to pay more to receive tech-driven hospitality services.

6.6 Emerging Technologies and Hybrid Models

The integration of robots with AI, IoT, predictive analytics, and Metaverse platforms is deepening exponentially (Chan et al., 2022; Yan et al., 2024). Future research should:

- Investigate the methods with which hybrid service models (human + robot) achieve maximum efficiency and personalisation.
- Explore the development of AI-enhanced emotional intelligence in robots and its impact on guest satisfaction.
- Identify the compatibility aspects between robotics and IoT ecosystems in smart hotels.
- Dream about far-out issues: what do you think about the synergy of digital twins or virtual concierge avatars with physical robots?

Table 9. Future Research Agenda

Gap Identified	Suggested Research Question	Proposed Method/Context
Lack of longitudinal evidence	How do guest and employee attitudes toward robots evolve over repeated interactions?	Longitudinal field studies in operational hotels
Limited cross-cultural analysis	How do cultural values (e.g., collectivism vs. individualism) affect robot acceptance?	Cross-country comparative surveys & experiments
Service quality measurement gap	How can SERVQUAL be adapted to robot-enabled services?	Scale development & validation studies
Ethical & economic challenges	What are the ROI, privacy, and well-being implications of robot adoption?	Mixed-method case studies; employee/guest interviews
Gen Z perspectives	How do Gen Z guests and employees shape the trajectory of hotel robotics adoption?	Generational comparative studies (Gen Z vs. others)
Hybrid & emerging tech	How do IoT, AI, and Metaverse platforms enhance robot–human collaboration?	Conceptual models & pilot implementations in smart hotels

6.7 Summary of Future Directions

To summarise, future research should concentrate more on the sustainable use of robots in hotel services rather than their mere novelty. Besides sustainability, researchers should also consider the aspects of inclusivity and generational readiness. Generation Z, being technologically savvy and digitally adept travellers, will probably become the main drivers of the robotisation process, not only as consumers but also as future employees of the hospitality sector. Consequently, the need for well-designed, long-term, intercultural, and morally sensitive studies that can ascertain that the use of robots in hotels will facilitate smooth operations and safeguard human values in hospitality cannot be overemphasised.

7. CONCLUSION

Thispaper presents a systematic review of 45 peer-reviewed studies that focus on the use of robotics and other related technologies in hotel services. The review employed the PRISMA 2020 framework to integrate descriptive trends and thematic insights across four domains: consumer perspectives, employee perspectives, technological attributes, and managerial implications. In essence, these themes reveal the benefits as well as the challenges that the hospitality industry encounters when using robotic solutions.

The consumers' point of view shows that their acceptance of service robots largely depends on the situation. As a rule, guests, and more specifically Generation Z, are found to be most receptive to the use of robots for the performance of standardized tasks, especially those related to hygiene, which is a trend that has been strongly supported during the COVID-19 pandemic. On the other hand, guests are still reluctant to replace human staff in areas where they have a high level of contact such as concierge assistance or reception. Research consistently points to a certain uneasiness with robots that are too humanlike, thus demonstrating the "uncanny valley" effect that is well known in the field of human–robot interaction. Consumers value efficiency and the newness of the service, however, they still prioritize empathy, personalization, and emotional labor as traditionally provided by human employees.

The insights into employees' views also bring to light a very complicated picture of the situation. It is true that robots can help in lessening the repetitive tasks and upgrading the efficiency of operations, yet, a considerable number of frontline workers express apprehension of losing their jobs, of losing the ability to connect with people on a personal level, and of having increased stress in case they are asked to resolve technical problems. When employees are in charge of handling the situations where robots fail even though these problems are out of their control, they are usually the ones held responsible. However, the younger workforce, most notably the Gen Z members, is generally more technologically savvy and therefore tends to get used to it quicker. This generational change signals that workforce planning in the future should not only consider training and reskilling, but also take into account the human elements that go hand-in-hand with automation.

Technological aspects are the main factors that influence the quality of service. Visitors are more concerned with the dependability, convenience, and cleanliness of the technology rather than the novelty or anthropomorphic aspects. The user experience is improved by polite communication styles, easy gesture recognition, and quick interaction flows; however, trust can be lost very quickly if, for example, there is poor voice recognition. The results of this study highlight the fundamental compromise made by robots: on the one hand, they are very good at providing consistent and efficient service, but on the other hand, they still have difficulty in areas that require empathy and personalization, which are two things that are very important in the hospitality industry.

Managerial implications are mainly about strategic, ethical, and communication considerations. The success of robot integration is not only dependent on the performance of the technology but also on the way hotels present and manage these changes. The introduction of robots as agents for the betterment of the service leads to the satisfaction of the guests, while the presentation of this innovation as a means of cutting down costs causes the colleagues and the guests to be sceptical of the technology. Hence, it is very important that management be effective, communication be open and staff be engaged in the process. The co-existence of humans and robots, in such a way that the robots take care of the mundane tasks while the humans provide the relational service, seems to be the most viable solution in the long run.

The review theoretical provides the extension of the Technology Acceptance Model (TAM) by adding situational moderators like pandemics and generational effects. Moreover, it supports the HRI theory by showing that guest perceptions are affected by micro-level communication strategies and deepens the discussions of job replacement and organizational dehumanization. In terms of practical implications, the findings serve as a roadmap for hotels to determine the areas where robots generate the most value, the ways to help staff adjust, and the strategies for attracting and engaging Gen Z as both future consumers and employees.

Future research can benefit from implementing longitudinal designs, cross-cultural comparisons, and ethical evaluations of automation. The changing expectations of Gen Z will be instrumental in determining what the next stage of robotic hospitality will look like. Ultimately, although robots possess transformative power, the future of hotel service lies in hybrid models that can strike a balance between technological efficiency and human empathy.

REFERENCES:

Acosta, L., González, E., Rodríguez, J., Hamilton, A., Méndez, J., Hernández, S., Sigut, M., & Marichal, G. (2006). Design and implementation of a service robot for a restaurant. *International Journal of Robotics and Automation, 21*(4). Advance online publication. DOI: 10.2316/Journal.206.2006.4.206-2909

Alsheibani, S., Messom, C., & Cheung, Y. (2020). Re-thinking the competitive landscape of artificial intelligence. *Proceedings of the . . . Annual Hawaii International Conference on System Sciences/Proceedings of the Annual Hawaii International Conference on System Sciences*. DOI: 10.24251/HICSS.2020.718

Belanche, D., Casaló, L. V., & Flavián, C. (2020). Frontline robots in tourism and hospitality: Service enhancement or cost reduction? *Electronic Markets, 31*(3), 477–492. DOI: 10.1007/s12525-020-00432-5

Belanche, D., Casaló, L. V., Flavián, C., & Schepers, J. (2019). Service robot implementation: A theoretical framework and research agenda. *Service Industries Journal, 40*(3–4), 203–225. DOI: 10.1080/02642069.2019.1672666

Bhardwaj, P. (2018, July 5). *Robots are replacing servers and delivering fresh seafood to people's tables at Alibaba's high-tech restaurant in Shanghai*. Business Insider. https://www.businessinsider.com/alibaba-shanghai-restaurant-robothe-robots-waiters-photos-2018-7

Bhimasta, R. A., & Kuo, P. (2019). What causes the adoption failure of service robots? *UBICOMP*, 1107–1112. DOI: 10.1145/3341162.3350843

Binesh, F., & Baloglu, S. (2023). Are we ready for hotel robots after the pandemic? A profile analysis. *Computers in Human Behavior, 147*, 107854. DOI: 10.1016/j.chb.2023.107854 PMID: 37389284

Blöcher, K., & Alt, R. (2020). AI and robotics in the European restaurant sector: Assessing potentials for process innovation in a high-contact service industry. *Electronic Markets, 31*(3), 529–551. DOI: 10.1007/s12525-020-00443-2

Bowen, J., & Morosan, C. (2018). Beware, hospitality industry: The robots are coming. *Worldwide Hospitality and Tourism Themes, 10*(6), 726–733. DOI: 10.1108/WHATT-07-2018-0045

Breazeal, C. (2003). Toward sociable robots. *Robotics and Autonomous Systems, 42*(3–4), 167–175. DOI: 10.1016/S0921-8890(02)00373-1

Cain, L. N., Thomas, J. H., & Alonso, M.Jr. (2019). From sci-fi to sci-fact: The state of robotics and AI in the hospitality industry. *Journal of Hospitality and Tourism Technology*, *10*(4), 624–650. DOI: 10.1108/JHTT-07-2018-0066

Canal, G., Escalera, S., & Angulo, C. (2016). A real-time Human-Robot Interaction system based on gestures for assistive scenarios. *Computer Vision and Image Understanding*, *149*, 65–77. DOI: 10.1016/j.cviu.2016.03.004

Chan, L., Hogaboam, L., & Cao, R. (2022). Artificial intelligence in tourism and hospitality. In *Applied innovation and technology management* (pp. 213–229). DOI: 10.1007/978-3-031-05740-3_14

Fu, S., Zheng, X., & Wong, I. A. (2022). The perils of hotel technology: The robot usage resistance model. *International Journal of Hospitality Management*, *102*, 103174. DOI: 10.1016/j.ijhm.2022.103174 PMID: 35095168

Garrell, A., Villamizar, M., Moreno-Noguer, F., & Sanfeliu, A. (2017). Teaching robots' proactive behaviour using human assistance. *International Journal of Social Robotics*, *9*(2), 231–249. DOI: 10.1007/s12369-016-0389-0

Ghosh, S., & Pherwani, J. (2015). Designing a natural voice assistant for mobile through a user-centred design approach. In *Lecture Notes in Computer Science* (pp. 320–331). DOI: 10.1007/978-3-319-20901-2_29

Gong, L., & Nass, C. (2007). When a Talking-Face Computer Agent is Half-Human and Half-Humanoid: Human Identity and Consistency Preference. *Human Communication Research*, *33*(2), 163–193. DOI: 10.1111/j.1468-2958.2007.00295.x

Huang, M., & Rust, R. T. (2018). Artificial intelligence in service. *Journal of Service Research*, *21*(2), 155–172. DOI: 10.1177/1094670517752459

Ivanov, S., Gretzel, U., Berezina, K., Sigala, M., & Webster, C. (2019). Progress on robotics in hospitality and tourism: A review of the literature. *Journal of Hospitality and Tourism Technology*, JHTT-08-2018-0087. Advance online publication. DOI: 10.1108/JHTT-08-2018-0087

Ivanov, S., & Webster, C. (2019). Economic fundamentals of using robots, artificial intelligence, and service automation in travel, tourism, and hospitality. In *Emerald Publishing Limited eBooks* (pp. 39–55). DOI: 10.1108/978-1-78756-687-320191002

Kabadayi, S., Ali, F., Choi, H., Joosten, H., & Lu, C. (2019). Smart service experience in hospitality and tourism services. *Journal of Service Management*, *30*(3), 326–348. DOI: 10.1108/JOSM-11-2018-0377

Kaushik, A. K., Agrawal, A. K., & Rahman, Z. (2015). Tourist behaviour towards self-service hotel technology adoption: Trust and subjective norm as key antecedents. *Tourism Management Perspectives*, *16*, 278–289. DOI: 10.1016/j.tmp.2015.09.002

Kim, S., Kim, J., Badu-Baiden, F., Giroux, M., & Choi, Y. (2020). Do you prefer robot service or human service in hotels? Impacts of the COVID-19 pandemic. *International Journal of Hospitality Management*, *93*, 102795. DOI: 10.1016/j.ijhm.2020.102795 PMID: 36919174

Kumar, S., Poudyal, A., Choudhury, S., & Jha, A. (2025). Analysing customer preferences for AI-Enabled service robots' physical appearance in the tourism and hospitality industry: Insights and Exploration. *Tourism Planning & Development*, 1–18. DOI: 10.1080/21568316.2025.2461008

Müller, C., & Kutzbach, N. (2022). World Robotics 2022 – Service Robots. In B. Graf & K. Pfeiffer (Eds.), *IFR Statistical Department, VDMA Services GmbH*. VDMA Services GmbH. https://ifr.org/img/worldrobotics/Executive_Summary_WR_Service_Robots_2022.pdf

Naumov, N. (2019). The impact of robots, artificial intelligence, and service automation on service quality and service experience in hospitality. In *Emerald Publishing Limited eBooks* (pp. 123–133). DOI: 10.1108/978-1-78756-687-320191007

Nomura, T., Kanda, T., Suzuki, T., & Kato, K. (2008). Prediction of human behaviour in human-robot interaction using psychological scales for anxiety and negative attitudes toward robots. *IEEE Transactions on Robotics*, *24*(2), 442–451. DOI: 10.1109/TRO.2007.914004

Palrão, T., Rodrigues, R. I., Madeira, A., Mendes, A. S., & Lopes, S. (2023). Robots in tourism and Hospitality: The perception of future professionals. *Human Behavior and Emerging Technologies*, *2023*, 1–12. DOI: 10.1155/2023/7172152

Qiu, H., Li, M., Shu, B., & Bai, B. (2019). Enhancing hospitality experience with service robots: The mediating role of rapport building. *Journal of Hospitality Marketing & Management*, *29*(3), 247–268. DOI: 10.1080/19368623.2019.1645073

Rodríguez-Lizundia, E., Marcos, S., Zalama, E., Gómez-García-Bermejo, J., & Gordaliza, A. (2015). A bellboy robot: Study of the effects of robot behaviour on user engagement and comfort. *International Journal of Human-Computer Studies*, *82*, 83–95. DOI: 10.1016/j.ijhcs.2015.06.001

Söderberg Granström, D. E. A. D. E. A., Pronk, A. M., & Criscione-Naylor, N.. (2023). Robotic services in the hotel industry: An examination of Henn-Na Hotels. In *International Journal of Gaming Hospitality and Tourism* (Vol. 1). https://stockton.edu/light/documents/ijght_vol.3no.1/robotic_services_henn_na_hotels-6.7.23.pdf

Torrey, C., Fussell, S. R., & Kiesler, S. (2013). How a robot should give advice. *2016 11th ACM/IEEE International Conference on Human-Robot Interaction (HRI)*, 275–282. DOI: 10.1109/HRI.2013.6483599

Tuomi, A., Tussyadiah, I. P., & Stienmetz, J. (2020). Applications and implications of service robots in hospitality. *Cornell Hospitality Quarterly*, *62*(2), 232–247. DOI: 10.1177/1938965520923961

Walters, M. L., Syrdal, D. S., Dautenhahn, K., Boekhorst, R. T., & Koay, K. L. (2007). Avoiding the uncanny valley: Robot appearance, personality and consistency of behaviour in an attention-seeking home scenario for a robot companion. *Autonomous Robots*, *24*(2), 159–178. DOI: 10.1007/s10514-007-9058-3

Xu, T., Zheng, Y., Zhang, J., & Wang, Z. (2025). "They" threaten my work: How AI service robots negatively impact frontline hotel employees through organisational dehumanisation. *International Journal of Hospitality Management*, *128*, 104162. DOI: 10.1016/j.ijhm.2025.104162

Yan, Z., Konar, R., & Balasubramanian, K. (2024). The future of hotel robotics, AI, and service automation in practice. In *Advances in e-business research series* (pp. 249–278). DOI: 10.4018/979-8-3693-9699-5.ch011

Ye, H., Sun, S., & Law, R. (2022). A review of Robotic applications in Hospitality and tourism research. *Sustainability (Basel)*, *14*(17), 10827. DOI: 10.3390/su141710827

Yu, C. (2019). Humanlike robots as employees in the hotel industry: Thematic content analysis of online reviews. *Journal of Hospitality Marketing & Management*, *29*(1), 22–38. DOI: 10.1080/19368623.2019.1592733

Zeng, Z., Chen, P., & Lew, A. A. (2020). From high-touch to high-tech: COVID-19 drives robotics adoption. *Tourism Geographies*, *22*(3), 724–734. DOI: 10.1080/14616688.2020.1762118

Zhang, J. (2018). Big data and tourism geographies – an emerging paradigm for future study? *Tourism Geographies*, *20*(5), 899–904. DOI: 10.1080/14616688.2018.1519719

KEY TERMS AND DEFINITIONS

Artificial Intelligence (AI): Systems that are enabled through computation to robots, to make them perceptive, capable of learning, and making decisions during guest interactions and hotel operations.

Contactless Service: A service approach that is primarily facilitated by automation and robotics to limit physical interaction, and is therefore, widely practiced during and after the COVID-19 pandemic.

Generation Z: A group of consumers and future hospitality professionals who are digitally native and show higher adaptability and openness toward robotic and automated service environments.

Human–Robot Interaction (HRI): The field that deals with the study of interaction between humans and robots. It mainly focuses on communication, acceptance, embodiment, and behavioural responses in hospitality contexts.

Robotics and Service Automation (RAISA): A comprehensive concept that explains the use of robotics, artificial intelligence, and automation in the service sector to improve operational efficiency and service delivery.

Service Robots: Autonomous or semi-autonomous machines that are capable of performing tasks that are normally done by hotel staff, such as delivery, cleaning, reception, and concierge functions.

Smart Hotels: Hotels that are enhanced by technology to be more efficient where robots, IoT systems, and intelligent devices collaborate to deliver seamless guest.

Technological Reliability: The degree of performance of a robot in a given task with respect to being trouble-free and accurate, thereby influencing the satisfaction of the guest and the workload of employees.

Chapter 3
The Convergence of Smart Hospitality and Robotics in Accommodation Operations

Gautam Tanwar
https://orcid.org/0009-0009-1815-6832
Bahra University

Vaibhav Verma
https://orcid.org/0000-0002-9041-7111
Bahra University, India

Ankit Shukla
https://orcid.org/0000-0002-7646-3517
Rajamangala University of Technology, Thanyaburi, Thailand

ABSTRACT

The convergence of robotics & smart hospitality in lodging operations is examined in this chapter, with a focus on how automation & artificial intelligence are changing housekeeping. It tracks the development of unsupervised service robots from manual & automated cleaning, focusing their operational responsibilities & contributions to consistency, efficiency & hygiene. It looks at the advantages & difficulties, as well as the technological, moral & social aspects of robotic implementation, using case studies from India & other countries. The chapter makes the case for a hybrid human & robot approach in which staff members may concentrate on client engagement &

DOI: 10.4018/979-8-3373-6541-1.ch003

providing high quality service while repetitive activities are automated. Data driven, AI-powered, multitasking robots that easily connect with IoT systems are expected to become common in the future. In the end it comes to the conclusion that human & robot partnership, in which technology boosts productivity while maintaining the empathy and adaptation is crucial for hospitality & will be the key to the future of housekeeping.

1 INTRODUCTION TO ROBOTICS IN ACCOMMODATION OPERATIONS

Hospitality has always been a fertile ground for technology adoption, and each cycle of technology innovation has redefined the guest experience and operations management. Robotics is the future of housekeeping. Robotics could be the next big leap after digitalisation, converting conventional housekeeping to labour rich service procedures, technology integrated functions in today's age. Academics characterise this transition as an element of a more general service automation continuum in which activities evolve from being manually performed, to machine assisted and, in the end, autonomous (Choi et al., 2020; Ivanov et al., 2020). This suggests that this acceleration and advent of service robots in hospitality operations should not be seen only as a result of technological progress, but also as a reaction to changing socio economic demands from the global shortage of cheap labour to higher hygiene expectations generated during COVID 19 (Fu et al., 2022). As hotels are changing their service model with automation and AI, housekeeping is where efficiency, cleanliness and innovation meet.

1.1 The Evolution of Automation

The road from manual housekeeping to robots in hospitality starts with mechanisation (e.g., vacuum cleaners, mechanised laundry), proceeds via digital automation IoT connected devices and ends with complete autonomous robots. Researchers suggest that the way in which the hospitality industry adopts robotics signals a paradigm shift from automation of tasks to autonomous agents (Ivanov, Seyitoğlu & Markova, 2020).

One key work observed this evolution: From early service kiosks and self check in machines to mobile robots doing back of house automation (Wan, Chan &Luo, 2020).. For example, the robotisation of hospitality has been expedited by, inter alia, labour cost driven pressures and health and hygiene considerations, especially after the COVID 19 pandemic, a trigger effect for robotic adoption in the hospitality industry (Wanet al., 2020). Therefore, the evolution can be considered in 3 stages:

- Manual human labour
- Mechanised/automated equipment
- Autonomous, connected and socially participating service robots.

From a theoretical viewpoint, this stream of literature is congruent with the "automation continuum" model in service operations research (Choi, Choi & Oh, 2019), as "a task can migrate from manual to semi automated or fully autonomous".

1.2 Defining the role of Robots

It is important to differentiate a full featured service robot (e.g., hotel housekeeping robots that are able to autonomously map, guide, detect obstacles and integrate with property systems as well as interact with guests or staff from an automation device such as a typical robotic vacuum, which will follow a fixed path. A service robot in hospitality research is described as a reprogrammable, multifunctional manipulator designed to move materials, parts, tools or specialised devices through various programmed motions for the performance of a variety of tasks (Yan, Konar & Balasubramanian, 2024).

This distinction finds empirical reaffirmation: *"Hotel managers considered robots to be fit for the 4D jobs, yet not appropriate for tasks demanding emotional intelligence or human rapport"* (Ivanov et al., 2020, Accordingly, the function of robotics is not to replace human labour, but to complement human processes by fulfilling predictable standardised work and empowering the human workforce to dedicate more time to complicated guest centric activity.

1.3 The Need for Robotics in Housekeeping

There are several converging factors which are pushing accommodation operations towards robotic housekeeping. These include: • Labour market pressures: Robotics is a desirable alternative in many hospitality areas due to labour shortages, high turnover, and rising wage alternatives (Ivanov et al., 2020). • Hygiene and guest expectations: Following the epidemic, there has been an increase in the need from visitors for frictionless operations and constant sanitation. According to research, visitors may favour robot assisted services because they feel safer (Wan et al., 2020). • Operational efficiency and consistency: By standardising cleaning procedures, lowering variability, and optimising room turnover times, robotics enables operators to increase output (Lee, 2022). • Technological developments and connectivity: The deployment of service robots is now more practical due to the increased accessibility of sensors, AI, and IoT. For instance, a review of 86 earlier studies on hospitality robotics revealed that robotics is becoming more feasible for front of house, housekeeping, and maintenance duties (Kim & Lee, 2023). When taken as a whole, these factors make a strong case for implementing robotics in housekeeping is not just as a novel idea, but also as a calculated reaction to changing operational, labour, and visitor expectations.

2 TYPES OF HOUSEKEEPING ROBOTS

The development of housekeeping robots includes a variety of machines that range from highly specialised to unique in terms of the mechanisms involved in bringing about a change for household tasks. As an example of the earliest and one of the most ubiquitous applications of automation in accommodations, consider floor cleaning robots. These robots use SLAM (simultaneous localisation and mapping), AI sensors, and adaptive cleaning patterns to ensure that they are able to clean the house in a shorter time frame, but consistently (Rosete et al., 2022). Besides inside operation: Here we heard also of the deployment of robot lawn mowers and garden helpers, a demonstration of how automation is moving outwards into property maintenance particularly in resorts (Lee 2022). On the specialist side, we see robots that fold laundry (never folding warm underwear again), wash windows and provide UVC disinfection services, reinventing the concept of service quality and hygiene security (Mukherjee et al., 2021). As a whole, these bots represent a layered ecosystem of automation in hospitality each one created not to replace human workers, but instead to boost operational dependability and quality of service.

2.1 Floor Cleaning Robots

Robotic floor cleaning is one of the most popular robotic applications in housekeeping. These are devices which use SLAM (simultaneous localisation and mapping), obstacle avoidance and programmatic cleaning cycles. Studies in hospitality suggest that cleaning time and variation of finish quality could be reduced by using such robots (Rosete et al., 2022). To illustrate, in some hotel applications, floor cleaning robots have been reported to save as much as 30–40% in cleaning time on corridors and public places versus exclusively manual cleaning (Hospitality Institute, 2024).

2.2 Lawn and Garden Robots

Hotels and resorts with extensive property have already started utilising the robots to mow lawns, control weeds, etc. While the outdoor aspect of robots has received less attention in hospitality literature, the same economics hold true automating the repetitive work in maintaining grounds will free up human labour for more guest facing activities, and maintain consistent aesthetics and safety on the property. The use of robots beyond guest area functions, including estate maintenance outside the confines, is an emerging frontier, as one review observes (Lee, 2022).

2.3 Laundry and Other Niche Robots

Outside of floor cleaning, hotels are looking into the use of robots for laundry room automation, window cleaning bots, minibar replenishment bots and room service trolley bots. Model driven studies on tidy up home robots (Taniguchi et al., 2020) lay the necessary groundwork for understanding how robots might engage with unstructured hotel room environments, finding objects, sorting linens, or organising supplies. The hospitality industry, too, has an example in India where robots are experimented upon for linen transportation, minibar restock and for delivery purposes (Raghuram, 2022).

2.4 Advanced Cleaning and Disinfection

At the top end of the market, roving robots are being set to work with UVC light disinfection systems, air purification modules, autonomous room sanitising functions and load balancing based on occupancy data. Studies indicate that these robots increasingly serve as guests perceived assurance of hygiene (Mukherjee et al., 2021). A literature review of hospitality robotics lamented the fact that one of, if not the fastest growing sub subdomain of COVID and post COVID disinfection robotics (Lee, 2022). Furthermore, the AI sensor inducing robots can adjust the cleaning frequency and path optimisation according to guest usage and surveillance data for environmental conditions (Kim & Lee, 2023). With the aid of this typology, you can position the broad range of robotic based products from Roombas to multi utility cleaning bots in the housekeeping services curve and add operational relevance for each.

3 OPERATIONAL IMPACT: PROS & CONS

Robotics in Housekeeping

The introduction of robotics in housekeeping offers a mix of operational efficiencies and organisational challenges. Evidence also implies that introducing robots in the workplace can help increase productivity increases up to 40%, standardise service standards and relieve workers from job related stress (Lee, 2022; Ivanov et al., 2022). These benefits are, however, moderated by technological, ethical and social issues. Challenges related zto the interoperability of systems, downtime and displacement workforce have become substantially important (Soares et al,2022). In addition, staff resistance (referred to as techno insecurity and role ambiguity in the literature) mostly prevents successful usage (Fu et al., 2022). Guest acceptance is also key; despite reliable and minimalistic robot performance, the friendliness or warmth offered by a person is valued by many customers related to top service quality (Seyitoğlu & Ivanov, 2023). So the future of housekeeping productivity will not be a matter of complete automation, but making that sweet spot complementarity between human flexibility and robot precision work for you, something which only strengthens your operational agility and hospitality factor.

3.1 Benefits for the Housekeeping Department

Housekeeping operations can benefit in a number of ways from robotic integration: • Efficiency and throughput: Robots can work before and after the typical human morning shift hours, maintain relatively consistent performance levels, and boost rooms cleaned per shift. This is supported by the promotion of the use of service robots leads to increased productivity (Lee, 2022). • Consistency and quality: As robots work through a programmed set of tasks, the variability introduced by fatigue or frustration among human workers is reduced. The end result can be that cleaning results are more consistent, reducing guest experience variability (Rosete et al., 2022). • Staff safety and ergonomics: physical tasks (such as corridor vacuuming, high rise window cleaning space constrained linen transportation) are handed to robots, allowing human staff to be reassigned to supervisory or guest facing roles (Ivanov et al., 2020). • Hygiene, security and Visitor confidence: Publicly exhibiting cleaning robots is not only found to be able to improve the guests' perception of hygiene important pandemic (Wan et al., 2020). Taken together, these advantages indicate that robotics can improve the effectiveness of an operation, the quality of services and enable human workers to concentrate on higher leveraged and value added activities.

3.2 Technical, Ethical and Social Implications

However, despite these benefits, there are many operational and social and ethical issues: • Technical obstacles: Hotels are, and I say this gently, the opposite of robots' native habitat, so to speak. Furniture gets moved around, people stand in the way, hallways can be narrow, and infrastructure is often outdated. Battery life, system downtime to perform maintenance, compatibility with property systems and risk reward trade-offs are other major issues (Kalsi 2023). • Cost and ROI issues: Robots have the potential to add efficiency, but with high upfront costs combined with maintenance, training, and risk of Obsolescence, return on investment may take years, especially in the sub segment properties (Soares et al., 2022). • Just what we need, a new attitude to workforce dislocation and employee resistance: one significant scholarly article offers a "Robot Usage Resistance Model" (RURM), noting that overload, techno insecurity and a lack of robot humanness can lead to staff resistance and less robot continuous usage (Ivanov et al., 2022). • Guest receptivity and human–robot interaction: Studies have indicated that guests would still choose humans over robots for emotional, social or interactive service activities, not able to withstand the efficiency of robots for standardised services (Seyitoğlu & Ivanov, 2023). • Ethical and privacy considerations: Guestroom collecting sensor data robots, navigating personal spaces, and autonomously performing tasks provoke questions of privacy, consent, data security and cultural acceptance, particularly when human interaction is and should be still key in hospitality (Humanoid Robots in Tourism and Hospitality, 2022). This limitation of robotic technology reinforces that robotics is not a plug and play solution, but rather the effective implementation requires addressing infrastructure, human factors, cultural fit and system design.

3.3 Human Robot Interaction and Integration

It will no longer be robots replacing human labour, but robots and human capabilities. Empirical evidence indicates that robots are most effective when they become part of the workflows, in which human employees are eligible to perform work with emotions, versatility and guest contact, whereas the tasks that are predictable and repetitive can be performed by robots (Ivanov et al., 2020). The literature advocates an augmented human model: robots augment staff, staff deal with exceptions, personalise service and provide a human touch. For example, a systematic review of service robotics for hospitality proposed the best solution to be a hybrid system where robots cooperate with human workers (Kim & Lee, 2023). Change management is vital: staff must be trained, their roles redefined, and internal resistance handled. Advances concerning guest robot HRI are another trend; for instance, analysing 50,000 reviews from TripAdvisor showed that robot sentiment was on the rise, albeit indicating preference towards human to staff ratios (Lee, 2022). In so many ways, in fact, that the human–robot interface is surely also a crucial frontier in making robotic housekeeping a reality yet not just when it comes to machine deployment and use, but also how people use, accept and supervise them.

4 CASE STUDIES IN HOUSEKEEPING ROBOTICS

Some of the international adoption of robotics in hospitality has led to mixed results, showing how automation can bring both promise as well as pain. A case in point that is especially rich in metaphorical use of humanisation for robots, etc., as well as utopian and dystopian future scenarios, can be found in the Henn-na Hotel in Japan: after much buzz about robotised service, shortcomings regarding adaptability, and to some extent maintenance brought them back to half personnel (Reis et al., 2020). Like minded experiments in Europe, the Middle East and India illustrate different motivations and challenges from a dearth of workers to cultural attitudes on automation (Kalsi, 2023). Among Indian hospitality, it is interesting to witness a pragmatic approach that selectively operates in the high end segment where robots are harnessed for linen transport, room delivery and UV ray disinfection; yet with a notable human vigil (Raghuram, 2022). The comparison study demonstrates that the hybrid system (robots do the monotonous work and human ones control emotional connection) achieves superior performance for guest satisfaction and cost effectiveness (Seyitoğlu & Ivanov, 2023). The case studies taken together reconfirm that robotic integration should be context bound, data driven, and culture sensitive.

4.1 Global Adoption of Robotic Housekeeping

From around the world, hotels are testing robotic housekeeping more and more but adopting slowly. For example, the famous case of Japan's Henn-na Hotel operated with robots also highlighted both potential and risks: while robots excelled at the basic standardised tasks, they failed at human interactions as well as adaptability and cost benefit (Reis et al., 2020) A larger literature 86 review of robotics in hospitality discovered that most research is currently around guest robot interaction or front desk services as opposed to back house cleaning (Lee, 2022).

4.2 Real world case studies: Operational changes, Return on Investment (ROI) and staff Feedback

Several studies report operational experiences: hotel managers describe that tasks most friendly towards robots can be described as repetitive and physical, responsibilities related to social capabilities remain human work (Ivanov et al., 2020). According to Return on Investment (ROI) case studies, some implementations do not perform as expected because of service and integration issues (Lee, 2022). One mixed methods study, examining housekeeping robotics at a regional hotel chain, reported resistance followed by increasing acceptance when staff roles were reframed

(Raghuram, 2022). The authors also reported that the return on investment was beyond 3–5 years, and its success depended on staff training and system reliability.

4.3 Robotics in the Indian Hotel Sector

In India, the uptake of housekeeping robots is at a nascent stage but rising rapidly. Managers at Indian hotels perceive robotics as a productivity lever, and identified Wi Fi connectivity, PMS (i.e. Property Management System) integration, lack of skills, seamless pollution cultural acceptance as major barriers in this context (Kalsi, 2023). Such a story has been exemplified in an Indian luxury hotel where robots that transport linen and replenish minibars have resulted in 25% time savings, but at the cost of extensive retraining and staff monitoring.

4.4 Comparative Analysis: Traditional versus Robotic Assisted Housekeeping

Traditional housework – by contrast, Traditional (i.e., in person) home cleaning has the aforementioned personal and human touch, but suffers from labour shortages, inconsistency and physical exhaustion. Robotic assisted devices offer reliability, scale and hygiene confidence, but can lack the human touch experience and are time consuming to set up and manage due to investment and change management required. Studies also indicate a hybrid model is best, with robots dealing with the day to day core work while staff take on nonstandard self focused guest activities, managing exceptions and emotional service delivery (Seyitoğlu & Ivanov, 2023). These case studies and comparative insights show that robotic housekeeping is possible, but success depends on operational design, human factors and context.

5 FUTURE DIRECTIONS AND CONCLUSION IN HOUSEKEEPING ROBOTICS

Over the next 10 years, isolated robot applications will mature into a connected system for a fully digitised factory. According to (Digital Transformation in Hospitality, 2023). Housekeeping of the future will be increasingly data driven, with sensors connected to IoT and predictive analytics determining what needs cleaning and where. Progress on AI and machine learning will enable robots to self optimise, adjusting routes, spotting anomalies, and becoming familiar with the preferences of guests (Shi et al., 2025). Robots that can multi task It will not only reframe the cost effectiveness equation, but also comprise a cleaning robot that can restock and maintain an environment autonomously (Kapur & Williams,2025). As this convergence advances, the success of issues relating to the ethical framework, the human robotic collaboration model and resilient workforce retraining will be ensured. In the end, this coming together of robotics and hospitality leads to a service model that meaningfully combines accuracy with empathy, technology with humanity and efficiency with experience.

5.1 Data Driven Cleaning

The next frontier in housekeeping is not just robotics but data driven robotics. Robots, occupancy sensors, guest flow analytics and IoT platforms deliver to dashboards that will enable predictive scheduling, prioritised cleaning (or not), dynamic workforce allocation and resource optimisation. Campaigns play an important role in imparting the right message digitally, but crafted content could make 2X impact, reverse engineering campaigns to cross publish and create digital org, event toolkits. Literature also confirms that hotels leveraging hotel tech or technology digitisation (AI, IoT) experienced a 30% improvement in efficiency during operations (Digital Transformation in Hospitality,2023). In housekeeping, this means robots that are trained on patterns: rooms of high or extended turnover get different cleaning; high foot traffic forces the robot to move more regularly; sensors in guest rooms provide data continuously so the robots and humans can optimise their schedule.

5.2 Advanced AI and Machine Learning

Robots will use more AI, machine learning and reinforcement learning. In home studies have shown that robots were able to grasp spatial sense and clean up effectively (Taniguchi et al., 2020). For the hospitality sector, this means robots that are able to: identify fallen objects, reconfigure themselves for different room configurations, utilise advanced sensors such as touch and force based detection in order to detect potential interference from hotel guests fine tune cleaning routes, collaborate with staff members, and display soft skills (Shi, Zhan &Lin, 2025). A further area of exciting research lies in human robot emotional interaction: robots equipped with anthropomorphic features, voice and expressive behaviour can potentially impact guest acceptance and staff trust (Ivanov et al., 2022).

5.3 The rise of the multi-tasking Robot

Isolated robots, vacuuming robot, laundry bot, concierge bot are only not the future; but multi-tasking with many arms to perform various housekeeping tasks, floor cleaning target, surface sanitising, disinfection, linen transport, etc., as well as minibar restocking and inventory monitoring all within one system is the actual

Future. This overlap simplifies the system and brings about a higher return on investment. The literature also suggests that robots could make up 25% of the hospitality labour force by 2030 in some instances, particularly for routine functions, Creating New Directions for Service Robots in Hospitality, 2024.

CONCLUSION

This chapter, therefore in briefly describes the migration of housekeeping from manual to robotics cleaning history and defines service robots, discusses the typology of housekeeping robots, evaluates the operational impact of these types, pros and cons, case studies with respect to the global scenario, Indian scenarios and outlines future projections. The takeaway for scholars and managers in hospitality: robotics is not just automation, it's a redesign of service delivery, demanding human robot synergy, data driven management and strategic deployment. In the era of smart hospitality, housekeeping will be characterised by the harmony between people and robots, facilitated by data. An assurance for efficiency, consistency, and guest satisfaction while retaining its core essence.

RECOMMENDATIONS:

- *Implement a hybrid service model*: Robotic in room amenities should be used for repetitive, risky tasks. Works by allowing staff to handle guest interaction and problem solving.
- *Invest in reskilling of the workforce*: Rather than investing together. Robot control and data analysis for improving human–robot collaboration.
- *Ethical and privacy considerations*: Prioritise the development of ethical and privacy data protection respect and provider policies, procedures that have been determined to be necessary. Guest authority to guarantee confidence in robotic procedures.
- *Interoperability Promotion*: Hoteliers should choose a robotics platform that is lawmakers and lobbyists to shape the continued development of drones; advocating for regulation and public support of sustainable, responsible drone use. Current property management systems (PMS) and Internet of Things (IoT).
- *Promote continuous assessment:* At the time of recruitment, managers should evaluate their employees through key performance profiles. Key Performance Indicators (KPIs), Return on Investment (ROI), guest satisfaction and staff adaptation to gauge success.

- *Encourage academic and industry collaboration*: Collaboration between academia, the robotics industry, and robotics stakeholders is key to developing a broad range of cutting edge robotic technologies. Companies and tourism service providers need to innovate according to local and operational contexts.
- *Automation to help meet sustainability objectives*: Schedulers to minimise environmental footprint and in line with green hospitality practices.

REFERENCES

Choi, Y., Choi, M., Oh, M., & Kim, S. S. (2020). Service robots in hotels: Understanding the service quality perceptions of human–robot interaction. *Journal of Hospitality Marketing & Management, 29*(6), 613–635. DOI: 10.1080/19368623.2020.1703871

Fu, S., Zheng, X., & Wong, I. A. (2022). The perils of hotel technology: The robot usage resistance model. *International Journal of Hospitality Management, 102*, 103174. DOI: 10.1016/j.ijhm.2022.103174 PMID: 35095168

Ivanov, S., Seyitoğlu, F., & Markova, M. (2020). Hotel managers' perceptions towards the use of robots: A mixed-methods approach. *Information Technology & Tourism, 22*(4), 505–535. DOI: 10.1007/s40558-020-00187-x

Ivanov, S., Seyitoğlu, F., & Webster, C. (2022). The economics of service robots in hospitality companies: Task automatability, operational impact and external drivers. *Service Industries Journal*. Advance online publication. DOI: 10.1177/13548166251362622

. Kalsi, M. (2023). Enhancing guest experience and operational efficiency in hotels through robotic technology—A comprehensive review. *International Journal for Multidimensional Research Perspectives.*

Kapur, P., & Williams, J. D. (2025). Balancing efficiency and human touch: The role of AI and robotics in hospitality. *Artificial Intelligence, Machine Learning, &. Robotics in Business, 1*(1), 138286. DOI: 10.32473/aimlrb.1.1.138286

Kim, H., & Lee, H. (2022). A review of robotic applications in hospitality and tourism research. *Sustainability (Basel), 14*(17), 10827. DOI: 10.3390/su141710827

Lee, K.-J. (2022). Effects of service robots in the hospitality and tourism sectors: A critical review and future research directions. *International Journal of Tourism & Hospitality Research, 36*(4), 21–33. DOI: 10.21298/IJTHR.2022.4.36.4.21

Lee, K.-J. (2024). Understanding service robots: A systematic literature review of hospitality employees' perspectives. *Tourism Management*. Advance online publication. DOI: 10.1016/j.tourman.2023.104456

Mukherjee, S., Baral, M. M., Venkataiah, C., Pal, S. K., & Nagariya, R. (2021). Service robots are an option for contactless services due to the COVID-19 pandemic in hotels. *Decision (Washington, D.C.), 48*(4), 445–460. DOI: 10.1007/s40622-021-00300-x

Raghuram, V. (2022). *A study on the functions and applications of robots within the housekeeping department of a hotel. International Journal of Advanced Research in Science, Communication and Technology (IJARSCT), 2*(4). Anjuman-I-Islam's Institute of Hospitality Management, Mumbai, India. https://ijarsct.co.in/A13483.pdf

Reis, J., Melão, N., Salvadorinho, J., Soares, B., & Rosete, A. (2020). Service robots in the hospitality industry: The case of Henn-na hotel, Japan. *Technology in Society, 63*, 101423. Advance online publication. DOI: 10.1016/j.techsoc.2020.101423

Rosete, A., Soares, B., Salvadorinho, J., & Reis, J. C. G. (2022). Service robots in the hospitality industry: An exploratory literature review. *Sustainability, 14*(17), 10827. DOI: 10.3390/su141710827

Seyitoğlu, F., & Ivanov, S. (2023). Humans and/or robots? Tourists' preferences toward the humans-robots mix in the service delivery systems of tourism and hospitality companies. *Frontiers in Psychology, 14*, 9786514. DOI: 10.3389/fpsyg.2023.9786514

Shi, Y., Zhan, L., & Lin, J. (2025). Adaptive learning for service robots in unstructured hospitality environments. *Artificial Intelligence in Tourism and Hospitality, 2*(1), 44–63. DOI: 10.1016/j.aihth.2025.01.004

Taniguchi, A., Isobe, S., El Hafi, L., Hagiwara, Y., & Taniguchi, T. (2020). Autonomous planning based on spatial concepts to tidy up home environments with service robots. *arXiv.* https://arxiv.org/abs/2002.03671

Wan, Y., Chan, E., & Luo, J. (2020). *The rise of service robots in the hospitality industry: Some actionable insights*. Boston University Hospitality Review.

Yan, Z., Konar, R., & Balasubramanian, K. (2024). The future of hotels: Robotics, AI, and service automation in practice. In *Robotics, AI and Service Automation in the Hotel Industry* (pp. 249–263). IGI Global. DOI: 10.4018/979-8-3693-9699-5.ch011

Chapter 4
Invisible Hands of Hospitality:
Robotics in Housekeeping, Reception, and Concierge Operations

Vipan Kumar
https://orcid.org/0009-0001-4765-1797
Lovely Professional University, India

Amrik Singh
https://orcid.org/0000-0003-3598-8787
Lovely Professional University, India

ABSTRACT

The dynamics of contemporary hospitality management have been redefined by the incorporation of robotics into hotel services, which has transformed operational workflows and guest experiences. The implementation of service robots in housekeeping, reception, and concierge tasks is examined in this chapter, emphasising how they can improve productivity, precision, and client satisfaction. Robots enable intelligent check-in procedures, personalised concierge interactions, and autonomous room cleaning by utilising developments in artificial intelligence, machine learning, and sensor-based navigation. The study also explains how robotics and Internet of Things (IoT) ecosystems can work together to facilitate real-time data sharing and predictive maintenance for smooth operations. It further focuses on workforce adaptation and ethical considerations, examining how the integration of robotics influences service quality and shapes overall brand perception. The transformative potential of robotics in technology-driven, contactless, and sustainable hotel service ecosystems.

DOI: 10.4018/979-8-3373-6541-1.ch004

INTRODUCTION:

The hospitality industry stands at a critical juncture in its evolution, shaped profoundly by technological disruption and changing consumer expectations. Hotels, once defined primarily by their physical spaces and human service interactions, are now reimagining themselves as hybrid environments where digital tools and human engagement converge to create seamless, personalized, and memorable experiences. The integration of artificial intelligence (AI), service automation, and social media into hotel operations and marketing is no longer an optional enhancement but rather a structural necessity for survival in an increasingly competitive global marketplace. This chapter builds upon recent research to examine how these forces are transforming both the operational foundations and strategic directions of modern hotels, situating technology not as a peripheral add-on but as a core driver of value creation and customer engagement.

One of the most significant areas of transformation is the adoption of robots, artificial intelligence, and service automation (RAISA), which are redefining how hotels deliver efficiency and personalization. As Lukanova and Ilieva (2019) note, RAISA technologies streamline key touchpoints of the service cycle such as check-in, payment, and concierge support, offering faster and more data-driven interactions. These systems allow hotels to minimize operational burdens while simultaneously personalizing offerings, creating the impression of a service model that is both highly efficient and tailored to individual needs. Yet, the authors are careful to stress that the hospitality sector's identity has always rested on its human-centered nature, raising questions about whether excessive reliance on machines could undermine the emotional engagement central to the guest experience. The challenge for hotels, therefore, is not simply adopting RAISA but doing so in a way that reinforces human warmth and authenticity.

Artificial intelligence in particular has emerged as a double-edged sword for hotels, promising radical improvements in personalization while also introducing significant ethical and trust-related challenges. Sharma et al. (2024) demonstrate how AI-powered systems on hotel websites are capable of analyzing browsing histories, prior purchases, and contextual signals such as location or device to generate tailored recommendations that enhance satisfaction and booking conversions. Such personalization can transform digital interfaces into intuitive companions for travelers, reducing choice overload and creating smoother customer journeys. Yet, as the authors caution, personalization carries risks if it appears intrusive or manipulative. Guests may perceive AI systems as violating privacy or unfairly steering their decisions, undermining trust in the brand. This tension illustrates the broader dilemma of digital transformation in hospitality: technological advances are only as effective as the ethical and transparent frameworks that support them.

Research further demonstrates that customer perceptions of AI are nuanced and contingent upon the contexts in which technologies are deployed. Banu et al. (2025) reveal that while many guests appreciate proactive recommendations, automated adjustments, or personalized dining suggestions, they continue to prefer human staff for services requiring empathy, creativity, or complex problem-solving. This reinforces the need for hybrid service models, where automation handles routine, efficiency-driven tasks while employees focus on cultivating emotional connections. Such models not only align with customer expectations but also offer sustainable paths for workforce integration, as staff roles evolve toward facilitators of technology rather than its competitors. The move toward "cobotics," where employees and intelligent systems work collaboratively, underscores this balancing act between efficiency and empathy, ensuring that hotels preserve their human essence even while embracing advanced digital tools.

Alongside automation and AI, social media has become an equally transformative force in shaping how hotels interact with their guests, manage reputations, and differentiate themselves in saturated markets. Saravanakumar and Narayanan (2018) emphasize that platforms like Facebook, Twitter, and Instagram have become indispensable for real-time communication, brand visibility, and customer loyalty cultivation. The power of these platforms lies in their capacity to amplify both positive and negative narratives, making active and authentic participation an operational necessity rather than a marketing luxury. Hotels that ignore this dimension risk reputational decline, while those that creatively engage audiences can secure competitive advantages by demonstrating responsiveness, authenticity, and cultural alignment.

As Ashton (2021) shows, the role of social media has evolved beyond promotion into a hybrid operational and experiential tool that supports storytelling, customer service, and crisis management. Guests increasingly expect immediate responses to queries or complaints raised online, making social media an extension of the hotel's front desk rather than a separate department. This integration demands not only technological infrastructure but also digital literacy among employees, ensuring that the brand voice remains consistent, engaging, and professional across diverse platforms. Moreover, as Bhinder, Kumari et al. (2025) highlight, consumer decision-making is now heavily mediated by digital reviews, influencer endorsements, and user-generated content, meaning that social trust rather than polished advertising often drives booking behavior. Hotels that embrace transparency and encourage authentic digital advocacy thus position themselves advantageously in the contemporary marketplace.

The literature further indicates that these transformations in technology and communication are not limited to consumer-facing interactions but extend into the very structure of hotel management and workforce development. Cobanoglu et al. (2021) stress the need for hospitality education to adapt curricula that prepare graduates with digital competencies in analytics, revenue management, and social media marketing alongside traditional service skills. Similarly, Stringam and Gerdes (2021) show that automation has altered staffing structures, reducing routine workloads but simultaneously requiring new roles focused on technological support and digital facilitation. The implication is clear: the hospitality workforce of the future must be digitally fluent, adaptive, and capable of working in tandem with intelligent systems to deliver consistent and humanized guest experiences.

Yet, as Chourasia (2024) observes, this digital transition has not been universally welcomed among employees. For many workers, the rapid adoption of technologies post-pandemic has generated anxieties about job security, workload intensification, and the psychological stress of continuous adaptation. Such concerns illustrate the dual nature of innovation, which can simultaneously empower organizations and strain their human foundations. Jaipuriar (2021) extends this argument by revealing gaps between the expectations of hospitality students and the realities of hotel internships, where inadequate training and managerial support hinder their ability to thrive in technologically transformed environments. These insights highlight that the success of digital transformation in hospitality depends not only on adopting new systems but also on fostering a supportive organizational culture that nurtures both staff morale and guest satisfaction.

The broader strategic implications of these technological shifts are further illustrated in recent research on digital transformation frameworks. Yan et al. (2025) argue that the integration of social media into hotel operations represents a structural evolution rather than a temporary adaptation. By embedding predictive insights from digital platforms into decision-making, hotels can anticipate guest preferences, personalize communications, and manage reputations more effectively, thereby fostering cycles of continuous improvement. Similarly, Kapogianni (2025) contends that social media adoption represents a cultural transformation in how hotels conceptualize service delivery, dissolving traditional boundaries between marketing, communication, and operations. In this sense, digital transformation is not a supplementary layer but a foundational recalibration of how hotels engage with customers, employees, and communities.

This technological evolution is accompanied by critical ethical and sustainability considerations. Moser (2022) emphasizes that while consumers appreciate personalization, they remain deeply concerned about data privacy and transparency. Hotels must therefore develop ethical frameworks that respect guest autonomy while still leveraging data to improve services. Štilić, Nicić, and Puška (2023) highlight sustainability as another defining factor, showing how AI can optimize energy use, reduce waste, and support environmentally responsible practices. These dimensions remind us that technological progress cannot be pursued in isolation but must be aligned with broader values of responsibility, inclusivity, and long-term trust-building.

Taken together, this body of research demonstrates that the integration of AI, RAISA, and social media into hotel operations and marketing is far more than a question of technological adoption. It represents a paradigm shift in the very fabric of hospitality, reshaping customer expectations, workforce dynamics, educational priorities, and ethical responsibilities. The challenge for managers and researchers alike is to navigate this complex terrain, balancing efficiency with empathy, personalization with privacy, and innovation with sustainability. As hotels recalibrate their strategies in an increasingly digital and socially connected marketplace, success will depend not only on mastering new tools but also on maintaining the essence of hospitality: creating meaningful, human-centered experiences that endure beyond the transaction.

RESEARCH METHODOLOGY:

Given the expansive scope of this research, which seeks to bridge theory and practice, a post-positivist paradigm is deemed suitable. At present, the study predominantly relies on secondary data to fulfil its objectives. These secondary sources include previous studies, research articles, industry reports, journals, magazines, and other relevant publications.

Literature Review:

Lukanova and Ilieva (2019) provide an in-depth exploration of the growing presence of robots, artificial intelligence, and service automation (RAISA) in modern hotel environments, focusing particularly on how these emerging technologies influence the customer assessment stage of the hospitality service cycle. Their research delves into both the operational and experiential implications of RAISA adoption, emphasizing how these tools can transform traditional service models through enhanced efficiency, personalization, and data-driven decision-making. For instance, automated check-in kiosks, robotic concierge services, and AI-enabled chatbots allow hotels to deliver faster, more consistent service while freeing human staff to focus on higher-value guest interactions. Moreover, the use of predictive analytics and intelligent systems helps hotels anticipate customer needs, tailor offers, and streamline back-end operations such as housekeeping scheduling or inventory management.

However, Lukanova and Ilieva caution that the benefits of automation must be weighed carefully against the potential loss of emotional connection - a defining characteristic of the hospitality experience. While guests may appreciate the novelty and convenience of robotic service agents, many still perceive them as gimmicks rather than genuine service enhancers. The absence of empathy, warmth, and spontaneous interaction can make hotel stays feel impersonal or mechanical. As such, the authors argue that the long-term success of RAISA depends not on technological sophistication alone, but on thoughtful integration that complements, rather than replaces, the human element. For hotel managers, this creates a strategic balancing act: leveraging automation to improve consistency, speed, and personalization, while ensuring that staff remain visible, approachable, and emotionally engaged in guest service. The future of hospitality, they conclude, lies in blending the precision of machines with the authenticity of human care, crafting experiences that are both efficient and emotionally resonant.

Sharma et al. (2024) expand on this conversation by shifting the focus to the digital frontiers of hospitality specifically, how artificial intelligence is revolutionizing hotel website personalization and reshaping the online travel decision-making process. Their study illustrates how AI-driven systems analyse vast datasets, including users' browsing patterns, past purchases, demographic information, and contextual cues such as device type or geolocation, to generate highly tailored content and recommendations. This intelligent personalization simplifies the booking journey by reducing information overload, helping travellers discover relevant room types, promotions, or add-on services aligned with their preferences. Consequently, hotels experience higher booking conversion rates, increased customer satisfaction, and stronger brand loyalty, as guests feel understood and valued on an individual level.

Nonetheless, Sharma et al. also highlight critical ethical and managerial challenges inherent in AI personalization. While the use of personal data enables superior targeting and customization, it simultaneously exposes hotels to privacy concerns, data breaches, and accusations of algorithmic bias or manipulation. If guests begin to perceive AI-driven suggestions as intrusive or exploitative, their trust in the brand can quickly erode. Therefore, the authors argue that transparency, consent, and ethical governance must underpin every aspect of AI deployment in hospitality. The study ultimately portrays AI as a double-edged sword: when implemented responsibly, it can enhance engagement, loyalty, and long-term profitability; when mismanaged, it can undermine brand credibility and alienate customers. For hotel leaders, the challenge lies in achieving a balance between innovation and accountability—embracing technology's potential to personalize experiences while upholding guest privacy, fairness, and emotional connection as central pillars of hospitality excellence.

Banu et al. (2025) take a comprehensive and forward-looking perspective on the transformative influence of artificial intelligence applications in shaping hospitality consumer preferences and behavior. Their study extends beyond specific service encounters to examine how AI impacts the broader guest experience, from pre-arrival planning to post-stay engagement. The findings suggest that AI not only enhances perceived service quality but also fundamentally alters customer expectations regarding speed, personalization, and convenience. Intelligent systems, for example, can anticipate guest needs by analyzing prior interactions, preferences, and contextual factors such as time of day or weather. This enables hotels to deliver proactive experiences like suggesting nearby dining options that match dietary habits, adjusting room temperature and lighting before check-in, or offering customized upgrades at optimal decision points. Such anticipatory service strengthens brand loyalty and improves operational efficiency simultaneously.

Nevertheless, Banu et al. caution that the success of AI adoption in hospitality depends heavily on maintaining trust, transparency, and meaningful human interaction. Many guests appreciate AI's ability to reduce waiting times and enhance precision, but they remain hesitant when automation encroaches on emotionally significant moments, such as resolving complaints or providing empathetic care. In these situations, human judgment, warmth, and problem-solving skills remain irreplaceable. The authors emphasize that excessive automation risks depersonalizing the hospitality experience—an outcome that contradicts the very essence of the industry. They therefore advocate for hybrid service models where human employees and intelligent systems collaborate seamlessly. In this approach, AI handles data-driven, repetitive, or predictive tasks, while staff focus on interpersonal engagement and creative problem-solving. This balanced integration, according to the authors, represents the most sustainable and ethical way forward one that merges technological precision with the authenticity and emotional intelligence that define true hospitality excellence.

Saravanakumar and Narayanan (2018) add another dimension to the discussion by investigating how social media marketing reshapes communication and branding strategies within the hotel industry. Their research underscores that platforms such as Facebook, Instagram, and Twitter have become indispensable tools for maintaining continuous engagement with guests, promoting services, and enhancing brand reputation. Through these platforms, hotels can showcase visual storytelling, share real-time updates, and create interactive campaigns that encourage user participation and word-of-mouth promotion. This dynamic digital environment offers vast opportunities for visibility and customer relationship building, especially among younger, tech-savvy travelers who rely heavily on online platforms for travel inspiration and decision-making.

However, the authors also highlight significant challenges accompanying this digital shift. Hotels must manage negative reviews, respond promptly to guest feedback, and maintain a consistent tone and message across multiple social platforms all of which require dedicated resources and strategic coordination. Furthermore, the intangible nature of social media success makes it difficult to measure direct returns on investment, compelling marketers to balance creative outreach with data-driven performance metrics. Saravanakumar and Narayanan warn that neglecting social media engagement can lead to reputational stagnation or decline, as today's digital audiences equate online visibility with credibility and trustworthiness. Conversely, hotels that adopt innovative, authentic, and responsive social media strategies are more likely to secure lasting competitive advantages. Their study concludes that success in this realm demands not only creativity and agility but also sustained organizational commitment, investment in skilled personnel, and a clear long-term vision aligning marketing goals with overall brand identity.

Cobanoglu et al. (2021) shift attention toward the educational and professional dimensions of technology integration within hospitality, emphasizing how digital transformation is reshaping both industry practices and academic preparation. Their research reveals that as hotels increasingly depend on digital platforms ranging from property management systems and mobile guest applications to AI-powered customer relationship tools; the competencies required of hospitality professionals are evolving rapidly. Future employees are expected to combine traditional service expertise with technical fluency in analytics, data interpretation, and technology-driven decision-making. The study illustrates that exposure to tools such as revenue management software, online booking systems, and digital marketing analytics during academic training can significantly enhance graduates' employability and readiness to meet industry demands.

Moreover, Cobanoglu et al. argue that hospitality education must undergo systemic reform to remain aligned with industry innovation. Universities should move beyond conventional, service-centric teaching models to adopt curricula that integrate emerging technologies, experiential learning, and interdisciplinary collaboration. This includes offering modules on automation ethics, AI in service design, and the strategic use of social media for brand management. By fostering digital literacy alongside emotional intelligence, educational institutions can cultivate professionals capable of navigating the dual demands of technological proficiency and guest-centered empathy. The authors conclude that aligning educational objectives with industry transformation not only prepares students for immediate employment but also empowers them to lead future innovations. In doing so, hospitality programs play a pivotal role in ensuring that the workforce of tomorrow can drive digital transformation responsibly - sustaining competitiveness while preserving the human spirit that lies at the heart of the hospitality experience.

Stringam and Gerdes (2021) conduct a comprehensive examination of how technological advancements are reshaping hotel staffing structures, redefining employee responsibilities, and altering the skill sets necessary for modern hospitality operations. Their research explores the integration of automation tools such as self-service kiosks, mobile check-in applications, digital ordering systems, and robotic service assistants—technologies that are rapidly becoming embedded within hotel ecosystems. These innovations have significantly reduced the routine and repetitive tasks traditionally handled by front desk agents, receptionists, and service attendants, allowing employees to focus more on strategic, guest-oriented functions. However, rather than completely replacing the human workforce, these technologies have reconfigured job roles, emphasizing collaboration between humans and machines. The authors introduce the concept of "cobots," or collaborative robots, designed to operate alongside employees to improve productivity and consistency in guest service delivery.

Stringam and Gerdes highlight that the implementation of automation has resulted in both staff reductions in certain operational areas and the creation of new, technologically specialized positions, particularly in system monitoring, maintenance, and digital guest relations. As a result, the hospitality industry is witnessing a structural shift toward a workforce that must balance traditional service expertise with technical literacy. The study underscores the rising importance of employee adaptability, digital fluency, and comfort with automated systems across all departments—from front office operations to housekeeping and maintenance. For instance, housekeeping staff now utilize smart room management tools, while concierges rely on AI-based recommendation systems to assist guests more effectively. The researchers conclude that technology in hospitality should not be viewed as a threat to employment but as a catalyst for upskilling and professional evolution. Success in this new era requires continuous employee training and an organizational culture that values technology as a partner in service delivery rather than as a replacement for human touch and empathy.

Uniyal and Sharma (2020) shift the discussion toward the practical and ethical challenges associated with adopting artificial intelligence (AI) and service automation in hotels. Their study identifies several critical barriers that impede successful technological integration, including interoperability issues between different systems, substantial installation and maintenance costs, and growing concerns over guest data privacy. They argue that although AI-driven tools enhance brand image and operational efficiency—through predictive analytics, chatbots, and personalized marketing—their widespread acceptance remains hindered by both technological limitations and cultural differences among global travelers. Guests from varying cultural and socio-economic backgrounds possess distinct expectations regarding personal service, emotional engagement, and data usage, resulting in diverse responses to automation.

Furthermore, Uniyal and Sharma raise important ethical questions surrounding the collection and storage of biometric data, such as facial or fingerprint recognition systems, which are often used for identity verification or customized experiences. The potential misuse or mishandling of such sensitive data could erode consumer trust and trigger reputational damage. Their analysis underscores that while artificial intelligence plays a central role in redefining the hospitality landscape through the Internet of Things (IoT), big data analytics, and machine learning, it simultaneously introduces new dilemmas regarding ethics, privacy, and emotional authenticity. The authors advocate for a measured, human-centric approach to AI adoption—one that integrates automation as a supplement to, rather than a substitute for, genuine interpersonal connection. They argue that the future of hotel technology lies in maintaining a delicate equilibrium between innovation and empathy, ensuring that digital systems enhance operational excellence without diminishing the human warmth that remains the cornerstone of hospitality.

Štilić, Nicić, and Puška (2023) adopt a forward-looking stance, envisioning the transformative potential of artificial intelligence to redefine the future of hotel operations in four key dimensions: personalization, automation, sustainability, and immersive technology. They propose that AI-powered systems will revolutionize the guest experience by curating hyper-personalized services—ranging from tailored room amenities and custom local recommendations to dynamic welcome messages that adapt to each guest's profile and preferences. Automation, in turn, will handle repetitive tasks such as check-ins, in-room service requests, and maintenance alerts, allowing human employees to concentrate on higher-order, emotionally engaging interactions. The authors note that innovations such as chatbots, cleaning robots, and smart-room control systems are already optimizing staff efficiency and enhancing convenience for guests.

A major contribution of their work is the emphasis on sustainability as an integral outcome of technological advancement. By leveraging AI to monitor energy consumption, manage waste, and regulate utilities, hotels can significantly reduce their environmental footprint while cutting operational costs. Moreover, they highlight the rising incorporation of voice-activated assistants and augmented reality (AR) features in hotel environments, creating immersive and interactive experiences that enhance guest engagement. Global hotel chains such as Marriott and Hilton serve as case studies in their analysis, showcasing how technology can serve as both an operational necessity and a strategic differentiator in a highly competitive marketplace. The authors ultimately conclude that AI represents not just a tool for efficiency but a transformative force capable of reshaping hospitality into a more sustainable, personalized, and emotionally resonant industry—one that balances digital sophistication with human creativity and service excellence.

Chourasia (2024) adds a valuable human perspective to this growing discourse by focusing on employee perceptions of technological change, particularly within concierge and guest services. Through surveys and qualitative interviews, the study captures a spectrum of employee attitudes toward automation and innovation. Some respondents expressed enthusiasm, viewing technology as a means of reducing tedious paperwork, streamlining communication, and improving response times. Others, however, voiced deep apprehension about job displacement, skill redundancy, and the psychological stress of adapting to an ever-evolving technological landscape. The study reveals that nearly one-third of employees reported heightened anxiety over employment stability, with many citing the post-COVID-19 surge in automation as a primary factor driving this insecurity.

Chourasia argues that while hotels adopt digital tools to enhance operational resilience and guest satisfaction, management must also consider the emotional and psychological impacts on employees. The rapid pace of digital transformation, though beneficial from a business standpoint, can contribute to burnout, fear of obsolescence, and resistance to change among staff. The author recommends that hospitality leaders implement comprehensive training programs, mental wellness initiatives, and transparent communication strategies to ensure smoother transitions into technology-enabled workflows. The study concludes that technology's role in hospitality is inherently dualistic: it serves as a powerful enabler of efficiency and innovation, yet also a potential source of disruption and stress within the workforce. The key to sustainable progress, Chourasia asserts, lies in cultivating an empathetic management approach that balances guest satisfaction with employee well-being—ensuring that technological modernization strengthens, rather than fractures, the human foundation of the hospitality industry.

Jaipuriar (2021) contributes by analyzing the role of technology in hotel accommodation operations from an employee perspective, with particular emphasis on internships and workforce readiness. The research underscores that while technology has transformed operations and heightened guest expectations, the industry's sustainability still depends on well-trained and motivated personnel. The study identifies a gap between student interns' expectations and the realities of hotel internships, revealing challenges such as lack of adequate training, insufficient managerial support, and issues of attrition. By calling for stronger collaboration between hospitality institutions, universities, and hotels, Jaipuriar advocates for structured internship modules, enhanced liaison with HR departments, and better experiential learning opportunities for interns. Importantly, the findings highlight that technological transformation in hospitality must be accompanied by parallel investment in human resource development, ensuring that the next generation of professionals can thrive in a digital-first yet human-centered service landscape

Yan et al. (2025) emphasize how the emergence of digital transformation strategies in hospitality has become a decisive factor in shaping competitive positioning. Their exploration highlights that hotels which adopt a systematic integration of social media tools within both operations and marketing gain deeper analytical capabilities, enabling them to understand customer behavior at levels of precision that were previously unattainable. By combining predictive insights from guest interactions with social platforms, hotels are not merely responding to real-time feedback but are actively anticipating guest desires and expectations. This approach builds a cycle of continuous improvement where each interaction refines service quality, fosters more effective engagement, and contributes directly to brand differentiation. The ability to personalize communication, manage reputation proactively, and integrate content strategies across multiple digital touchpoints demonstrates that social media, when strategically embedded, is not just a promotional accessory but a driver of sustainable business growth. The study makes it clear that digital transformation in hospitality must be understood as a structural evolution rather than a temporary adaptation, one where hotels recalibrate their core strategies around the realities of a data-driven, socially connected marketplace.

Ashton (2021) builds upon this perspective by illustrating the nuanced role of social media in bridging the operational aspects of hotel management with the experiential dimensions of guest interaction. The research demonstrates that platforms like Facebook, Instagram, and Twitter are no longer limited to promotional spaces but have evolved into multi-functional tools supporting customer service, brand storytelling, and crisis management. Through examples of hotels that successfully leveraged these platforms during reputational challenges, Ashton shows that immediacy and authenticity in digital communication are now essential traits of effective hospitality leadership. Hotels that fail to integrate social media deeply into their structures risk falling behind in responsiveness, particularly as guests increasingly expect direct and transparent communication in online environments. By conceptualizing social media as a hybrid operational and marketing tool, Ashton reframes its role from being an auxiliary department function into a fundamental part of the hospitality business model. The study also underscores the importance of equipping staff with the digital literacy necessary to maintain consistency, creativity, and professionalism across social platforms, ensuring that the brand voice remains coherent while adapting to the diverse expectations of a global audience.

Bhinder, Kumari et al. (2025) extend the discussion by focusing on how social media platforms actively shape consumer decision-making processes in hospitality. Their findings reveal that the persuasive power of digital reviews, user-generated content, and influencer partnerships fundamentally alters how guests perceive value and trust in hotel brands. By analyzing patterns in consumer behavior, the study underscores that purchasing choices are now heavily mediated by the credibility of online endorsements rather than traditional advertising. Hotels are therefore compelled to rethink their marketing mix, prioritizing interactive strategies that highlight authenticity and community engagement over polished promotional messaging. The research stresses that transparency in communication and responsiveness to digital feedback are vital not only for retaining customer loyalty but also for attracting new segments who rely extensively on social signals before making reservations. Moreover, the study points out that the immersive nature of platforms such as TikTok and Instagram reels has created new pathways for storytelling where visual narratives significantly influence emotional connection with brands. Hotels that adopt these strategies effectively can elevate guest perception, reinforce positive brand associations, and create a stronger cycle of digital advocacy that feeds back into revenue growth and market positioning.

Kapogianni (2025) ties these developments together by arguing that the adoption of social media strategies in hospitality represents not just a technological shift but also a cultural transformation in how hotels conceptualize service delivery. The study emphasizes that the boundaries between marketing, communication, and operations are becoming increasingly blurred as social media becomes embedded in every stage of the guest journey. From pre-booking interactions shaped by targeted digital campaigns to on-property engagement enhanced by real-time social responses, and even post-stay loyalty nurtured through personalized digital follow-ups, social media constructs a continuous loop of brand presence. Kapogianni stresses that this holistic cycle challenges traditional hierarchies within hotel management, urging leadership to embrace flatter, more adaptive structures that can respond swiftly to the dynamic demands of digital engagement. Importantly, the research also raises awareness of the ethical and sustainability considerations tied to social media use, such as the need to avoid manipulative marketing practices and to ensure inclusivity in online representation. This reflection broadens the debate by acknowledging that while social media delivers operational and marketing advantages, it must also be wielded with responsibility, transparency, and cultural sensitivity to secure long-term guest trust and brand resilience in an increasingly globalized hospitality landscape.

Mukherjee et al. (2021) provide insightful information about how social media platforms can be used as structural facilitators of visitor engagement and loyalty, rather than merely as marketing accessories. According to their research, social media sites like Facebook and Instagram offer more than just advertising visibility; they also build vibrant ecosystems where interactive campaigns, user-generated content, and storytelling all work together to strengthen brand authenticity. Because happy customers willingly spread the brand message, hotels benefit from both increased organic reach and decreased reliance on expensive advertising. Social media should be seen as a component of a comprehensive digital strategy that links back to operational delivery, the study stresses. If the experience promised online is not mirrored in service quality on-site, the entire cycle of credibility collapses. This relational aspect of social media underscores its role in co-creating value alongside customers rather than dictating it from a managerial perspective.

The discussion continues with Rather (2021), who shows how consumer interaction on digital platforms goes well beyond light conversation and directly influences the development of emotional ties with hospitality brands. Digital relationships are now just as important as face-to-face service encounters, according to Rather, who analyses how social exchanges foster affective commitment, trust, and satisfaction. According to his research, hotels encourage advocacy behavior where customers willingly spread the word about the brand to their networks as well as repeat business when they actively respond to online reviews, resolve grievances, and highlight guest stories. This dynamic demonstrates how client relationships in the hospitality sector are developing virtually rather than being limited to the lobby or front desk. Such conclusions challenge managers to rethink service boundaries, recognizing that hospitality now operates in a hybrid environment where digital and physical touchpoints must be seamlessly integrated.

Hoq (2022) endorses this hybrid paradigm by situating social media within the broader framework of digital transformation. His research indicates that hotels that use social media in their customer service, marketing, and revenue management strategies see measurable gains in performance. More importantly, Hoq emphasises the cultural shift required for digital adoption: in addition to investing in technology, hotels must cultivate organisational mindsets that value innovation and flexibility. The study found that resistance to change, particularly in traditional family-run businesses, frequently stalls the integration of social platforms into regular operations. On the other hand, progressive properties that incorporate analytics-based decision-making into their digital procedures put themselves in a better position than rivals. This creates a widening gap between adopters and laggards, suggesting that the digital divide in hospitality is not merely technological but strategic in nature.

Tiberghien et al. (2020), who study how destinations use digital platforms to craft cultural narratives and influence visitor perceptions, advance the global perspective of social media's role in tourism. Despite being focused on heritage and cultural tourism, their research has direct relevance to hotel marketing tactics. In addition to being places to stay, hotels are positioned as cultural bridges that influence travellers' experiences before, during, and after their journey. Hotels can stand out in a crowded market by emphasising sustainability, community connections, and authenticity through carefully chosen online storytelling. Tiberghien et al. also caution, however, that such narratives must align with on-ground realities; misrepresentation or superficial branding risks alienating the increasingly discerning and digitally literate traveler. This illustrates the delicate balance hotels must strike between persuasive digital marketing and authentic operational delivery, where overstatement can undermine long-term credibility.

By emphasising that the hospitality industry's reliance on digital engagement generates both opportunities and vulnerabilities, Moser (2022) expands this conversation into the areas of ethics and consumer trust. His results demonstrate that although consumers value tailored offers and carefully chosen content, they are also sensitive to concerns about data misuse, privacy, and transparency. Moser stresses the significance of implementing ethical frameworks that safeguard consumer rights without stifling innovation in a time when customer relationship management is incorporating predictive analytics and artificial intelligence. For hotels, this entails walking a tightrope between using visitor data to tailor services and protecting autonomy by upholding transparency and explicit consent. Ultimately, the argument put forth by Moser resonates with the broader trajectory of hospitality's digital evolution: success will not be defined solely by technological adoption but by the ability to sustain trust and authenticity in a rapidly shifting landscape

RECOMMENDATIONS

1. **Balance Technology with Human Touch**: AI and automation should be used in hotels to supplement human interaction, not to replace it. Hospitality still requires emotional connections, and employees must play a key role in providing empathy in addition to technological efficiency.
2. **Invest in Responsible Data Practices**: To maintain customer trust, robust procedures for data security, privacy, and transparency must be created. Explicit consent procedures should be included with personalised services to prevent impressions of manipulation.
3. **Adopt Hybrid Service Models**: Efficiency and authenticity are guaranteed when intelligent systems are combined with human expertise. This method enables staff to concentrate on high-value, intricate, and emotionally delicate services while routine tasks can be automated.
4. **Strengthen Social Media Integration**: Social media platforms ought to be viewed as essential to business operations rather than as external marketing avenues. To stay competitive, hotels need to use them for storytelling, reputation management, and real-time engagement.
5. **Embed Digital Competencies in Education**: In order to prepare graduates for success in a service industry driven by technology, hospitality schools must broaden their curricula to include analytics, artificial intelligence, social media, and digital revenue management.
6. **Support Workforce Adaptation**: Hotels must provide staff with ongoing training and assistance as they adjust to new technologies. This entails resolving concerns regarding job stability and developing new positions that combine technical and human skills.
7. **Promote Ethical and Inclusive Practices**: Fairness, inclusivity, and transparency must be given top priority in the adoption of digital technology. Hotels can gain the trust of a global clientele by avoiding deceptive advertising and honouring cultural sensitivities.

8. **Align Technology with Sustainability Goals**: By implementing AI and automation in ways that minimise waste and energy use, hotels can fulfil their sustainability pledges while providing effective guest experiences.

FUTURE WORK:

In order to develop sustainable, well-balanced guest experience models, future hospitality technology research should delve deeper into the investigation of the intersection between digital innovation and human service. The increasing use of automation, artificial intelligence (AI), and data analytics in hotels necessitates thorough research into how these technologies can coexist with the human and emotional aspects of hospitality. Future research should concentrate on creating and assessing hybrid service models that preserve the spontaneity, warmth, and empathy of human interaction while optimising operational efficiency through automation. Research of this kind may reveal the most effective ways to combine employee intuition and machine intelligence, leading to consistent and emotionally impactful service experiences. Furthermore, to investigate the long-term impacts of AI-driven personalisation on visitor retention and loyalty, empirical research is required. Although personalisation boosts customer satisfaction in the short term, it is unclear if customers will continue to trust companies that mainly use data tracking and predictive analytics. Whether repeated exposure to automated personalisation fosters stronger bonds or causes weariness and privacy concerns could be investigated through longitudinal studies. In data-intensive settings, researchers should also look into how transparency helps maintain consumer trust, especially how sharing information about data usage affects perceptions of security and fairness. Another crucial area of study is cross-cultural comparison. Travellers from Europe or North America may perceive automation and digital engagement as impersonal or intrusive, while visitors from South Korea or Japan may see them as creative and efficient. Future research should therefore look at how attitudes towards automation and AI interaction are influenced by generational differences, socioeconomic backgrounds, and cultural norms. Global hotel brands can better adapt their digital transformation strategies to a variety of clientele by being aware of these distinctions. The workforce dimension is equally significant. Researchers ought to investigate how hotels can assist staff in adjusting to ongoing technological advancements without sacrificing morale or job satisfaction. To ensure workforce resilience, research into digital literacy initiatives, training frameworks, and leadership techniques that promote flexibility will be essential. Future studies could also examine the psychological effects of adopting technology, with an emphasis on ways to reduce stress, avoid burnout, and encourage a sense of empowerment in workers who collaborate with automation and artificial intelligence.

Last but not least, interdisciplinary and longitudinal research ought to evaluate the ways in which the hospitality sector's use of technology supports environmental sustainability. The long-term effects of AI-driven systems on waste reduction, water conservation, and energy management, as well as the total carbon footprint of digital operations, could be research topics. Future studies can provide a more comprehensive understanding of how digital transformation can concurrently enhance worldwide sustainability goals, business performance, and guest satisfaction by tying technological advancement to environmental responsibility.

CONCLUSION:

All of the reviewed literature emphasises how complicated and multifaceted the digital transformation of the hospitality industry is, including automation, artificial intelligence, social media engagement, and changing workforce dynamics. A common finding throughout all of the research is that, although technology is changing almost every facet of hotel operations and the visitor experience, it cannot take the place of the human element that is fundamental to hospitality. Operations have been completely transformed by innovations like RAISA (Robotics, Artificial Intelligence, and Service Automation), AI-driven personalisation, and data analytics platforms, which have increased productivity, improved personalisation, and given visitors never-before-seen convenience. However, in an increasingly automated workplace, these same technologies present serious obstacles to preserving genuine human connection, guaranteeing ethical data governance, and safeguarding employee wellbeing. One defining theme is still the fundamental conflict between empathy and efficiency. The future of the hospitality sector hinges on striking a balance between emotional authenticity and technological prowess. Hotels that ignore innovation may find it difficult to meet the growing demands of speed, accuracy, and personalisation, while those that rely too heavily on automation run the risk of offending visitors who are looking for warmth and spontaneity. Therefore, integration integrating technology into service frameworks that respect individuality, trust, and compassion rather than substitution is the way forward. Another structural pillar of hospitality strategy is social media. Social media sites like Facebook, Instagram, and X (formerly Twitter) have developed into essential parts of customer relationship management and brand management, going beyond simple promotional tools. Hotels can influence opinions, react to criticism, and cultivate loyalty throughout the client journey by using storytelling and real-time engagement. But the need for genuineness and regular communication emphasises how important innovation, openness, and long-term organisational dedication are to the success of digital marketing. In this dynamic ecosystem, workforce preparedness and education are equally important pillars. In order to prepare future professionals with both traditional service excellence and advanced digital fluency, hospitality education must change as hotels embrace new technologies. To prepare graduates to spearhead the upcoming wave of responsible transformation, curricula must incorporate instruction in analytics, automation ethics, and sustainable innovation. Opportunities for ongoing education that foster emotional intelligence and flexibility in addition to technical proficiency must also be provided to employees.

Lastly, one important factor that determines long-term resilience is the ethical aspect of technological advancement. In an interconnected, socially conscious marketplace, trust must be upheld through transparency in data usage, inclusivity in service design, and a dedication to sustainability. The reviewed studies collectively demonstrate that the success of the hospitality sector in the digital age will rely on enhancing the human touch through careful, ethically grounded, and culturally sensitive technology integration rather than replacing it. In a competitive global environment where guest expectations, technological capabilities, and social values continue to evolve at an unprecedented pace, hotels that adopt this holistic approach where innovation coexists with empathy and sustainability are best positioned to thrive.

REFERENCES

Ashton, M. T. (2021). *Robotics, artificial intelligence and service automation in hotel service process design: Can customer dominant logic provide improved value?* (Doctoral dissertation, Bournemouth University).

Banu, S., Samantaray, R. R., Ayesha, N., Fathima, A., & Tabassum, Z. (2025). Robotics and automation in hospitality services. In *Cutting-edge technologies for business sectors* (pp. 367–400). IGI Global.

Bhinder, H. S. (n.d.). Emerging technology trends adopted in hotel guest cycle for enhancing guest stay experiences. In *Recent trends in tourism and hospitality* (p. 18).

Chourasia, B. (2024, June). To study the technological impact and innovation on the hotel industry in relation to the employees working in the concierge section of front office. In *International Conference on Innovation and Regenerative Trends in Tourism and Hospitality Industry (IRTTHI 2024)* (pp. 158–176). Atlantis Press. DOI: 10.2991/978-94-6463-437-2_11

Cobanoglu, C., Dogan, S., Berezina, K., Collins, G., Nanu, L., Shahtakhtinskaya, K., & Parvez, O. (2021). *Hospitality and tourism information technology* (Vol. 17). University of South Florida M3 Center Publishing.

Jaipuriar, S. (2021). Role of technology in accommodation operation department of hotel: An employee's perspective. In *Hospitality and tourism: Emerging practices in human resource management* (p. 70).

Kapogianni, I. (2025). Enhancing the guest experience in smart hotels through technology: A glimpse into the future.

Kumari, M., Kumar, S., & Devi, N. (2025). The impact of service automation and robotics on hospitality business models. In *Metaverse and sustainable business models in SMEs* (pp. 317–338). IGI Global Scientific Publishing. DOI: 10.4018/979-8-3693-9005-4.ch013

Lukanova, G., & Ilieva, G. (2019). Robots, artificial intelligence, and service automation in hotels. In *Robots, artificial intelligence, and service automation in travel, tourism and hospitality* (pp. 157–183). Emerald Publishing Limited. DOI: 10.1108/978-1-78756-687-320191009

Moser, J. (2022). Look but don't touch: Contactless solutions to enhance the customer journey in hotels (Master's thesis, Universidade Catolica Portuguesa, Portugal).

Mukherjee, S., Baral, M. M., Venkataiah, C., Pal, S. K., & Nagariya, R. (2021). Service robots are an option for contactless services due to the COVID-19 pandemic in the hotels. *Decision (Washington, D.C.)*, *48*(4), 445–460.

Saravanakumar, S., & Narayanan, M. B. (2018). The service automation and robotics in hospitality industry: A study on business implications. *International Journal of Mechanical and Production Engineering Research and Development*, *8*(6), 91–100. DOI: 10.24247/ijmperddec201810

Sharma, M., Tyagi, P. K., Semwal, R., Tyagi, P., Singh, V. J., & Kumari, N. (2024, June). Robotic concierges and smart spaces: A vision for the future of hospitality services. In *International Conference on Innovation and Regenerative Trends in Tourism and Hospitality Industry (IRTTHI 2024)* (pp. 511–525). Atlantis Press. DOI: 10.2991/978-94-6463-437-2_34

Štilić, A., Nicić, M., & Puška, A. (2023). Check-in to the future: Exploring the impact of contemporary information technologies and artificial intelligence on the hotel industry. *Turističko poslovanje, (31)*.

Stringam, B. B., & Gerdes, J. H. (2021). *Hotel and guest room technology* (Vol. 17). University of South Florida M3 Publishing.

Uniyal, R. S. D. M., & Sharma, N. (2020). Emerging use of artificial intelligence, robots & service automation in hotel industry. *International Journal of Future Generation Communication and Networking*, *13*(3), 2785–2790.

Yan, Z., Konar, R., & Balasubramanian, K. (2025). The future of hotels: Robotics, AI, and service automation in practice. In *Sustainability, innovation, and consumer preference* (pp. 249–278). IGI Global Scientific Publishing.

Zhong, L., Coca-Stefaniak, J. A., Morrison, A. M., Yang, L., & Deng, B. (2022). Technology acceptance before and after COVID-19: No-touch service from hotel robots. *Tourism Review*, *77*(4), 1062–1080. DOI: 10.1108/TR-06-2021-0276

Chapter 5
Balancing Technology and Human Expertise: An Industry Perspective on AI Integration in Housekeeping Operations

Rachna Chandan
https://orcid.org/0000-0002-5144-4336
Banarsidas Chadiwala Institute of Hotel Management, Delhi, India & Voice of Housekeepers, India

Dharna Shukla
https://orcid.org/0000-0001-5857-2130
Amity University, Lucknow, India

Sandhya Anilkumar
M.S. Ramaiah University of Applied Sciences, India

Madhu Chandhok
https://orcid.org/0009-0000-0785-2116
Independent Researcher, India

ABSTRACT

The rapid growth of Artificial Intelligence (AI) and automation has transformed housekeeping operations in hospitality. Technology streamlines routine tasks, enhances efficiency, and promotes sustainability, yet the human element remains vital for service quality and guest personalization. This study, based on a survey of 67 Indian hospitality organizations, explores how AI and human expertise complement each other. Findings show AI improves process optimization, speed, and decision-

DOI: 10.4018/979-8-3373-6541-1.ch005

making, while humans ensure empathy, ethics, and cultural sensitivity. Challenges include data privacy, workforce adaptability, and continuous upskilling. Balancing innovation with human values is essential for sustainable growth and superior guest experiences.

INTRODUCTION

The hospitality industry is undergoing a profound technological transformation driven by rapid advancements in Artificial Intelligence (AI) and automation. Among all operational areas, housekeeping has experienced one of the most significant shifts, moving from a predominantly labour-intensive function to a technology-assisted, data-driven system. AI-enabled tools such as autonomous cleaning robots, predictive analytics, smart inventory systems along with real-time monitoring platforms are increasingly deployed to enhance cleanliness standards, operational efficiency as well as resource optimization (Chandhok et al., 2025). These developments have positioned housekeeping not merely as a support function but as a strategic contributor to service quality and operational performance.

Beyond efficiency gains, AI adoption in housekeeping is closely linked to sustainability objectives. Smart systems enable optimized energy usage, reduced chemical consumption, efficient linen management with waste minimisation—aligning housekeeping operations with broader environmental goals (Alemede, 2025). As sustainability becomes integral to brand reputation and guest decision-making, hotels increasingly rely on AI-driven housekeeping solutions to meet expectations of regulatory, environmental and ethical nature. However, these technological advancements also introduce structural and human challenges that demand careful management.

Historical parallels can be drawn between the current AI revolution and earlier technological disruptions such as the rise of the internet. While early predictions anticipated large-scale job displacement and organizational collapse, digital transformation ultimately enhanced productivity, transparency raising accountability across industries (Manoharan & Ashtikar, 2025). Similarly, AI adoption in hospitality has generated apprehension among employees, including concerns related to job security, technostress along with resistance to change. Research indicates that technological support alone may not immediately eliminate such stress; nevertheless, reducing support can have detrimental effects, whereas sustained investment in training and assistance produces long-term positive outcomes (Bausch, Kraemer, & Mauroner, 2024).

Despite the increasing sophistication of AI systems, hospitality remains fundamentally human-centric. Housekeeping employees contribute emotional intelligence, ethical judgment, cultural sensitivity, situational adaptability—qualities that remain beyond the scope of automation (Nira, 2025; Mujtaba, 2025). While AI excels in standardization, prediction, and efficiency, human expertise is indispensable in ensuring personalised service delivery, service recovery, and emotionally meaningful guest interactions. This highlights the necessity of viewing AI not as a replacement for labour but as an enabler that amplifies human potential.

The coexistence of AI and human expertise has consequently reshaped managerial and strategic priorities within hospitality organizations. Hotels are increasingly required to adopt human-centred AI implementation models that encourage employee participation, open communication with continuous upskilling. The emergence of new professional roles—such as AI Hospitality Managers, data interpreters, analytics specialists—illustrates how technological integration can generate employment transformation rather than displacement. These roles act as intermediaries between intelligent systems and human teams, ensuring alignment with organizational values, service standards, and ethical principles.

In addition to operational and human considerations, AI integration in housekeeping raises critical ethical and governance issues. Data privacy, algorithmic transparency, workforce equity and responsible use of guest information have emerged as central concerns (Watts & Munir, 2025). Addressing these challenges requires structured implementation frameworks, clear policies and institutional collaboration between industry and academia to build AI literacy and ethical competence among employees (Talukdar & Kumar, 2024). Without such frameworks, the benefits of AI may be undermined by resistance, mistrust leading to service inconsistencies.

Although existing literature acknowledges the efficiency and sustainability advantages of AI in housekeeping, there remains a lack of empirical clarity regarding how organizations can achieve an optimal balance between automation and human expertise. Limited attention has been paid to recognizing the specific value added by human contributions in AI-driven environments and to identifying best practices for integrated implementation. Addressing these gaps is essential for ensuring that AI adoption enhances service quality, employee well-being as well asbrand integrity rather than eroding hospitality's emotional core.

This chapter therefore examines the coexistence of Artificial Intelligence and human expertise in housekeeping operations across selected service sectors. By focusing on balance, recognition, and implementation strategies, the study aims to provide a structured understanding of how AI-human integration can enhance operational performance and guest satisfaction while preserving the values of warmth, empathy, and ethical service that define hospitality.

RESEARCH GAPS AND PROBLEM STATEMENT

Evidence from the industry suggests three primary gaps that this study addresses:

- The Balance Gap: There is a limited understanding of the optimal equilibrium between AI and human expertise. This study investigates how a harmonized approach ensures that automation serves as a supportive tool rather than a total replacement for human potential.
- The Recognition Gap: There is insufficient recognition of the specific value human expertise adds to AI-driven operations. This study argues for the irreplaceable role of human judgment and creativity in maintaining the emotional core of the guest experience.
- The Implementation Gap: There is a lack of established best practices for AI-human integration. Effective strategies for workforce development, such as AI literacy and the creation of roles like "AI Hospitality Managers," remain largely undefined.

RESEARCH OBJECTIVES

Utilizing a mixed-methods research design—incorporating quantitative data from 67 organizations (including hotels, hospitals, and aviation) and qualitative thematic analysis—this study is guided by the following objectives:

Objective 1- To establish the impact of creating a balance between AI and human expertise in Housekeeping: Evaluating how a strategic balance streamlines operations and facilitates new operational opportunities.
Objective 2- To examine the significance of human expertise in AI-driven Housekeeping operations: Analyzing how human involvement significantly improves "Quality Services" and upholds brand reputation and sustainability.
Objective 3- To identify best practices for integrating AI and human expertise in Housekeeping: Determining how AI-powered tools can enhance individual productivity while being complemented by the human touch to reshape guest experiences.

By harmonizing technology with human expertise, organizations can achieve operational excellence while offering memorable, emotionally intelligent guest experiences that technology alone can never replicate.

LITERATURE REVIEW

The rise of Artificial Intelligence (AI) in housekeeping has become one of the most groundbreaking changes in the hospitality sector. Researchers point out that AI-powered tools—like automated cleaning robots, smart sensors, advanced linen inspection systems—are reshaping how hotels operate and setting new standards for service (Yang J., 2024; Thakur A. et. al., 2025). These innovations boost accuracy, ensure consistent cleanliness thus reducing human error, which is crucial for maintaining high hygiene standards, especially in the wake of the pandemic. In addition to, AI-driven predictive maintenance tools help spot potential equipment issues before they affect guests, ensuring everything runs smoothly and comfortably (Manoharan G. & Ashtikar S.P., 2025). On another front, predictive analytics in AI systems streamline scheduling and task management by examining real-time room occupancy and guest information, which significantly cuts down on turnaround times and operational expenses (Kaul D., 2023; Soofastaei A., 2025). Smart room technologies—like voice-activated assistants and AI-managed environmental controls—have revolutionized how quickly staff can respond to guest needs, allowing them to focus on more complex interactions that require a personal touch (Mbarushimana, S., 2024). Together, these advancements signal a shift towards a more data-driven, responsive and a guest-focused approach to housekeeping management. While AI certainly boosts efficiency, experts consistently emphasize that technology should enhance, not replace, human skills. The current view is that AI serves as a tool to elevate human performance, taking care of routine tasks so that staff can concentrate on the more emotionally nuanced and creative aspects of their roles that require empathy and sound judgment (Gill A. & Mathur A., 2024; Challoumis C., 2024).

The use of AI in hospitality services can be explained by well-established theories of innovation and technology acceptance. AI adoption is explained by the Technology Acceptance Model (TAM) (Davis, 1989) based on perceived utility and usability, both of which are crucial in housekeeping settings where accuracy and efficiency are crucial. In a similar vein, Diffusion of Innovation Theory (Rogers, 2003) emphasizes how AI technologies progressively transition from early use to widespread use as their relative benefit, compatibility, and observability become apparent in hotel operations. The Technology–Organization–Environment (TOE) framework offers additional explanatory power at the organizational level by showing how decisions about AI integration are influenced by organizational culture, technological readiness, and external factors like expectations for post-pandemic hygiene (Tornatzky & Fleischer, 1990). Together, these frameworks provide an explanation for the quick adoption of AI in housekeeping, especially in labor-intensive and high-occupancy hotel settings.

The integration of AI into housekeeping also signifies a broader transformation in organizational culture within the hospitality industry. Sectors such as that of hotels are analyzing guest feedback and metrics of performance through use of AI, identifying trends and areas for improvement and aiding them in the decision making process. This transition encourages a proactive rather than reactive approach to service delivery, ensuring that potential issues are resolved before they affect the satisfaction of the guest. Also, AI's role in sustainability—such as monitoring water and energy use, optimizing cleaning chemical quantities or reducing waste—contributes to eco-friendly operations. AI is redefining housekeeping as a strategic, technology-driven function vital to overall hotel excellence.

The "human touch" is absolutely essential for guest satisfaction, especially in luxury and boutique hotels where personalized interactions really shape brand loyalty (Lei S.S.I. et al., 2025). Relying solely on automated service models can strip away the warmth, flexibility and spontaneity that guests expect from genuine hospitality. Plus, certain tasks in housekeeping and maintenance—like removing stains, arranging décor and customizing requests—demand a level of sensory judgment and adaptability that AI just can't match right now. Research shows that while AI can handle cleaning and scheduling with impressive accuracy, the more complex aspects of service still depend on human intuition and emotional intelligence (Huang M.H. & Rust R.T., 2018). As a result, hybrid operational models that blend human and technological efforts are becoming the go-to approach for the future of hotel operations. However, the impact of AI on the workforce brings some significant challenges. Studies suggest that automation could put low-skilled, entry-level jobs at risk, particularly in housekeeping, where tasks are routine and can be easily automated (Appelbaum, E., Bernhardt, A., & Murnane, R. J., 2003). This raises valid concerns about job loss and economic disparities in communities that rely on hospitality jobs. Employee pushback against automation—often referred to as "techno-insecurity"—is a common theme in discussions, fueled by fears of being replaced and a lack of digital skills (Bausch, D., Kraemer, T., & Mauroner, O., 2024).

The enduring importance of human connection in hospitality cannot be overstated, as it forms the emotional core of guest satisfaction and loyalty. Guests often remember not the technology that served them, but the warmth and attentiveness of the people who made their stay special. In an age of increasing automation, human staff bring creativity, empathy, adding a personalized touch that technology cannot emulate. Their ability to interpret non-verbal cues, anticipate unspoken needs and adaptation to unexpected situations ensures that service feels authentic and meaningful. Therefore, while AI enhances efficiency, the essence of hospitality continues to rest on genuine human engagement and care.

To tackle these disruptions, experts suggest that we focus on upskilling the workforce, redefining job roles, and creating hybrid training programs that enable employees to work seamlessly with AI systems. These strategies aim to evolve traditional housekeeping jobs into more supervisory or analytical roles that highlight oversight, creativity, and tech-savviness. Research on multigenerational workforces also emphasizes the importance of customizing AI implementation to fit the varied values and communication styles of different employees (Lakshmikanth R. et al., 2024). On a positive note, recent studies indicate that collaborating with AI can boost employee engagement and creativity by alleviating the emotional fatigue that comes from repetitive tasks (Liu X. & Li Y., 2025). The robot-employee interdependence theory suggests that when AI systems are seen as supportive partners instead of threats, they can enhance motivation and innovation among staff (Le, K. B. Q., Sajtos, L., & Fernandez, K. V., 2023). The researchers suggested that positive views of AI as a helpful partner can boost employee enthusiasm, creativity, and engagement. Additionally, new research on emotionally intelligent cobots—robots designed to read and respond to emotional signals—shows a growing interest in human-robot collaboration that combines efficiency with empathy (Toichoa Eyam, A., Mohammed, W. M., & Martinez Lastra, J. L., 2021). Modern literature promotes human-machine collaboration models, in which AI enhances human talents rather than taking the place of human labor. According to Davenport and Kirby (2016), the Augmentation Theory of AI views technology as a facilitator that improves human performance by lowering cognitive and physical effort. This translates into AI managing data-driven, standardized duties like housekeeping, while staff members concentrate on problem-solving, guest contact, and supervisory responsibilities.

In addition to training, organizational culture plays a critical role in ensuring successful AI integration. Managements' encouragement of an environment of trust, open communication, and continuous learning helps employees view technology as an enabler rather than a competitor. Encouraging staff participation in decision-making about AI adoption can reduce resistance and increase confidence in new systems. In addition, companies should emphasize lifelong learning through workshops, certifications, cross-functional exposure to building of digital competences. Recognizing and rewarding adaptability and innovation will further motivate employees to embrace change. Aligning human growth with technological progress ensures a harmonious, future-ready workforce in hospitality.

From the guests' viewpoint, the arrival of AI in housekeeping and related services stirs up a mix of excitement and concern. AI-driven personalization—through behavior tracking, predictive housekeeping, and tailored room environments—makes things more convenient but also brings up issues of privacy and trust (Nira R.A., 2025; Chandhok M. et al., 2025). Research shows that how satisfied guests feel with AI services largely hinges on their trust in the technology's reliability and the ethical handling of their personal data (Makivić, R. et al., 2024). If there's no clarity about data collection, personalization can come off as intrusive rather than beneficial (Karwatzki, S. et al., 2017).

Guests are increasingly becoming aware of how their data is collected and used these days, making transparency a crucial aspect of AI-driven hospitality. The purpose of data collection should be clearly communicated along with the benefits to the guests, and the measures being taken to protect their privacy. Establishing ethical guidelines and implementing secure data management systems can strengthen guest confidence and brand reputation. Also, allowing guests to customize their level of interaction with AI—such as opting in or out of automated features—can enhance their sense of control and comfort. Building this trust ensures that technology enhances, rather than disrupts, the overall guest experience.

Generational differences definitely come into play here: younger guests, especially those from Generation Z, tend to embrace AI-driven service robots and automation, seeing them as both efficient and fun (Peter R. et al., 2025). On the flip side, older travelers usually lean towards human interaction, which they associate with warmth and reliability (Praharaj, Swagat & Mishra et. al., 2023). The COVID-19 pandemic shook things up a bit, as guests started prioritizing health and safety over personal contact. During that time, AI and robotics stepped in to provide "touchless hospitality," which really helped improve perceptions of hygiene and safety in hotels (Gaur L. et al., 2021, Trihas, N., Zozonakis, G., & Kourgiantakis, M., 2022). Technologies like UV-C disinfection robots and autonomous cleaning systems became crucial rather than just experimental, tackling safety concerns and the labor shortages that the pandemic made worse (Zaman A. et al., 2022). AI-integrated housekeeping systems boosted efficiency in busy properties while also promoting eco-friendly practices, like cutting down on chemical use and optimizing energy consumption (Ayyappan V. & Ayyappan I., 2025).

As the hospitality industry continues to evolve, understanding generational preferences has become key to designing meaningful guest experiences. Younger travelers appreciate speed, digital convenience with seamless automation, while older generations value authenticity, reassurance and personalized services. Hotels adopt a blended service model that caters to both ends of the spectrum, integrating AI for efficiency without losing the warmth of human engagement. The pandemic accelerated digital adoption, but it also reminded the industry of the emotional importance of trust and safety. Striking a balance between innovation and empathy will be essential to meeting diverse guest expectations worldwide.

While the benefits of automation are well-documented, experts warn against becoming too reliant on it. Leaning too heavily on AI tools can stifle human creativity and cut down on those valuable interactions between employees and guests, which are so vital in the hospitality industry (Makivić, R. et al., 2024). Also, current AI models still struggle with emotional communication, highlighting just how crucial personal engagement is. Guests often find that services led by humans feel more genuine and emotionally satisfying, particularly in situations that require discretion, empathy, or problem-solving (Steinhoff L. & Martin K., 2023). As a result, many researchers suggest a balanced hybrid approach, where AI takes care of back-end tasks like predictive maintenance, smart inventory management, and workflow optimization, while human staff concentrate on crafting unforgettable guest experiences. This collaboration not only boosts productivity and guest satisfaction but also fosters sustainable and inclusive growth in the hospitality sector.

Maintaining equilibrium between automation and human involvement warrants that technology remains a supportive tool rather than a replacement for human potential. True hospitality thrives on warmth, adaptability along with emotional intelligence—qualities that machines are yet to fully emulate. Overdependence on AI could risk diluting the essence of service excellence and the spontaneity that defines memorable guest experiences. Therefore, continuous employee engagement, training as well as their empowerment are essential to complement technological efficiency with human sensitivity. By preserving this synergy, hotels can achieve operational excellence while ensuring that the soul of hospitality—genuine care and connection—remains at the heart of every interaction.

Summarizing, the literature shows that integrating AI into housekeeping is transforming the hospitality landscape by improving operational efficiency, sustainability, and guest personalization. However, human expertise is still essential for maintaining the emotional and experiential elements of service. Scholars agree that the future of housekeeping will thrive on collaboration between humans and machines rather than competition. As AI continues to advance, hotels need to focus on ethical technology use, empowering employees, and fostering emotional connections to create a well-rounded model of excellence—one that leverages the accuracy of technology while preserving the warmth of human hospitality.

RESEARCH METHODOLOGY

This descriptive research employed a mixed-methods approach, combining both quantitative and qualitative data analysis to thoroughly investigate the balance between AI integration and human expertise within the housekeeping sector. The research was guided by the following objectives:

1. Objective 1- To establish the impact of creating a balance between AI and human expertise in Housekeeping: Evaluating how a strategic balance streamlines operations and facilitates new operational opportunities.

Objective 2- To examine the significance of human expertise in AI-driven Housekeeping operations: Analyzing how human involvement significantly improves "Quality Services" and upholds brand reputation and sustainability.

Objective 3- To identify best practices for integrating AI and human expertise in Housekeeping: Determining how AI-powered tools can enhance individual productivity while being complemented by the human touch to reshape guest experiences.

Quantitative Data Collection and Analysis

Data Collection

Primary quantitative data were collected using a structured questionnaire. A calculated sample of 67 organizations across India was targeted. The sample included diverse housekeeping verticals, such as hotels, hospitals, facility management service providers, and the aviation sector.

To collect data for the study, a convenience sampling method was used. The questionnaire, which was closed-ended and Likert scale-based, was distributed via mail and followed up with phone calls to the departmental heads and managers specifically dealing with automation.

Items for the questionnaire focused on several key areas, including service enhancement, operational efficiency and enhanced productivity, guest experiences, impact of human expertise on AI operations, brand reputation, and energy consumption and sustainability

Data Analysis

The data were analyzed using SPSS 20 software. Descriptive statistics (e.g., frequency distribution, measures of central tendency, and variability) were utilized to summarize the data and identify trends related to organizational efforts in AI adoption.

The reliability and adequacy of the collected data and sample were rigorously assessed using Cronbach's alpha (confirmed as .881)

Qualitative Data Collection and Analysis

Secondary data analysis was conducted to contextualize the findings. Relevant existing data correlated to the research topic were sourced from various mediums, including the internet/websites, books, journals, magazines, newspapers, reports, old records, and feedback forms.

Reliability Test Results

A total of 92 expert responses were obtained from 67 organizations across the different housekeeping verticals (hotel, hospital, facility management and aviation). Table 1 shows that the reliability statistics yielded a Cronbach's alpha value of more than 0.7. As noted by Sekaran (2003), a Cronbach's alpha value exceeding 0.7 indicates that the data collection instrument is reliable. Cronbach's Alpha is calculated using the following formula:

$\alpha=(K)/(K-1) * Sy^2-\text{Sum } Si^2/Sy^2$

Where: α = Cronbach's Alpha

K = the number of items in the scale =16

Si = the sum of the item scores for the each item (13.92379)

S = the sum of the total scores for all items (80.21094)

Table 1. Reliability statistics exhibiting Cronbach's alpha value of the study (Source: Authors' interpretation)

Cronbach's Alpha	N of Items
.881	16

Data Analysis

- **Demographic Profile**

Response rate This study's sample population consists of employees from the supervisory level and managerial level.

Table 2. Frequency Distribution of Respondents' Demographic Profile(Source: Authors' interpretation)

SECTORS	FREQUENCY	PERCENTAGE	VALID PERCENTAGE	CUMULATIVE PERCENTAGE
HOTEL	48	52.17	52.17	52.17
HOSPITAL	7	7.61	7.61	59.78
FACILITY MANAGEMENT	28	30.43	30.43	90.22
AVIATION	9	9.78	9.78	100.00
TOTAL	**92**	**100**	**100**	

Table 2 shows total of 92 responses were received from the experts in different sectors from housekeeping. Mugenda & Mugenda (2003) quantified that 50 % or more than that is acceptable for analysis. Babbie (2004) also proclaimed in his study that 50 % return rates are satisfactory, 60 % is good and 70 % is perfect to analyze. The achieved response rate was 100 % implying that the response rate was excellent.

The study sought to establish the demographic profile of the organizations responding to the research questionnaire, ensuring the data reflects a dedicated industry focus. A total of 67 organizations (Annexure 1) participated in this research work, contributing the primary data for the quantitative analysis. All participating entities were drawn from the four housekeeping verticals, including Hotels, Hospitals, Facility Management Service Providers, and Aviation Sector.

- AI adoption rate

Responding to the organizational willingness to invest in AI-powered housekeeping technologies, most of the respondents agreed that in the coming two to three years their organizations will be adopting technologies in Housekeeping processes as shown in Table 3.

Table 3. Frequency Distribution of AI adoption (Source: Authors' interpretation)

RESPONSES	FREQUENCY	PERCENT	VALID PERCENT	CUMULATIVE PERCENT
Moderately likely	21	22.83	22.83	22.83
Very likely	34	36.96	36.96	59.78
Most likely	21	22.83	22.83	82.61
Not likely at all	4	4.35	4.35	86.96
Slightly likely	12	13.04	13.04	100.00
TOTAL	92	100	100	

Substantial responding verticals agreed that their organization, while adopting technology will definitely align with ethical & sustainability standards as exhibted in Table 4.

Table 4. Adoption of technology aligns with ethical & sustainability standards (Source: Authors' interpretation)

Housekeeping verticals	Yes	No	May be	
Aviation	6	0	3	9
Facility Management	19	0	9	28
Hospital	6	0	1	7
Hotel	28	1	19	48
TOTAL	59	1	32	**92**

- **Objective-wise Analysis**

  Objective 1: To establish the impact of creating a balance between AI and human expertise in Housekeeping: Evaluating how a strategic balance streamlines operations and facilitates new operational opportunities.

Figure 1. Impact of balancing AI and Human expertise in Housekeeping Department (Source: Authors' interpretation)

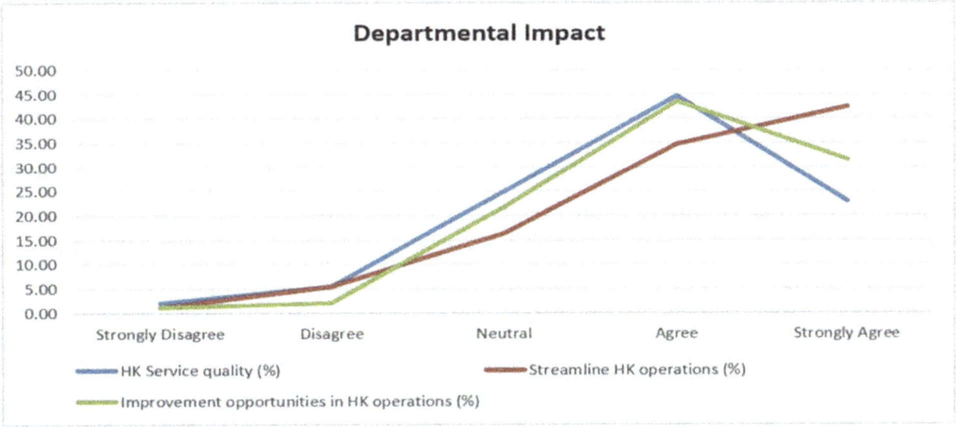

Many of the respondents (67%) agreed that the balanced use of automation in housekeeping contributes to improved service quality, while most also emphasized (77 %) that a strategic equilibrium between AI and human expertise is crucial for streamlining housekeeping operations, as shown in Figure 1. Additionally, majority of the respondents (75%) believed that the combined integration of AI and human expertise enhances operational opportunities by promoting innovation and efficiency. This balance allows AI to handle repetitive, data-driven tasks while empowering staff to focus on personalized guest experiences, problem-solving, and emotional engagement. Thus, maintaining harmony between technology and human insight ensures not only productivity and cost-effectiveness but also preserves the core values of hospitality—warmth, attentiveness, and authenticity in service delivery.

Figure 2. Impact of AI powered tools on optimizing Housekeepers performance (Source: Authors' interpretation)

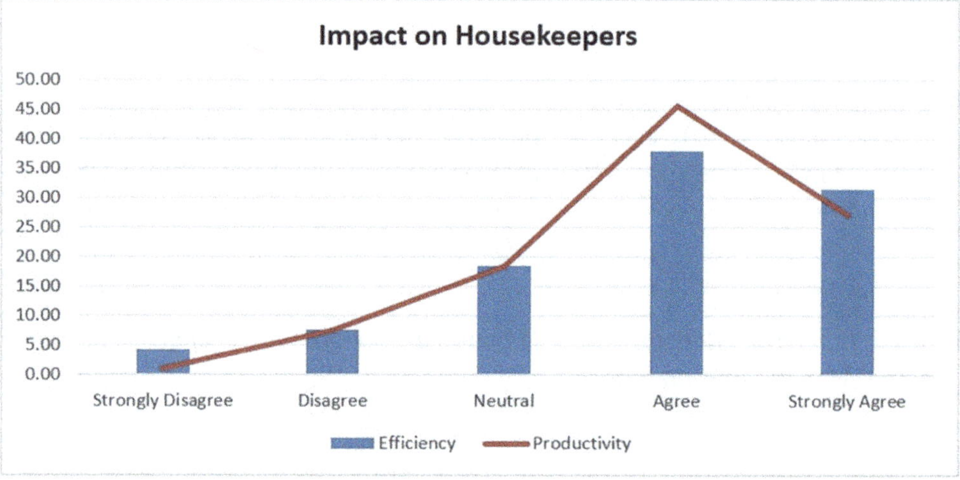

A significant majority of respondents (69%) agreed that AI-powered tools will substantially enhance the overall efficiency of the Housekeeping Department, as shown in Figure 2. Meanwhile, most (73%0 of the respondents acknowledged that the adoption of AI tools would positively impact the individual productivity of housekeepers. These findings highlight that technology-driven solutions, such as automated scheduling, predictive cleaning and smart inventory systems, streamline routine operations and reduce manual workload. Consequently, housekeepers can focus more on guest-centric tasks, improving service quality and operational accuracy. The data suggests that AI adoption not only improves time management and workflow efficiency but also supports staff morale and performance by reducing repetitive strain and enhancing task precision.

Figure 3. Impact of Artificial intelligence in reshaping guest experiences (Source: Authors' interpretation)

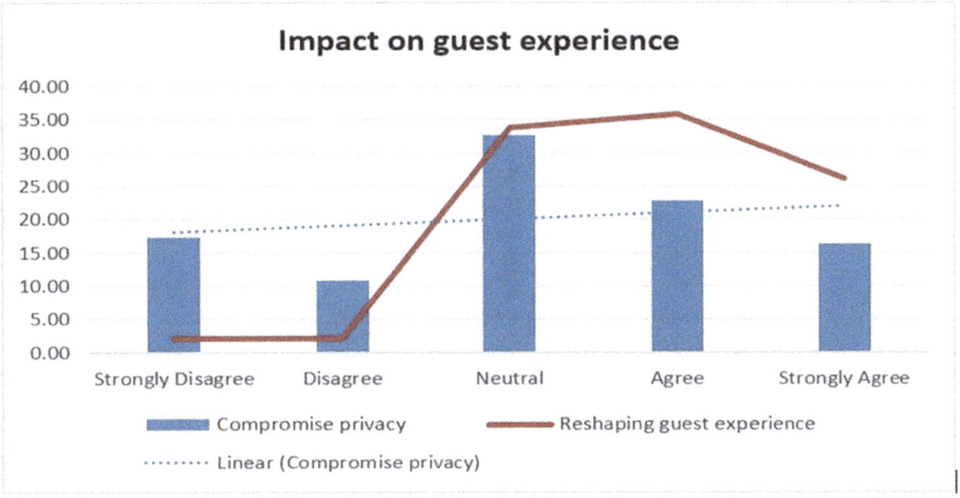

There was an agreement (67% of the respondents) that the integration of AI complemented by a human touch will significantly reshape guest experiences as shown in Figure 3. This highlights a belief that successful AI implementation in service delivery requires balancing technology with human interaction. In contrast, the level of agreement regarding privacy was mixed where 39% of respondents agreed that AI is not compromising with guest privacy as exhibited in Figure 3. This indicates that while many see the experiential benefits of AI, a significant portion (61%) either believe AI does compromise guest privacy or are uncertain about its impact on data security.

Objective 2: To examine the significance of human expertise in AI-driven Housekeeping operations: Analyzing how human involvement significantly improves "Quality Services" and upholds brand reputation and sustainability.

Figure 4. Significance of Human expertise in AI driven Housekeeping Operations (Source: Authors' interpretation)

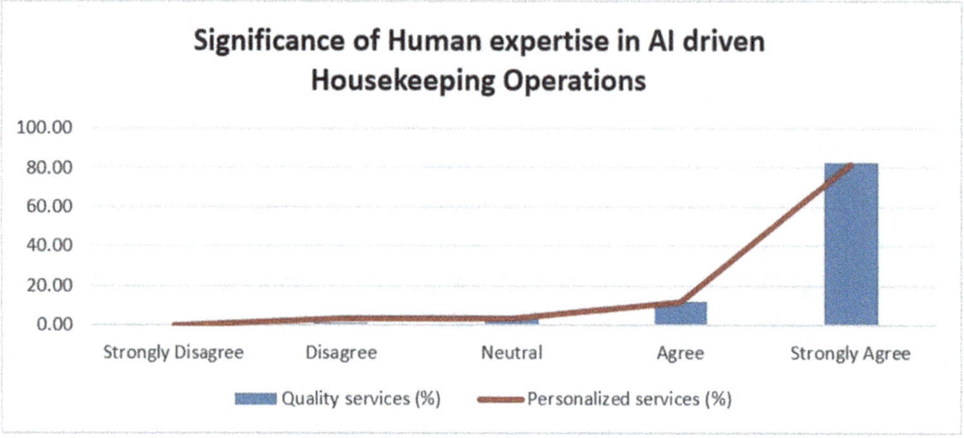

Figure 4 shows that majority of the respondents (93%) agreed that human expertise has a significant impact in human-driven housekeeping operations, while most of them (82%) believed that human expertise in AI-driven housekeeping operations significantly improves service quality. This finding highlights that despite automation, the human role remains central in maintaining personalization, emotional engagement, achieving and maintaining quality assurance. Employees' judgment, adaptability, and empathy ensure that service standards align with guest expectations. Human oversight in AI-supported systems also minimizes operational errors, improves communication, and ensures ethical handling of guest data, thereby reinforcing the need for a balanced human-technology collaboration in modern hospitality.

Objective 3: To identify best practices for integrating AI and human expertise in Housekeeping: Determining how AI-powered tools can enhance individual productivity while being complemented by the human touch to reshape guest experiences.

Figure 5. AI and Human expertise integrating Best practices in Housekeeping (Source: Authors' interpretation)

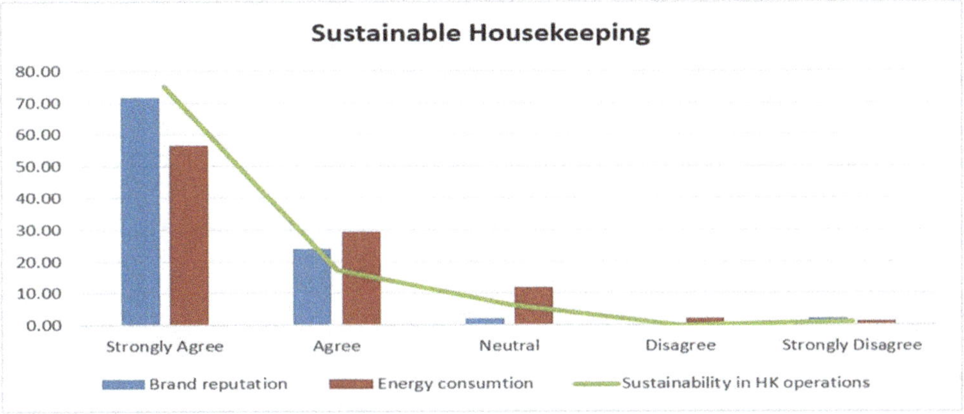

The majority of respondents agreed that AI significantly contributes to strengthening brand reputation, as shown in Figure 5. Additionally, majority of respondents also acknowledged that AI effectively integrates sustainability into housekeeping operations, supporting eco-friendly initiatives such as waste reduction and efficient resource management. Further, most of them (85%) agreed that AI plays a key role in monitoring and optimizing energy consumption across properties. These findings suggest that integrating AI with human expertise not only enhances operational efficiency but also aligns with broader organizational goals of sustainability, brand value and responsible business practices. Combining human intuition with AI-driven insights ensures consistency with adaptability ensuring continuous improvement in housekeeping standards.

FINDINGS

Outcome of data analysis, and findings includes:

- Organizations are eager to invest and adopt AI-powered housekeeping technologies.
- Organization are ready to accept and follow ethical & sustainability standards while adopting technology,
- Balanced use of automation in housekeeping significantly improve service quality,
- Strategic balance between AI and human expertise positively streamline Housekeeping operations,

- Collective integration of AI and human expertise would facilitate the improvement of Housekeeping operational opportunities,
- AI-powered tools will significantly enhance the efficiency of the Housekeeping Department,
- AI-powered tools will positively enhance the productivity of individual Housekeepers,
- Integration of AI complemented by a human touch will significantly reshape guest experiences,
- Human expertise has significant impact on AI driven Housekeeping Operations,
- Human expertise in AI driven Housekeeping Operations significantly improve Quality services,
- Human expertise and AI significantly helps in integrating brand reputation,
- Human expertise and AI integrates sustainability into Housekeeping operations,
- Human expertise and AI signifies the energy consumption in an organization.

DISCUSSIONS AND CONCLUSION

The study set out to examine the demographic profile of respondents and their perceptions regarding the integration of Artificial Intelligence in housekeeping operations. The findings indicate that a majority of respondents (52%, n=48) belonged to the hotel industry, followed by 30% from Facility Management services, reflecting strong representation from service-intensive sectors. To remain competitive, organizations must carefully manage the balance between investment costs and potential employee resistance to technological change (Valtonen & Holopainen, 2025). While frequency analysis reveals that most of the respondents (82.61 %) anticipate the adoption of technology in housekeeping processes within the next two to three years, while only 64 % expressed confidence that ethical and sustainability standards would be fully adhered to. This gap reinforces the argument by Omar et al. (2024) that widespread acceptance of AI in hospitality depends on balancing operational innovation with ethical responsibility. Responsible digital transformation, therefore, must be firmly embedded within sustainability frameworks (Pappas et al., 2023).

Amidst persistent skill shortages and escalating labour costs, respondents perceive AI integration as a strategic mechanism for standardising Standard Operating Procedures (SOPs) and enhancing service quality. The findings demonstrate that respondents (67%) believed a balanced application of automation directly improves service quality, while most (75 %) acknowledged that the synergy between AI and human expertise expands operational opportunities. Further, majority of the respondents (77 %) emphasise the necessity of a strategic balance, indicating a shift from total automation towards collaborative human–machine models. Consistent with Mandić, Panić, and Čičin-Šain (2023), the study confirms that AI streamlines operations, enhances service delivery, and reduces physical workload. The results further establish that the integration of AI and human expertise significantly influences housekeeping performance by improving service standards, operational consistency, and future growth prospects. Supporting this, Zahidi et al. (2024) advocate that balanced AI–human engagement strengthens organizational culture and positions AI as an enabler of customer satisfaction.

The study also reveals that AI-powered tools substantially enhance housekeeping efficiency and employee productivity. These findings align with Tan et al. (2024), who highlight that balanced AI adoption positively affects employee competitiveness and productivity. Moreover, while AI optimises processes and performance, the complementary role of human expertise is critical in fostering repeat visitation and customer loyalty (Kadagidze & Ugrelidze, 2023). The integration of AI with human touch is thus shown to significantly reshape guest experiences, corroborating Al-Hyari et al. (2023), who assert that automation enhances guest satisfaction only when it supports—not replaces—personalised human interaction.

Ethical concerns, particularly relating to guest privacy and data security, emerge as a critical challenge. Consistent with Devaraj (2023) and Zahidi, Kaluvilla, and Mulla (2024), respondents expressed apprehensions regarding ethical governance and technical vulnerabilities (Taufiqur Rahman & Waked, 2025). Supporting these concerns, only 39% of housekeeping experts agreed that AI does not compromise guest privacy, highlighting a persistent trust deficit. Despite these challenges, the findings confirm the continued significance of human expertise in AI-driven housekeeping operations, demonstrating a positive and significant impact on service quality. This supports Srivastava and Tyagi (2024), who argue that automation enhances competitiveness and profitability when supported by skilled human oversight.

The study further identifies best practices for effective AI–human integration, with respondents emphasising brand reputation, sustainability initiatives, and energy optimisation as key outcomes of balanced adoption. Mustikasari et al. (2025) similarly establish that ethical AI practices confer competitive advantage by strengthening customer relationships and enhancing brand perception, thereby supporting long-term loyalty. Empirical evidence from the study reinforces this view, with majority (94%) of experts associating AI adoption with stronger brand reputation, and a high number (92%) recognising its role in advancing sustainability. An equally high number of respondents (85%) acknowledged its effectiveness in monitoring and optimising energy consumption. Nevertheless, organizational resistance persists, largely driven by fears of job displacement and ethical lapses in data management.

A notable "trust gap" continues to impede full-scale AI adoption. Although personalization opportunities generate optimism, stakeholders (61%) remain uncertain or concerned about data security and privacy. Addressing this paradox necessitates robust risk assessment, transparent governance mechanisms, and a firm commitment to ethical data stewardship to ensure that technological progress does not erode guest trust.

Ultimately, the transition from traditional housekeeping to AI-integrated operations is fundamentally a human journey. The industry is currently navigating a "rippling phase," where the primary challenge lies not in technological capability but in workforce skill gaps—particularly among frontline employees with limited technical exposure. The most viable path forward lies in Strategic Stewardship, requiring organizations to actively guide workforce evolution by investing in continuous upskilling and soft-skill development, empowering staff through supportive AI tools that reduce physical strain, and preserving the human touch that defines hospitality. By positioning AI as a collaborative partner rather than a substitute, the housekeeping sector can bridge the trust gap and cultivate an operational environment that is both technologically advanced and profoundly human.

FUTURE IMPLICATIONS

The findings reinforce that balancing technology with Human expertise will help in enhancing the Housekeeping department. Results identified that implementation of ethical and sustainable AI standards are required for adopting technology. With the help of this finding, Housekeeping managers are recommended to implement ethical and sustainable standards in order to enhance customer experience and brand loyalty.

Results also identified that strategic balance between AI and human expertise positively streamline Housekeeping operations. In order to reshape guest experiences, improve quality services and maintain brand reputation, the finding of this research work recommends Housekeeping managers to integrate AI complemented with human expertise.

SCOPE FOR FUTURE RESEARCH

- The study is conducted at the national level; future research can be carried out at international level.
- A sample size of the study employee from 67 organizations from different verticals of Housekeeping, which may increase in future research work.

REFERENCES

Al-Hyari, H.S., Al-Smadi, H.M., & Weshah, S.R. (2023). The impact of Artificial Intelligence (AI) on guest satisfaction in Hotel Management: an empirical study of luxury hotels. GeoJournal of Tourism and Geosites, 48(2spl), 810–819. https://doi.org/10.30892/gtg.482spl15-1081

Alemede, V. O. (2025). Innovative process technologies: Advancing efficiency and sustainability through optimization and control. *Int J Res Publ Rev*, *6*(2), 1941–1955. DOI: 10.55248/gengpi.6.0225.0904

Appelbaum, E., Bernhardt, A., & Murnane, R. J. (2003). *Low-wage America: An overview. Low Wage America: How Employers Are Reshaping Opportunity in the Workplace*. Russell Sage Foundation.

Ayyappan, V., & Ayyappan, I. (2025).Eds.Das J. & Eugene J., Predictive Analytics and IoT for Cleanliness Optimization in Indian Hotels in Sustainable Cleanliness, Teeneu Publications, ISBN 978-81-968444-3-1

Bausch, D., Kraemer, T., & Mauroner, O. (2024). Technology-induced stress and employee resistance in the context of digital transformation and identification of countermeasures. *International Journal on Management of Innovation & Technology*, *21*(04), 2450029. DOI: 10.1142/S0219877024500299

Challoumis, C. (2024). The dawn of artificial intelligence. In *XIX International Scientific Conference*. London. Great Britain(pp. 169-205).

Davenport, T. H., & Kirby, J. (2016). *Only humans need apply: Winners and losers in the age of smart machines*. Harper Business.

Davis, F. D. (1989). Perceived usefulness, perceived ease of use, and user acceptance of information technology. *Management Information Systems Quarterly*, *13*(3), 319–340. DOI: 10.2307/249008

Gaur, L., Afaq, A., Singh, G., & Dwivedi, Y. K. (2021). Role of artificial intelligence and robotics to foster the touchless travel during a pandemic: A review and research agenda. *International Journal of Contemporary Hospitality Management*, *33*(11), 4079–4098. DOI: 10.1108/IJCHM-11-2020-1246

Gill, A., & Mathur, A. (2024). Emotional Intelligence in the Age of AI: Enhancing Workforce Development for Human-Machine Collaboration. In Sustainable Innovation for Industry 6.0 (pp. 263-285). IGI Global.

Huang, M. H., & Rust, R. T. (2018). Artificial intelligence in service. *Journal of Service Research, 21*(2), 155–172. DOI: 10.1177/1094670517752459

Kadagidze, L., & Ugrelidze, E. (2023). AI in Hospitality industry: a comprehensive study on its impact on operations, customer experience, and revenue management. Conference: VII International European Conference on Interdisciplinary Scientific Research

Karwatzki, S., Dytynko, O., Trenz, M., & Veit, D. (2017). Beyond the personalization–privacy paradox: Privacy valuation, transparency features, and service personalization. *Journal of Management Information Systems, 34*(2), 369–400. DOI: 10.1080/07421222.2017.1334467

Kaul, D. (2023). Ai-driven real-time inventory management in hotel reservation systems: Predictive analytics, dynamic pricing, and integration for operational efficiency. *Emerging Trends in Machine Intelligence and Big Data, 15*(10), 66–80.

Lakshmikanth, R., Vijai, C., Krishna, S. H., Alla, P. C. R., Lourens, M., & Soni, M. (2024, March). Utilizing AI to tailor leadership styles and interventions for optimal employee engagement and performance. In *2024 International Conference on Trends in Quantum Computing and Emerging Business Technologies* (pp. 1-7). IEEE.

Le, K. B. Q., Sajtos, L., & Fernandez, K. V. (2023). Employee-(ro) bot collaboration in service: An interdependence perspective. *Journal of Service Management, 34*(2), 176–207. DOI: 10.1108/JOSM-06-2021-0232

Lei, S. S. I., Wu, D., Liu, G. G., & Law, R. (2025). Customizing customer journey in hotels: A focus on human touch. *International Journal of Hospitality Management, 124*, 103956. DOI: 10.1016/j.ijhm.2024.103956

Liu, X., & Li, Y. (2025). Examining the double-edged sword effect of AI usage on work engagement: The moderating role of core task characteristics substitution. *Behavioral Sciences (Basel, Switzerland), 15*(2), 206. DOI: 10.3390/bs15020206 PMID: 40001838

Makivić, R., Vukolić, D., Veljović, S., Bolesnikov, M., Dávid, L. D., Ivanišević, A., Silić, M., & Gajić, T. (2024). AI Impact on Hotel Guest Satisfaction via Tailor-Made Services: A Case Study of Serbia and Hungary. *Information (Basel), 15*(11), 700. https://doi.org/10.3390/info15110700DOI: 10.3390/info15110700

Mandić, D., Panić, A., & Čičin-Šain, M. (2023). Automated service delivery in hotels: balancing efficiency and human interaction for optimal guest satisfaction. *International Scientific Conference on Economy, Management and Information Technologies*, Vol. 1: 167-174. DOI - DOI: 10.46793/ICEMIT23.137M

Manoharan, G., & Ashtikar, S. P. (2025). AI Era in Hospitality: The New Frontier in Guest Satisfaction and Operations. In Digital Disruption in Hospitality, AI, and Emerging Technologies: A Roadmap to Personalized Experiences, Enhanced Operations, and Revenue Growth (pp. 59-77). Emerald Publishing Limited.

Mbarushimana, S. (2024). The Impact of Smart Room Technology on Guest Satisfaction and Operational Efficiency. *Journal of Modern Hospitality.*, *3*(2), 1–13. DOI: 10.47941/jmh.1956

Mujtaba, B. G. (2025). Human-AI Intersection: Understanding the Ethical Challenges, Opportunities, and Governance Protocols for a Changing Data-Driven Digital World. *Business Ethics and Leadership*, *9*(1), 109–126. DOI: 10.61093/bel.9(1).109-126.2025

Mustikasari, A., Hurriyati, R., Dirgantari, D. P., Sultan, A. M., & Sugiana, S. N. (2025). The Role of Artificial Intelligence in Brand Experience: Shaping Consumer Behavior and Driving Repurchase Decisions. *International Journal of Advanced Computer Science and Applications*, *16*(4), •••. DOI: 10.14569/IJACSA.2025.0160432

Nira, R. A. (2025). AI-Driven Hyper-Personalization in Hospitality: Application, Present and Future Opportunities, Challenges, and Guest Trust Issues. *International Journal of Research and Innovation in Social Science*, *9*(4), 5562–5573. DOI: 10.47772/IJRISS.2025.90400397

Pappas, I., Mikalef, P., Dwivedi, K. Y., Jaccheri, L., & Krogstie, J. (2023). Responsible Digital Transformation for a Sustainable Society. *Information Systems Frontiers*, *25*(3), 945–953. https://doi.org/10.1007/s10796-023-10406-5DOI: 10.1007/s10796-023-10406-5 PMID: 37287710

Peter, R., Roshith, V., Lawrence, S., Mona, A. E., Narayanan, K. B., & Yusaira, F. (2025). Gen AI – Gen Z: Understanding Gen Z's emotional responses and brand experiences with Gen AI-driven, hyper-personalized advertising. *Frontiers in Communication*, *10*, 1554551. DOI: 10.3389/fcomm.2025.1554551

Praharaj, S., Mishra, B., Mishra, U., Panigrahi, R. R., & Mishra, P. C. (2023). Role of service automation on guest experience of hotel industry. *Tourism and Hospitality Management*, *29*(2), 265–278. DOI: 10.20867/thm.29.2.11

Rogers, E. M. (2003). *Diffusion of innovations* (5th ed.). Free Press.

Omar, S. I., Aliah, N., & Kwong, T. A. (2024). The Ethical Implications of AI-Powered Personalization in Hospitality and Tourism. Impact of AI and Tech- Driven Solutions in Hospitality and Tourism. IGI Global, 103-122. DOI:10.4018/979-8-3693-6755-1.ch006

Soofastaei, A. (2025). Intelligent scheduling: how AI and advanced analytics are revolutionizing time optimization. In *Mastering Time-Innovative Solutions to Complex Scheduling Problems*. IntechOpen. DOI: 10.5772/intechopen.1008531

Srivastava, P. & Tyagi, K. P. (2024). Implications of AI in Hotel Housekeeping Operations. Hotel and Travel Management in the AI Era, IGI Global. DOI - DOI: 10.4018/979-8-3693-7898-4.ch016

Steinhoff, L., & Martin, K. D. (2023). Putting data privacy regulation into action: The differential capabilities of service frontline interfaces. *Journal of Service Research*, *26*(3), 330–350. DOI: 10.1177/10946705221141925

Tan, K. L., Hofman, P. S., Noor, N., Tan, S. R., Hii, I. S. H., & Cham, T. H. (2024). Does artificial intelligence improve hospitality employees' individual competitive productivity? A time-lagged moderated-mediation model involving job crafting and meaningful work. *Current Issues in Tourism*, *28*(20), 3284–3301. DOI: 10.1080/13683500.2024.2391114

Taufiqur Rahman, B. S., & Waked, N. H. (2025). The impact of artificial intelligence on enhancing guest experience in luxury hotels in Kuala Lumpur, Malaysia. [DOI: 10.22271/27069583.2025.v7.i1c.141]. *International Journal of Tourism Hotel Management*, *7*(1), 208–216.

Thakur, A., Kaipa, K., Banerjee, A. G., Cappelleri, D. J., Krovi, V. N., & Gupta, S. K. (2025). Physical AI for Powering the Next Revolution in Robotics. *Journal of Computing and Information Science in Engineering*, •••, 1–21.

Toichoa Eyam, A., Mohammed, W. M., & Martinez Lastra, J. L. (2021). Emotion-driven analysis and control of human-robot interactions in collaborative applications. *Sensors (Basel)*, *21*(14), 4626. DOI: 10.3390/s21144626 PMID: 34300366

Tornatzky, L. G., & Fleischer, M. (1990). *The processes of technological innovation*. Lexington Books.

Trihas, N., Zozonakis, G., & Kourgiantakis, M. (2022, May). Contactless Tourism in the Post-COVID-19 Era: How Ready Are Greek Tourists? In International Conference on Modern Trends in Business Hospitality and Tourism (pp. 103-118). Cham: Springer Nature Switzerland.

Valtonen, A. & Holopainen, M.,(2025). Mitigating employee resistance and achieving well-being in digital transformation. Information Technology and People, Vol. 38 (8): 42-72. DOI-DOI: 10.1108/ITP-05-2024-0701

Watts, S., & Munir, T. (2025). Bridging the gap: Exploring innovation enablers, challenges and AI adoption for enhanced workforce productivity. *International Journal of Productivity and Performance Management*, •••, 1–17. DOI: 10.1108/IJPPM-01-2025-0001

Yang, J. (2024). The investigation of bionic-companionship framework for Humanoid Service Robotics (HSR) with deep learning image captioning (Doctoral dissertation, Cardiff Metropolitan University).

Zahidi, F., Kaluvilla, B. B., & Mulla, T. (2024). Embracing the new era: Artificial intelligence and its multifaceted impact on the hospitality industry. Journal of Open Innovation: Technology, Market, and Complexity. DOI. Vol. 10(4 https://doi.org/10.1016/j.joitmc.2024.100390

Zaman, A., Majib, M. S., Tanjim, S. A., Siddique, S. M. A., Islam, S., Aadeeb, M. S., & Islam, M. N. (2022). UVC-PURGE: A novel cost-effective disinfection robot for combating COVID-19 pandemic. *IEEE Access : Practical Innovations, Open Solutions*, *10*, 37613–37634. DOI: 10.1109/ACCESS.2022.3163243 PMID: 35582495

APPENDIX

Table 5. Organizations participated in the research

S. No.	Organization (respondents)
1	Accenture
2	Acciona Facility Services
3	Ibis, Kolkata (Accor Group)
4	Akasa Air
5	Angsana Oasis Spa & Resort, Bengaluru
6	Apollo Hospital
7	Aramark-Food Services and Facilities Management
8	Aster RV Hospital
9	Bangalore International Airport Ltd
10	BGS Institute of Hotel Management, Bengaluru
11	Bhartiya Skill Development University (BSDU), Jaipur
12	British Council
13	CBRE South Asia Pvt Ltd
14	Cleantech Management Services Pvt. Ltd.
15	Conrad, Bengaluru
16	Denali Management Services, Bengaluru
17	Duster Total Solutions Services Pvt Ltd
18	Fernandez Hospital, Stork Home
19	FMCO facilities (KSA)
20	Grand Hyatt,
21	Grand Mercure Bengaluru at Gopalan Mall
22	Green park Hotels and Resorts Limited
23	Guru Nanak Institute of Hotel Management (JIS Group), Kolkata
24	Haldiram Snacks Pvt.Ltd.
25	Holiday Inn Express Kolkata Airport
26	Hotel Pride Executive, Solapur
27	Taj Puri Resort & Spa, Odisha
28	IHM, Kolkata
29	IHM-PUSA
30	IHM Shri Shakti, Hyderabad
31	Indian Hotels Company Limited (Taj Group of hotels)

continued on following page

Table 5. Continued

S. No.	Organization (respondents)
32	Indian School of Hospitality
33	Insight Mall Developer Private Limited
34	Institute of Hotel Management, Chennai
35	International Real Estate Partners (IREP)-Facility Management Solutions
36	ITC Kakatiya, Hyderabad
37	JLL, Gurgaon
38	Kony India Pvt. Ltd.
39	Lemon Tree Hotels
40	M3M Broadway
41	Manipal Hospitals, Kolkata
42	Noah Sky Suites, Kochi
43	Medcare Hospitals & Medical Centers
44	Management & Entrepreneurship and Professional Skills Council (MEPSC)
45	Merlin Group, Kolkata
46	Naipunnya Institute of Management and Information Technology, Kerala
47	NIS Management Limited, Kolkata
48	Noida International Airport
49	Park Hyatt, Chennai
50	Prestige Group Hotels
51	S.E.A. college of Science Commerce Arts-Department of Hotel management
52	Sheraton Grand Bangalore Hotel at Brigade Gateway
53	St Joseph's Institute of Hotel Management & Catering Technology, Palai (SJIHMCT)
54	Sodexo India Services Private Limited
55	Sula Vineyards
56	TAFE services
57	Taj City Centre, Gurugram
58	O by Tamara Coimbatore (a unit by Tamara Leisure Experiences)
59	TGI Grand Fortuna, Hosur
60	The Aviyaan, Lucknow
61	The Oterra Hotel
62	The Phoenix Mills Co. Ltd.
63	The Sonnet Hotel, Kolkata
64	The Umrao Hotel, New Delhi
65	Vivanta Kolkata EM Bypass

continued on following page

Table 5. Continued

S. No.	Organization (respondents)
66	Yamuna International Airport Private Limited
67	Zolostays Property Solution Pvt Ltd.

Chapter 6
Reception Automation and Guest Experience:
Balancing Efficiency and Personalization in Hotel Front Offices

Rathindra Nath Samanta
https://orcid.org/0009-0004-8157-8871
Welcomgroup Graduate School of Hotel Administration, Manipal Academy of Higher Education, Manipal, India

Vishwajeet P. Mathad
https://orcid.org/0009-0002-3144-9719
Welcomgroup Graduate School of Hotel Administration, Manipal Academy of Higher Education, Manipal, India

Rakshith Kamath
https://orcid.org/0009-0007-0801-4086
Welcomgroup Graduate School of Hotel Administration, Manipal Academy of Higher Education, Manipal, India

ABSTRACT

This chapter explores how hotel front offices are being reshaped by reception automation while guest expectations for personalized service continue to rise. Technologies such as self-service kiosks, mobile check-in, AI chatbots, and integrated CRM systems improve speed, accuracy, and operational efficiency, yet may reduce the emotional warmth central to hospitality. Using established theories and industry evidence, it analyses this tension and proposes a high-tech, high-touch model in which auto-

DOI: 10.4018/979-8-3373-6541-1.ch006

mation manages routine tasks and staff focuses on empathy, problem-solving, and meaningful guest engagement. This balanced approach enables hotels to deliver both efficiency and richer, more personalized experiences.

1. INTRODUCTION - THE MODERN FRONT OFFICE DILEMMA

The hotel front office has been referred to as the "nerve centre" of hotel operations, a critical nexus where revenue management, security, and most prominently guest relations intersect. It is a crucial revenue center that handles reservations, check-in/check-out, and payment processing. At the same time, it is the main stage for creating the important first and last impressions that guests have of the establishment (S. Ivanov et al., 2017). Historically, this field was exclusively human-centred, relying intrinsically on the soft skills, emotional intelligence, and nuanced expertise of front desk agents to provide warm welcomes, recognize returning patrons, and deliver bespoke service that formed the bedrock of hospitality.

However, the contemporary hospitality industry is navigating an unprecedented landscape. Soaring operational costs, volatile staffing levels post-pandemic, and the pressing demand for 24/7 service availability have intensified the pursuit of operational efficiency (Ivanov & Webster, 2019). Simultaneously, the profile of modern travellers has evolved. Empowered by digital technology and saturated with choice, today's guests exhibit a complex set of expectations: they demand the speed, seamlessness, and autonomy of digital transactions alongside the thoughtful, memorable, and highly personalized experiences that were the hallmarks of traditional hospitality (I. Tussyadiah, 2020a).

The convergence of internal operational pressures and external guest expectations has resulted in the rapid adoption of reception automation technologies. From self-service kiosks and mobile check-in apps to AI-powered chatbots, these tools promise a definitive solution to the efficiency dilemma. However, their integration introduces a new, critical question for both scholars and practitioners: can algorithms and machines deliver the genuine warmth, discretionary effort, and intuitive personalization that lie at the very heart of the hospitality concept? This study delves into this complex interplay, arguing that the future of the front office does not reside in a binary choice between technology and humanity but in a deliberate, strategically designed synergy. The path to a sustainable competitive advantage lies in leveraging technology to handle standardized transactions with flawless efficiency, thus enabling human capital to focus on creating rich, emotionally resonant interactions. This chapter analyses this tension, assesses the effects of automation, and proposes a robust "high-tech, high touch" framework to achieve this important balance.

2. LITERATURE REVIEW

Reconciling Technological Efficiency with the Demands of Personalized Hospitality

To fully understand the modern issue facing hotel front offices, one must embark on an investigative journey through their historical evolution, theoretical foundations, and competing demands that describe their modern functions. This review synthesizes a wide range of literature to contextualize the current paradigm shift, highlighting it not as a sudden disruption but as the latest chapter in the ongoing automation and dehumanization of hospitality.

2.1. The Historical Evolution of the Front Office: From Gatekeeper to Experience Gateway

The history of the front office is a story of transforming societal expectations of service, power dynamics, and technology adoption. Three separate eras can be distinguished in its evolution.

2.1.1 The Pre-Industrial Era: The Front Desk as Fortress and Gatekeeper

In the inns and coaching houses of the 18th and 19th centuries, the front desk often had a simple counter or a private office as a symbol of authority and authority. The interaction was transactional and paternalistic, with minimal expectations of personalized services beyond basic shelter and sustenance. The "guest record" was often a ledger book, and its primary function was administrative control and revenue collection. This era established the front office's foundational role as the operational and financial hub of the establishment. It has been noted that pre-industrial hospitality systems were shaped by rigid social hierarchies and limited service diversity, where innkeepers controlled guest interactions and standardized offerings prevailed due to resource and labor constraints. This early model reflected a goods-dominant logic in which service value was embedded in basic functional delivery rather than experiential differentiation (Lashley, 2008). As such, personalization was minimal because hospitality was perceived primarily as shelter provision, not an orchestrated guest experience.

2.1.2 The Grand Hotel Era: The Concierge as Cultural Intermediary and the Rise of Theatrical Hospitality

The 19th century saw the rise of the grand hotel in Europe and America, epitomized by establishments like The Savoy in London and The Ritz in Paris. This period marked a significant shift from the mere provision of services to performance. The front desk has become a stage for the "theatre of hospitality" (King, 1995). The role of the staff, particularly the concierge, evolved into that of a cultural intermediary and an excellent problem solver. Armed with a vast network of contacts and an intuitive understanding of aristocratic and later high-society tastes, the concierge could secure impossible theatre tickets, recommend tailors, or arrange private tours. This was the genesis of high-touch anticipatory personalization. This relationship was deeply human-centric, relying on discretion, memory, and social acuity. The rise of grand hotels also marked the emergence of hospitality as an experience-centric service, where luxury properties began constructing symbolic value through architectural grandeur, personalized attention, and social exclusivity. Historical analyses show that this era laid the foundations for experiential differentiation, as guest satisfaction increasingly depended on emotional engagement, status recognition, and the concierge's cultural expertise(Ottenbacher & Harrington, 2007). This shift represented one of the earliest forms of curated, high-touch guest experience in modern hospitality.

2.1.3 The Modern Corporate Era: Systematization, Brand Standards, and the Efficiency Mandate

The post-war expansion of corporate hotel chains (e.g., Holiday Inn, Hilton) introduced a new imperative: standardized, scalable, and efficient service delivery. The front office was systematized through detailed operating procedures, brand standards, and initial computerization of operations. Efficiency, accuracy, and adherence to brand scripts became key performance indicators. While this era brought unprecedented consistency and scale, it also began to subtly de-skill the front desk role, prioritizing process over personal intuition.

Therefore, the current wave of automation is not an isolated event but an acceleration of the corporatization trend. However, as (S. Ivanov et al., 2017) argue, its impact is fundamentally different because it moves beyond supporting the human agent to directly mediating core guest-host interactions. This challenges a dynamic that has been central to hospitality for centuries, forcing a redefinition of service encounters. Studies emphasizes that the historical evolution of the front office has mirrored broader transformations in service management and consumer expectations. As digital technologies matured, guest–firm interactions shifted from purely transactional exchanges to experience-driven encounters, where technology increasingly became a mediator of value creation (Buhalis & Sinarta, 2019). This evolution reflects a long-term progression from manual administrative tasks toward sophisticated, data-enabled service ecosystems. The modern front office, therefore, represents the culmination of decades of technological, cultural, and organizational changes that have gradually reoriented hospitality toward personalization, immediacy, and seamless service delivery. Automation today should thus be viewed as the latest phase in an ongoing continuum rather than a disruptive break from the past.

2.2. Theoretical Lenses for Understanding Automation Adoption

To analyse reception automation and guest experience in a rigorous manner, this chapter draws on complementary theoretical perspectives that operate at individual, organisational, and service-system levels. Combining these lenses helps to explain (1) why guests adopt (or reject) automation, (2) how hotels transform capabilities to derive competitive advantage, and (3) how value is co-created or co-destroyed when automation interacts with human service.

2.2.1 The Individual Guest: Technology Acceptance Model (TAM) and Beyond

The Technology Acceptance Model (TAM), developed by (Davis, 1989), remains the dominant framework for understanding guest adoption of self-service technologies. TAM posits that a user's intention to use a system is determined by two primary factors: *Perceived Usefulness (PU)* the degree to which a person believes that using a system will enhance their job performance or, in this context, their experience and *Perceived Ease of Use (PEOU)* the degree to which a person believes that using a system will be free of effort. Numerous studies in hospitality have validated TAM, showing that guests are more likely to use mobile check-in or kiosks if they find them useful for saving time and are easy to navigate (Morosan & DeFranco, 2016). Recent evidence also shows that empathic chatbot interfaces can enhance customers' perceptions of social presence and information quality, which significantly strengthens satisfaction with automated service encounters. At the same time, these systems may become counterproductive when guests experience time pressure, demonstrating that perceptions of usefulness and ease of use are context-dependent rather than stable across situations (Juquelier et al., 2025).

However, contemporary research has expanded TAM to include other critical variables in the hospitality context. Perceived Risk (concerns about data security and payment fraud), Personal Innovativeness (an individual's willingness to try out new technologies), and the influence of Social Factors (seeing other guests use technology) have been shown to significantly influence adoption (Morosan, 2016). In addition to TAM, the Unified Theory of Acceptance and Use of Technology (UTAUT) highlights the role of social influence and facilitating conditions in shaping technology adoption decisions (Venkatesh et al., 2003). Furthermore, the concept of Technology Readiness (Parasuraman, 2000), which segments consumers into pioneers, skeptics, and sluggards based on their overall propensity to embrace technology, provides a crucial psychographic lens for understanding why a one-size-fits-all automation strategy is destined to fail (Parasuraman & Colby, 2015). UTAUT2 further extends adoption theory by introducing the role of habit and consumer context, emphasising that repeated exposure to digital tools and prior technology experience strongly shape guests' willingness to use automated reception systems (Venkatesh et al., 2012).

A further application of TAM in the reception automation context lies in understanding how guests evaluate the usefulness and ease of automated check-in systems compared to human service. For instance, perceived usefulness increases when kiosks or mobile apps demonstrably reduce waiting time or allow guests to bypass queues. If the process is equally slow or inconvenient, usefulness is judged to be low, which reduces the likelihood of adoption. Likewise, perceived ease of use becomes critical because hotel guests may be fatigued or stressed upon arrival. Any friction, such as multiple steps, scanning failures, or complicated screens, quickly diminishes PEOU. Extensions such as TAM2 (Venkatesh & Davis, 2000) and TAM3 further emphasize the role of external variables, including social influence and technology readiness. Supporting this view, research on AI-driven chatbot interactions confirms that user trust, perceived competence, and socially oriented communication styles strongly enhance satisfaction with automated services. These findings reinforce the TAM2 perspective that social influence and cognitive appraisals meaningfully shape guests' willingness to adopt front-office automation (Al-Oraini, 2025). For example, a business traveler with high digital confidence may find kiosks highly useful, whereas a leisure guest seeking personalized service may perceive them as less useful due to their desire for human interaction.

2.2.2 The Organizational Perspective: Resource-Based View (RBV) and Strategic Capability

From a strategic management standpoint, the Resource-Based View (RBV) of the firm provides a compelling rationale for investment in automation. RBV argues that firms gain sustainable competitive advantage by possessing valuable, rare, inimitable, and non-substitutable (VRIN) resources and capabilities (Barney, 1991). A hotel's technological infrastructure- its integrated network of kiosks, mobile platforms,

PMS, CRM, and data analytics can be viewed as a tangible resource. This "digital hospitality capability" allows a hotel to differentiate itself in a crowded market, improve profitability, and build guest loyalty in ways that competitors cannot easily replicate, as it is embedded in the organization's unique processes and culture.

In the context of reception automation, RBV helps explain why technology alone seldom creates sustainable competitive advantage. Basic automation such as standard kiosks or generic mobile check-in platforms may be valuable but is not rare, inimitable, or non-substitutable, since most hotels can acquire similar systems. Such tools, therefore, lead only to competitive parity (Wade & Hulland, 2004). The true strategic advantage arises when automation is deeply integrated with proprietary data systems and organizational routines. For example, a kiosk that recognizes returning guests, prepopulates their preferences, automatically adjusts room assignments, and completes check-in within seconds represents a capability that is difficult for competitors to imitate (Mithas et al., 2011). Moreover, RBV supports the argument that a human technology hybrid capability, where automation handles routine tasks, and employees focus on empathy driven interactions, constitutes a unique service asset. This integrated capability becomes embedded in culture, processes, and staff skills, meeting RBV's VRIN criteria more effectively than technology alone.

2.2.3 The Service System Perspective: Service-Dominant Logic (SDL) and Value Co-Creation

Perhaps the most profound theoretical lens for understanding the current shift is Service-Dominant Logic (SDL). Pioneered by (Vargo & Lusch, 2004), SDL reframes all exchanges as service-for-service, where value is not embedded in a tangible good but is co-created with the consumer through the application of knowledge and skills. In a traditional front office, value is co-created through interactive dialogue between the guest and agent. Automation reconfigures the value co-creation process (Neuhofer et al., 2015).

In an automated model, the guest becomes a co-producer of their own experience, performing tasks once done by staff (e.g., data entry, key issuance). This can enhance the value of some guests by providing them with a sense of autonomy and control. However, it also carries the risk of the term "value co-destruction (Gebauer et al., 2010). The positive, relational interaction that is the foundation of the service is destroyed if the visitor is made to perform unpaid labor, the technology is difficult to use, or the procedure seems impersonal and transactional. Therefore, SDL emphasizes that automation is not just a matter of functional performance, but also of designing service systems that facilitate, rather than hinder, mutually beneficial value collaboration (Vargo & Lusch, 2008).

A key extension of Service Dominant Logic is the distinction between operand and operant resources. In hotel reception automation, kiosks, mobile apps, and chatbots represent operand resources, tangible tools, while employees' knowledge, empathy, and problem -solving capabilities represent operant resources (Lusch & Nambisan, 2015). SDL asserts that meaningful value emerges when these resources interact, and guests actively participate in the service encounter (McColl-Kennedy et al., 2012). However, recent evidence suggests that the effect of social presence on satisfaction in automated service encounters is highly contextual, indicating that technology-enabled co-creation does not uniformly translate into higher value for all guests or situations (Al-Oraini, 2025). For example, when guests use a kiosk, they are not merely receiving service; they are co-creating their check-in experience by integrating their own data, preferences, and devices with the hotel's systems. This shift in automation away from the Goods Dominant perspective of cost reduction and toward a value-in-context perspective enables automation to free human staff to engage in higher-order, personalized tasks. Through this lens, reception automation becomes a mechanism that enhances rather than replaces human contribution, reinforcing the high-tech, high-touch model proposed in this chapter.

2.3. The Drivers of Automation: A Multi-faceted Imperative

The push towards automation is driven by a combination of powerful operational, economic, and strategic factors that make it an almost irresistible force for hoteliers.

2.3.1 Operational and Economic Drivers: The Bottom Line

The most immediate drivers were operational. Technologies such as self-service kiosks have been empirically shown to reduce average check-in times by over 50%, directly addressing a major pain point—queues during peak arrival times (S. Ivanov et al., 2017)This leads to tangible labor cost optimizations, not necessarily through mass layoffs, but by allowing existing staff to be redeployed from repetitive transactional duties to more complex, revenue-generating roles, such as concierge services or proactive guest engagement (Ivanov & Webster, 2019). Furthermore, automation enhances data accuracy, reduces errors in billing and room allocation, and provides 24/7 service capability for routine inquiries, all of which contribute to a leaner and more resilient operation.

2.3.2 Strategic and Competitive Drivers: Beyond Cost-Cutting

Beyond mere cost reduction, automation is a strategic tool for capacity management and revenue optimization. In a world of fixed room inventory and perishable services (an unsold room night is revenue lost forever), maximizing human resource productivity is paramount. Automation allows hotels to handle higher volumes of guests without a proportional increase in front-desk staff. Moreover, these technologies open new avenues for revenue management research. Targeted automated upselling prompts during digital check-in (e.g., for room upgrades, spa treatments, or dinner reservations) have been shown to significantly increase ancillary revenue (Marriott-2024-Annual-Report, n.d.) .In this light, automation is not an expense but an investment in operational scalability and revenue generation.

2.4. The Countervailing Imperative: The Enduring Power of Personalization

Despite the compelling case for automation, a vast body of literature affirms that the soul of hospitality lies in personalization, a domain where humans retain a decisive advantage.

2.4.1 Deconstructing Personalization: Transactional vs. Experiential

It is critical to distinguish between the different layers of personalization. Transactional personalization involves using data to customize a specific interaction, such as addressing a guest by name or acknowledging their membership status. This level is increasingly within the reach of sophisticated CRM-driven automation systems. However, experiential personalization is a more profound concept. It involves anticipating needs, offering bespoke recommendations, demonstrating genuine empathy during service recovery, and engaging in spontaneous acts of hospitality that make guests feel uniquely understood and valued (Kandampully et al., 2015). This form of personalization is deeply relational and contextual, requiring emotional intelligence, intuition, and the ability to read nonverbal cues, capabilities that remain firmly in the human domain of care.

2.4.2 The Psychology and Economics of Emotional Connection

The value of experiential personalization is rooted in consumer psychology. Studies based on attachment theory and neuroscience suggest that positive personalized social interactions can trigger the release of oxytocin, fostering trust and emotional bonds (Post, 2019).In a commercial context, this emotional connection is the primary driver of loyalty, which transcends price sensitivity. The economic value is clear: it is far more profitable to retain loyal guests than to acquire new ones. As (I. Tussyadiah, 2020b) aptly concludes, while automation excels at "mass personalization" (using data to offer standardized customized options), it fundamentally lacks the capacity for the "individual personalization" that forges these deep, emotional, and lucrative loyalties.

2.5. Identifying the Research Gap: The Integration Paradox

The literature presents a clear yet unresolved tension. On the one hand, there is robust evidence supporting the operational and economic necessity of automation. Conversely, there is an equally compelling body of work underscoring the irreplaceable value of human-delivered experiential personalization for building sustainable competitive advantage. Therefore, the critical gap in the existing research is not in understanding these forces in isolation but in comprehending how to integrate them synergistically.

Much of the extant literature inadvertently positions technology and human touch as being in a zero-sum game, leading to what can be termed the "integration paradox": the very systems implemented to improve service efficiency can, through poor design or misapplication, systematically erode the emotional value and personal connection that constitute the core of the hospitality product. This paradox creates a strategic blind spot for hotel managers. The question is no longer whether to automate but how to do so in a way that augments rather than replaces the human element.

This chapter addresses this critical gap. It shifts beyond a dichotomous view that pits efficiency against personalization and proposes a nuanced, practical framework for a "high-tech, high touch" approach. In order to resolve the integration paradox and achieve a sustainable balance, the following sections outline the strategic roles of technology, the changed function of human staff, and the critical importance of integrated data systems as the glue holding this new service paradigm together.

3. HARNESSING AUTOMATION IN HOTEL FRONT OFFICES: BALANCING EFFICIENCY, INNOVATION, AND GUEST EXPERIENCE

3.1 The Drive for Efficiency: Automation Technologies and Their Impacts

Front office automation is not a monumental concept, but a comprehensive range of technologies, each designed to intercept and streamline specific touchpoints along the guest journey. A comprehensive typology reveals the distinct functionalities, operational impacts, and limitations of each system as follows.

3.1.1 Self-Service Kiosks: The Automated Portal

Self-service kiosks are the most visible form of front-office automation, providing physical proxies to traditional front desk agents. Modern kiosks are much more sophisticated than basic check-in terminals; they are integrated systems with key encoders to create digital or physical keycards, credit card readers to process secure payments, and passport scanners to verify identity. From an operational point of view, kiosks provide significant advantages in terms of data accuracy and staff allocation. By regulating the input process, they reduced the error rate in manual data entry. This allows hotels to provide front-desk staff with more value-added connections. However, their implementation is not without its shortcomings. The initial capital expenditure is substantial, and dedicated floor space is required. Furthermore, they can present accessibility issues for guests with disabilities or those who are technologically averse, potentially creating a negative first impression if not supported by readily available human assistance. Therefore, the success of kiosks is contingent on thoughtful placement, intuitive user interface (UI) design, and their role within a broader hybrid service model.

3.1.2 Mobile-Centric Technologies: The Decentralized Concierge

The widespread adoption of smartphones has facilitated a fundamental transition in service delivery, moving from centralized systems to decentralized networks. Mobile applications facilitate a comprehensive "contactless" journey, beginning with pre-arrival online check-in and room selection from a digital floor plan. The most significant innovation is the mobile key, which uses Bluetooth Low Energy (BLE) or Near Field Communication (NFC) to allow guests to bypass the front desk entirely and proceed directly to their rooms (Morosan & DeFranco, 2016).

- Beyond access, these apps have become powerful communication and service delivery platforms. Integrated messaging systems allow guests to make requests (e.g., towel delivery, wake-up calls) that are logged directly into the Property Management System (PMS), creating an efficient and auditable workflow. For hotels, this generates a rich stream of data on guest preferences and behaviors. Challenges include ensuring a robust and secure IT infrastructure, managing cybersecurity risks associated with mobile access, and overcoming the "digital divide" to accommodate all guest segments. The success of mobile platforms hinges on their reliability, seamless integration with other hotel systems, and ability to offer genuine value that incentivizes downloads and use.

3.1.3 AI-Powered Conversational Agents: The Cognitive Front Line

AI-powered chatbots and virtual assistants represent a strategic shift from mechanical to cognitive automation. Embedded on hotel websites, messaging apps (e.g., WhatsApp, WeChat), or in-room smart devices, these agents use Natural Language Processing (NLP) and Machine Learning (ML) to understand and respond to a vast array of guest inquiries. While their foundational role has been to handle frequently asked questions and provide instant, accurate, and multilingual 24/7 support (I. P. Tussyadiah & Park, 2018) their capabilities are rapidly expanding. Modern systems are increasingly deployed for complex, task-oriented functions such as order placement and tracking, reservation management, and personalized menu composition in restaurants, moving far beyond simple Q&A, (Sam & Jasim, 2025)

- The sophistication of these systems is advancing from simple, rule-based responders to more complex, AI-driven agents capable of handling multi-intent queries and understanding context. For instance, an advanced chatbot can process a complex request like, "Can I have a late checkout and also book a taxi for noon tomorrow?" The underlying technology now enables more sophisticated interactions, with advanced NLP allowing chatbots to perform semantic search, sentiment analysis, and emotion detection, pushing the boundaries towards more intuitive and responsive services (Sam & Jasim, 2025).
- The primary limitation, as previously noted, is their current inability to manage truly empathetic or unscripted situations. A guest expressing frustration may receive a scripted apology from a chatbot, but only a human can de-escalate the situation with genuine empathy and discretionary problem-solving. The future development of this technology lies in improving emotional intelligence (Affective Computing) and creating seamless handoff protocols to human agents when a query exceeds the bot's capabilities.

3.1.4 Robotic Process Automation (RPA): The Digital Backbone

While less visible to the guest, Robotic Process Automation (RPA) is a critical driver of back-office efficiency. RPA involves the use of "software robots" to automate high-volume, repetitive, rule-based digital tasks. In the front office context, this includes automating processes such as generating and sending pre-arrival confirmation emails, reconciling folios for no-show charges, updating guest profiles across the PMS, CRM, and loyalty platform, and automating night audit reports (S. Ivanov & Webster, 2019).

- The benefits of RPA are profound; it virtually eliminates human error in these tasks, ensures 100% consistency, and operates 24/7 without interruption. By automating these administrative burdens, RPA frees managerial and supervisory staff to focus on more analytical, strategic, and guest-facing tasks. The implementation challenge lies in process mapping to identify suitable tasks for automation and ensuring that the RPA software can be integrated with legacy systems. However, the return on investment is often clear and rapid, making RPA a foundational element for creating operational efficiency that enables frontline personalization.

3.1.5 CRM Systems and Data Analytics: The Central Nervous System

Customer relationship management (CRM) platforms and their integration with property management systems (PMS) function as the central nervous system of the modern automated front office. These systems aggregate and synthesize guest data from demographic profiles and booking history to stated and observed preferences sourced from online bookings, loyalty programs, and all guest-staff interactions (Ivanov & Webster, 2019). An effective, integrated CRM creates a unified guest profile, which is the basis for both automated and human-driven personalization. For example, a guest's pre-arrival response requesting a high floor can automatically generate a work order for the housekeeping team, while at check-in stage, the same system can prompt a staff member's device with the guest's name, photo, and their preference for extra pillows, which enables a genuinely personalized welcome.

- Additionally, sophisticated CRM engines can generate targeted sales by automatically emailing spa promotions to guests who have frequently used spa services in the past. These systems enable hotels to segment their customers with precision and automate personalized marketing at a rapid pace. The efficacy of such tactics is highlighted by research, which confirms that specific CRM elements like personalization and a targeted customer approach are the most significant factors in improving guest satisfaction, while knowledge management greatly drive loyalty (Sofi et al., 2025). Moreover, the study found that guest engagement acts as a powerful amplifier; high levels of engagement significantly strengthen the positive effects of CRM on both satisfaction and loyalty

3.2 The Human Dimension of Personalization: The Enduring Value of the Human Element

Despite clear efficiency gains, hospitality is fundamentally people centric. Today's guests still crave genuine human care and individualized attentiveness. Personalization, which involves anticipating a guest's unique preferences, remains a proven driver of satisfaction, loyalty, and positive reviews. For example, doing something as simple as remembering a returning guest's name, pillow choice, or special occasion can create an emotional "wow" moment that automated channels alone cannot provide. Indeed, personalized gestures (welcoming Mr. Sharma with his favorite drink or an unexpected room upgrade) engender goodwill and repeat business. As hospitality scholar Kandampully notes, "superior customer experiences" built on personal attention are "essential" for loyalty (Kandampully et al., 2015).

Automated systems can support personalization at scale by analyzing data and making suggestions, but current AI still struggles with spontaneous empathy. As (I. Tussyadiah, 2020b) observes, technology excels at "mass personalization" (applying general preferences en masse) but falls short on "individual personalization," which requires emotional intelligence and creativity. In practice, if a guest has a complex complaint or a last-minute special request (e.g., hosting an unexpected guest), only a trained staff member can respond flexibly and empathetically. Overreliance on automation risks creating a sterile, purely transactional experience. If all interactions are automated, hotels lose the "magic" that differentiates one brand from another; guests may feel alienated even if every task is technically handled.

Studies confirm this balance: guests want convenience for simple tasks but human contact for special ones. For instance, when asked, most travellers said they appreciated faster check-in, but expected a friendly human at check-out or for hospitality gesture (*Booking.Com-Sustainable-Travel-Report2023*, n.d.) .Modern hospitality literature thus stresses that technology should enhance human services, not replace them. A recent analysis of hotel tech readiness concluded that in the industry, "human aspect cannot be completely taken out" of the guest journey (Ivanov & Webster, 2019). In short, the essence of hospitality – warmth, attentiveness, and creativity – must still be delivered by people, even as machines speed up routine parts of the stay.

3.3 The Central Tension: Framing the Efficiency-Personalization Paradox

Hoteliers face a classic paradox: efficiency versus personalization of services. Automation thrives on standardization, consistency, and scale, whereas personalization thrives on flexibility and variance. This conflict is evident in the fundamental design of the service technologies. For instance, while automation excels at "mass personalization" by applying general preferences at scale, it falls short on "individual personalization," which requires empathy, emotional intelligence, and spontaneous problem-solving that are uniquely human traits (I. Tussyadiah, 2020b). Recent research highlights that automation can increase perceptions of convenience while simultaneously reducing perceived relational value, creating a dual effect on satisfaction. Studies show that when service tasks become overly standardized through automation, guests may perceive a loss of warmth and authenticity, which in turn weakens their emotional attachment to the service provider (Belanche et al., 2021). Conversely, automation improves functional quality, especially in contexts where speed and accuracy are prioritised. This reinforces that guest satisfaction emerges from both functional and emotional dimensions, and an imbalance in either direction can reduce overall service evaluations. At an extreme, forcing all guests through self-serve channels maximizes throughput but annihilates the personal warmth. Conversely, insisting on staffed check-in for every guest ensures a human touch but can be inefficient and costly. This challenge is illustrated by simple trade-offs: an automated kiosk can process 3–4 check-ins simultaneously, whereas a human agent can only handle one at a time. However, the same agent can spot a special request or build rapport in ways a kiosk cannot.

This tension is widely acknowledged in hospitality practices. A recent industry viewpoint notes that pushing 100% of guests through machines would "maximize efficiency but sacrifice personal touch" (*Balancing Automation and Personalization in Service Industries*, n.d.). Indeed, Marriott's annual report notes that hotels must "balance standardization and customization" to meet business and leisure needs (*Marriott International Annual Report 2024*, n.d.). Research on user acceptance reveals that this balance is influenced by the design of the technology. Features that enhance anthropomorphism, such as perceived warmth, competence, and social presence cues (e.g., using a personalized tone, greetings, and emojis), can significantly increase user trust and acceptance of automated services like chatbots, helping to bridge the impersonal nature of technology (Sam & Jasim, 2025). Guest research reinforces that most travellers prefer hybrid options: using apps for convenience but expecting human service for concierge tasks or problem resolution (Morosan & DeFranco, 2016). For example, younger "digital natives" happily check in on a phone yet still want a friendly face when they have a complaint or question.

Looking forward, the next frontier in mitigating this tension lies in the development of artificial emotional intelligence. The future evolution of this technology points toward "emotion recognition" capabilities, which would allow systems to provide more empathetic and context-aware responses (Sam & Jasim, 2025). However, this remains a nascent field, and the essence of hospitality warmth, attentiveness, and creative problem-solving must still be delivered by people, even as machines speed up the routine parts of the stay.

Importantly, guest attitudes vary by segments. A 2025 survey found that Gen Z leads in tech adoption: 82% preferred app/kiosk check-in over the desk, versus ~55% of Boomers (*70% of Travelers Would Skip the Front Desk, Mews Survey Reveals: The Rise of Self-Check-In Hotels*, n.d.). In general, business travellers may prioritize speed, whereas luxury resort guests may prioritize a personalized arrival. Hoteliers must therefore segment and identify which guest types or situations call for automation and which call for human attention. Bridging this gap requires integrating technology with service design so that technology frees staff to focus on guest-centric interactions rather than simply replacing them (Ivanov & Webster, 2019).

3.4 Striking the Balance: A Framework for a High-Tech, High-Touch Model

The high-tech, high-touch model offers a framework for reconciling efficiency with empathy. In this hybrid approach, routine transactional tasks are handled by technology ("high-tech"), while staff concentrate on higher-level, personalized service ("high-touch") (Ivanov & Webster, 2019). Crucially, this does not mean that some guests must be fully automated and others fully human; rather, each guest experience can blend both. A critical design principle for this blend is creating seamless handoff protocols from technology to human agents, ensuring the guest feels continuously supported when a query becomes too complex or emotionally charged (Sam & Jasim, 2025).

The key components of the hybrid model are as follows.

3.4.1 High-Tech Layer (Automation of Routine Tasks):

Technology adopts standardized processes. Self-serve check-in kiosks and mobile check-in manage room allocations and key issuances. AI-powered chatbots handle common FAQs and increasingly more complex, task-oriented functions, such as order placement and reservation management(Sam & Jasim, 2025). CRM systems automate targeted emails (e.g., welcome messages or pre-arrival upsell offers). Back-office tasks, such as data entry, billing, and routine reporting, are automated. These systems operate continuously and at scale, ensuring consistency and speed. By shifting these laborious tasks to machines, staff can be redeployed for creative and discretionary work.

3.4.2 High-Touch Layer (Human-Centered Service):

Employees focus on tasks requiring judgment, creativity, and emotional intelligence. This includes greeting guests by name, addressing special requests, problem solving, and providing local knowledge and empathy. Human hosts excel at reading guests' moods and tailoring their interactions. For example, if a family arrives with a sick child, a staff member might expedite room cleaning or bring extra pillows – something an app cannot do alone. In luxury or boutique hotels, high-touch might involve one-on-one concierge service, personal greetings, or concierge recommendations that delight the guests. Even in economy hotels, staff can ensure a warm welcome or handle exceptions (e.g., overbooking) personally.

3.4.3 Data-Driven Personalization (Integrating Technology and Touch):

The linchpin is data and CRM. Recent hospitality studies show that hybrid service models outperform fully automated or fully human models because they leverage complementary strengths. Automation enhances reliability, accuracy, and consistency, while humans provide adaptability and emotional intelligence, which remain critical for service recovery and complex interactions (Wirtz et al., 2018). Furthermore, hybrid models support dynamic personalization, where AI-driven insights inform staff behavior, allowing employees to tailor greetings, recommendations, and recovery efforts with greater precision. This form of augmented intelligence enables hotels to scale personalized service without proportionally increasing labour costs. Here, information flows from technology to people and back. For instance, when a guest uses mobile check-in or interacts with a chatbot, the data (including preferences and inquiries) is fed into the CRM. This makes the chatbot a crucial data conduit, where information collected (e.g., a guest asking about spa hours via a chatbot) is instantly available to staff, empowering them to make informed, personalized follow-ups (Sam & Jasim, 2025). The next staff member sees a consolidated profile (guest name, loyalty status, prior interactions) on a tablet, enabling a truly informed greeting. Similarly, after their stay, guest feedback can be captured digitally and used to train staff or refine AI recommendations. Thus, CRM acts as an integrator, ensuring that technology-generated data inform human services, and human observations update technology systems. Staff use technology-enabled profiles to anticipate needs, and technology learns from staff feedback.

Table 1. High-Tech vs. High-Touch Division of Tasks

Aspect	High-Tech (Automation)	High-Touch (Human)
Check-in Process	Guests self-check in via kiosk or app (ID scan, payment, keys).	Concierge greets guest, handles luggage, and assists with personal needs.
Routine Info	Chatbot/app answers FAQs (Wi-Fi, hours, directions).	The front desk explains policies and offers recommendations by name.
Special Requests	Pre-arrival app collects preferences (room setup, dietary needs).	Staff proactively fulfill unique requests (room upgrade, occasion setups).
Upselling	Automated offers pushed via app or email (spa discounts, upgrades), e.g., NOR1 upgrades, FPG upsell	Personalized upsells: staff suggest services at check-in or concierge.
Problem Resolution	Automated alerts flag issues (late check-out auto-charge warning).	Human staff handle complaints/ emergencies with empathy.
Guest Data Use	CRM aggregates profiles/preferences for segmentation and mailing.	Staff interpret data to surprise guests (e.g., remembering name/preference).

The concept can be summarized in the table above, which contrasts tasks handled by automation with those handled by humans in the high-tech/high-touch paradigm. In this model, automation tackles scalable, rule-based tasks, whereas humans handle tailored, discretionary services.

By intelligently assigning tasks in this manner, hotels can maximize both efficiency and personalization. Automation smooths routine touchpoints, while people ensure that the stay feels special. As one hospitality authority concluded, "technologies and human employees should work together" – leveraging each other's strengths to improve service quality and efficiency (Ivanov & Webster, 2019). In practice, managers must carefully design processes to ensure that technology amplifies human care. For example, digital check-in frees a receptionist to spend more time greeting late arrivals or sending personalized thank-you notes to guests than before. This philosophy has been echoed across the industry: leading hotel executives stress that technology is not an end in itself but a tool to allow "staff to really welcome their guests and provide remarkable experiences" (*Marriott International Annual Report 2024*, n.d.).

3.5 Implementation Challenges and Strategic Considerations

Transitioning to a high-tech, high-touch model extends beyond technological procurement and requires comprehensive organizational change management. The principal implementation challenges include the following:

3.5.1 Technological Integration:

Many hospitality organizations operate with legacy systems, including property management, reservation, and point-of-sale platforms, that lack inherent interoperability. The introduction of self-service kiosks and mobile applications necessitates seamless integration with core systems to enable real-time data synchronization. Implementers frequently encounter challenges with disparate databases, incompatible data formats, and organizational data silos, which can create operational inefficiencies and compromise data integrity (Ivanov & Webster, 2019). For instance, a mobile check-in application requires integration with both room-assignment algorithms and electronic locking systems. Industry analysis consistently shows that "integration capabilities represent the most significant technological hurdle, with fragmented systems leading to disrupted guest experiences and operational bottlenecks"(*Hospitality Industry in 2023: Responding to New Trends*, n.d.)

3.5.2 Staff Training and Change Management:

The implementation of automation technologies often generates employee apprehension regarding job security and role transformation. Strategic change management must emphasize technological augmentation rather than replacement, requiring comprehensive retraining programs focused on new digital tools and elevated service competencies (I. Tussyadiah, 2020b). Front-line staff must transition from transactional processors to experience curators, utilizing customer relationship management dashboards to deliver personalized services. Research indicates that "successful digital transformation in hospitality hinges on transparent communication about role evolution and investment in continuous skill development to build employee confidence and competence with new technologies" (Ivanov & Webster, 2019).

3.5.3 Guest Segmentation and Education:

A uniform approach to technology implementation risks alienating guest segments with varying technological aptitudes and service expectations. Effective implementation requires sophisticated guest segmentation strategies that align service delivery channels with demographic preferences and behavioural patterns (Morosan & DeFranco, 2016). Contemporary industry research reveals pronounced generational disparities in technology adoption, with "over 80% of Gen Z travelers preferring self-service options compared to approximately 60% of Baby Boomers, necessitating flexible service delivery systems" (*70% of Travelers Would Skip the Front Desk, Mews Survey Reveals: The Rise of Self-Check-In Hotels*, n.d.). In addition, Generation Z shows a stronger preference and more positive attitude towards chatbots than Generation X, who may value more human-like, socially interactive bots. This underscores the need for a segmented service approach, not a one-size-fits-all technology rollout (Sam & Jasim, 2025). Strategic signage, digital communication, and staff guidance are essential for facilitating guest adoption and ensuring seamless service experiences across all touchpoints.

3.5.4 Organizational and Cultural Transformation

Successful implementation demands a fundamental cultural shift toward digital innovation while preserving the core hospitality values. This transformation extends beyond technology to a fundamental redefinition of staff roles and responsibilities. As identified by (Sam & Jasim, 2025), successful implementation requires exploring the "potential collaboration between human and artificial intelligence to create innovative and improved customer experiences", moving staff from manual task doers to experience enhancers and problem solvers. To support this evolution, organizations must revise their performance metrics to balance operational efficiency with experiential quality indicators. Organizational leadership must champion this holistic evolution through what (Chi et al., 2020) describe as "a phased, iterative approach that prioritizes cross-departmental collaboration and continuous improvement, transforming technological adoption from a discrete project into an ongoing strategic priority".

These implementation challenges necessitate systematic planning and execution. Progressive organizations typically adopt staged implementation methodologies, beginning with controlled pilot deployments that allow performance assessment and procedural refinement before an organization-wide rollout. The interconnected nature of front-office automation requires coordinated planning across multiple departments, including operations, information technology, marketing, and human resources, to ensure alignment and operational coherence.

3.5.5 Privacy, Data Ethics, and Consent

As automation and CRM systems collect more guest information, hotels must manage privacy and data ethics carefully. Automated check-in, chatbots and mobile apps often gather personal, behavioural and sometimes biometric data, making transparent consent essential. Guests must clearly understand what data are collected, why, and how they can opt out. Lack of clarity reduces trust and lowers adoption of automated services.

Hotels should practice data minimization, collecting only information needed for service delivery, and ensure strong cybersecurity across integrated systems through encryption, secure APIs and regular audits. Ethical safeguards are also required to prevent algorithmic bias in personalization or automated decision-making.

Key best practices include:

1. Clear opt-in consent on all digital channels.
2. Transparent data-use statements and retention policies.
3. Guest control over stored data.
4. Staff training on ethical data handling.

Strong privacy governance increases guest trust and supports wider acceptance of automation.

3.6 Post-COVID Trends and Evolving Guest Expectations

The COVID-19 pandemic served as a profound catalyst, accelerating the adoption of automation technologies from a value-added feature to a health and safety imperative in the food industry. As guests became accustomed to digital-only interactions during the lockdowns, their baseline expectations permanently shifted. Contemporary travelers routinely assume the availability of self-service technologies, with industry data from (*Oracle's Top Hospitality Industry Predictions for 2023*, n.d.) indicating that 60% of travelers rank mobile check-in as a top expectation. This period also fundamentally altered perceptions of automation; This heightened comfort with automation has led to the rapid deployment of cleaning robots, contactless kiosks, and digital menus, many of which have been retained as permanent fixtures.

Industry leaders have confirmed that frictionless convenience is now the standard. (*Beyond the Front Desk: How Cloud, Mobility, and AI Are Redefining Hospitality*, n.d.) of Mews observed that the modern traveler demands "seamless, digital-first experiences," and that "the traditional front desk check-in is increasingly perceived as an obstacle rather than a welcome". Technologies such as contactless payments, mobile keys, and QR-code menus have become commonplace, and their absence can now be a source of guest dissatisfaction. Paradoxically, while technology has become essential for safety, the pandemic has underscored the irreplaceable value of human empathy in times of crisis. Staff who provide compassionate, flexible service such as handling anxious guests or facilitating contactless curb side amenities forge powerful emotional connections, demonstrating that the human touch remains a critical differentiator (S. H. Ivanov et al., n.d.).

The pandemic has expanded the scope of data-driven personalization. Hotels have begun collecting new categories of guest data related to health and safety preferences, which can now be leveraged to tailor the guest experience more precisely. Guests' willingness to share data for a better experience has increased; a

(*Hospitality Industry in 2023: Responding to New Trends*, n.d.) survey found that 93% of travellers are willing to provide personal preferences to enhance their stay. This creates an unprecedented opportunity for hotels to build sophisticated profiles of their guests. Furthermore, the pandemic's economic pressures have made operational efficiency non-negotiable, forcing many hotels to adopt automation out of necessity. As noted by (Chi et al., 2020), "the pandemic acted as a forced experiment in digital transformation, compressing a decade's worth of innovation into a matter of months and establishing a new, elevated baseline for technological capability in the industry". In summary, COVID-19 irrevocably accelerated the integration of technology, cementing dual expectations among guests for both flawless, contactless efficiency and meaningful, human-driven care.

4. SCOPE FOR FUTURE RESEARCH

This chapter outlines a framework for balancing automation and personalization and reveals several promising avenues for future scholarly inquiry such as:

4.1 Competitive Differentiation and Service Culture

Future success in hospitality will hinge on how effectively properties balance technology and human interaction. Brands that seamlessly integrate data and digital tools while delivering genuine and heartfelt service will secure stronger guest loyalty. For instance, a midsize hotel chain might pair a smooth mobile check-in process with a welcome amenity tailored to guest preferences, while another might differentiate itself with a personal concierge service triggered by smart CRM alerts. Nevertheless, investing in technology alone is inadequate; nurturing a service-oriented ethos capable of leveraging these tools is equally critical.

4.2 Evolving Metrics and Performance Measurement

Conventional hospitality metrics, such as occupancy and average daily rate (ADR), are no longer capturing the full picture of guests. Hotels should implement multidimensional performance indicators that can measure the satisfaction of both automation and human interaction. For example, Net Promoter Scores (NPS) can be segmented to assess guest feedback on self-service check-in kiosks and staff-led communications. Collecting and analyzing such data allows managers to continuously calibrate the ideal technology–human service mix, improving both efficiency and personalization.

4.3 Decision-support and Segmentation Models

One of the most promising future research areas involves the development of intelligent decision-support tools to determine when human intervention is preferable to an automated service. Predictive analytics and machine learning models can help hotels decide, in real time, whether a particular guest or situation warrants staff involvement or is better suited to a self-service channel. Such systems would enhance service quality, optimize staffing, and personalize the guest journey more effectively.

4.4 Advancing Personalization

Through AIAs artificial intelligence and machine learning continue to evolve, the next frontier of personalization lies in their predictive capabilities. Future research should explore how AI can infer unspoken guest needs and proactively guide staff actions. If a guest browses wellness amenities on a hotel app, the system may notify the front desk to provide a spa voucher when they arrive. This kind of human AI partnership, like "AI plus doctor" models in healthcare, might modify service characters and guest engagement in hotels.

4.5 Ethical Dimensions of Data Usage

The growing dependence on guest data for personalization grants multifaceted moral implications. Future research should examine the influence of privacy, transparency, and consent on trust and satisfaction in data-intensive fields. The importance of understanding guest perceptions is essential for establishing personalization strategies that are both effective and ethically sound.

4.6 Cross-Cultural and Demographic Insights

Various cultures, geographical regions, and demographic groups have different expectations for automation and human interaction from guests. Global hotel chains can improve their strategies to a variety of audiences in the future by conducting comparative studies that examine how many market segment's view and value this balance. As hospitality brands enter new markets and cater to a larger number of customers, these types of insights will become increasingly important.

4.7 Workforce Evolution and Education

As technology becomes more integrated into front office operations, the nature of work in hospitality will shift significantly. Research is required to determine how positions such as "Experience Curators" evolve over time, how job satisfaction is impacted, and how hospitality education needs to be changed to prepare future professionals for these positions. By being aware of these changes, workforce development will be able to keep up with technological innovations.

4.8 ROI and Longitudinal Impact Studies

The importance of determining the return on investment (ROI) of hybrid service models is another important research matter. Significant empirical insights into the long-term business impact of high-tech/high-touch properties will be provided by longitudinal studies that monitor metrics like the Net Promoter Score (NPS), guest lifetime value (LTV), and revenue per available guest (RevPAG).

4.9 Beyond the Front Desk

The introduction of Visitor Experience Check-in and concierge services should not be the only examples of innovation. New technologies such as AI-powered virtual concierges, augmented reality property tours, and in-room voice assistants offer new opportunities for personalization. The guest experience could be further expanded outside of the hotel by incorporating other platforms, such as ride-sharing applications or smart city infrastructure. Hotels can improve their services and stay on top of guest expectations by keeping an eye on and trying with these trends.

4.10 Strategic Roadmaps and Leadership Vision

Finally, strong leadership and strategic planning are essential for successful implementation. Hotels should create unambiguous technology plans that complement their target audience and brand promises. Depending on performance results, pilot projects, such as testing chatbots or AI concierge services, may be expanded. The leadership must also convey a compelling vision, such as "We utilize technology to make stays seamless, so our people can make them memorable. This tenet guarantees that cultural development, training, and investment all contribute to the ultimate goal of providing guests with memorable experiences.

CONCLUSION

As a result of changing guest expectations, technological advancements, and competitive pressures, hotel front offices are becoming increasingly a priority rather than a matter of choice. The primary conflict between guest personalization and operational efficiency is a dynamic equilibrium that should be prioritized and wisely managed rather than a conflict to be defeated by one side. The "high-tech, high touch" model, which views automation and human service as complementary forces that, when combined, produce better visitor experiences rather than as alternatives, offers a strong framework for achieving this balance.

Through self-service kiosks, mobile platforms, AI chatbots, and CRM-driven personalization systems, automation serves as a robust and effective operational backbone in this integrated paradigm, handling predictable transactional tasks like check-ins, payments, and routine inquiries with ease. In addition to improving convenience, which is now a standard expectation in the post-COVID era, this technological layer frees up human employees to concentrate on qualities that machines cannot match, such as empathy, creativity, emotional intelligence, and providing genuinely memorable service moments.

It is important that the connection between these two dimensions is determined. Hotels can anticipate needs, tailor offerings, and foster loyalty in earlier unattainable ways by changing distinct guest interactions into an ongoing, personalized narrative. When automation is utilized carefully, it enables staff members to move beyond transactional positions and become "experience curators," concentrating their efforts on valuable touchpoints that enhance their connections with visitors.

The most prosperous hotels in the future will be those that utilize technology to provide genuine hospitality rather than those that embrace it for its own benefit. They generate service atmospheres that feel both natural and incredibly human by automating transactions, allowing their employees to master the process.

In the end, combining guest personalization with reception automation is a crucial requirement that ensures that the essence of hospitality is not only present but flourishes in the digital age. Future studies should keep improving this hybrid model, examine the best ways to divide the roles of humans and technology, and examine fresh approaches to personalization. At present, hoteliers need to provide a comprehensive approach to change management, strategic planning, focused employee training, and guest-centric design at the moment. The hospitality sector can combine human empathy and machine agility to create front desk experiences that are both emotionally and operationally compelling, which is the real significance of the next-generation hospitality.

"The most competitive hotels of the future will not be those that automate fastest, but those that design guest journeys where technology amplifies, rather than replaces, the warmth, creativity, and emotional resonance that define true hospitality."

REFERENCES

Al-Oraini, B. S. (2025). Chatbot dynamics: Trust, social presence and customer satisfaction in AI-driven services. *Journal of Innovative Digital Transformation, 2*(2), 109–130. DOI: 10.1108/JIDT-08-2024-0022

Balancing Automation and Personalization in Service Industries. (n.d.). Retrieved October 7, 2025, from https://www.hospitalitynet.org/opinion/4123063.html

Barney, J. (1991). Firm Resources and Sustained Competitive Advantage. *Journal of Management, 17*(1), 99–120. DOI: 10.1177/014920639101700108

Belanche, D., Casaló, L. V., & Flavián, C. (2021). Frontline robots in tourism and hospitality: Service enhancement or cost reduction? *Electronic Markets, 31*(3), 477–492. DOI: 10.1007/s12525-020-00432-5

Beyond the Front Desk: How Cloud, Mobility, and AI Are Redefining Hospitality. (n.d.). Retrieved October 7, 2025, from https://www.hospitalitynet.org/opinion/4129074.html

Booking.com-sustainable-travel-report2023. (n.d.).

Buhalis, D., & Sinarta, Y. (2019). Real-time co-creation and nowness service: Lessons from tourism and hospitality. *Journal of Travel & Tourism Marketing, 36*(5), 563–582. DOI: 10.1080/10548408.2019.1592059

Chi, O. H., Denton, G., & Gursoy, D. (2020). Artificially intelligent device use in service delivery: A systematic review, synthesis, and research agenda. *Journal of Hospitality Marketing & Management, 29*(7), 757–786. DOI: 10.1080/19368623.2020.1721394

Davis, F. D. (1989). Perceived usefulness, perceived ease of use, and user acceptance of information technology. *Management Information Systems Research Center, University of Minnesota, Vol. 13,*(No. 3 (Sep., 1989),), 319-340 (22 pages).

Gebauer, H., Johnson, M., & Enquist, B. (2010). Value co-creation as a determinant of success in public transport services: A study of the Swiss Federal Railway operator (SBB). *Managing Service Quality, 20*(6), 511–530. DOI: 10.1108/09604521011092866

Hospitality Industry in 2023: Responding to new trends. (n.d.). Retrieved October 7, 2025, from https://www.hospitalitynet.org/opinion/4114325.html

Ivanov, S., & Webster, C. (Eds.). (2019). *Robots, Artificial Intelligence, and Service Automation in Travel, Tourism and Hospitality*. Emerald Publishing Limited., DOI: 10.1108/9781787566873

Ivanov, S., Webster, C., & Berezina, K. (2017). *Adoption of robots and service automation by tourism and hospitality companies.* https://papers.ssrn.com/sol3/papers.cfm?abstract_id=2964308

Ivanov, S. H., Webster, C., & Seyyedi, P. (n.d.). Consumers' attitudes towards the introduction of robots in accommodation establishments. *ResearchGate, 66/ No. 3/ 2018/ 302-317*, 302–317.

Juquelier, A., Poncin, I., & Hazée, S. (2025). Empathic chatbots: A double-edged sword in customer experiences. *Journal of Business Research, 188*, 115074. DOI: 10.1016/j.jbusres.2024.115074

Kandampully, J., Zhang, T., & Bilgihan, A. (2015). Customer loyalty: A review and future directions with a special focus on the hospitality industry. *International Journal of Contemporary Hospitality Management, 27*(3), 379–414. DOI: 10.1108/IJCHM-03-2014-0151

King, C. A. (1995). What is hospitality? *International Journal of Hospitality Management, 14*(3–4), 219–234. DOI: 10.1016/0278-4319(95)00045-3

Lashley, C. (2008). Studying Hospitality: Insights from Social Sciences1. *Scandinavian Journal of Hospitality and Tourism, 8*(1), 69–84. DOI: 10.1080/15022250701880745

Lusch, R. F., & Nambisan, S. (2015). Service Innovation: A Service-Dominant Logic Perspective1. *Management Information Systems Quarterly, 39*(1), 155–175. DOI: 10.25300/MISQ/2015/39.1.07

Marriott-2024-Annual-Report. (n.d.). Marriott Investor Relations. https://marriott.gcs-web.com/financial-information/annual-reports

Marriott International Annual Report 2024.pdf. (n.d.).

McColl-Kennedy, J. R., Vargo, S. L., Dagger, T. S., Sweeney, J. C., & Kasteren, Y. V. (2012). Health Care Customer Value Cocreation Practice Styles. *Journal of Service Research, 15*(4), 370–389. DOI: 10.1177/1094670512442806

Mithas, S., Ramasubbu, N., & Sambamurthy, V. (2011). How Information Management Capability Influences Firm Performance1. *Management Information Systems Quarterly, 35*(1), 237–256. DOI: 10.2307/23043496

Morosan, C. (2016). An empirical examination of U.S. travelers' intentions to use biometric e-gates in airports. *Journal of Air Transport Management, 55*, 120–128. DOI: 10.1016/j.jairtraman.2016.05.005

Morosan, C., & DeFranco, A. (2016). Modeling guests' intentions to use mobile apps in hotels: The roles of personalization, privacy, and involvement. *International Journal of Contemporary Hospitality Management, 28*(9), 1968–1991. DOI: 10.1108/IJCHM-07-2015-0349

Neuhofer, B., Buhalis, D., & Ladkin, A. (2015). Technology as a Catalyst of Change: Enablers and Barriers of the Tourist Experience and Their Consequences. In Tussyadiah, I., & Inversini, A. (Eds.), *Information and Communication Technologies in Tourism 2015* (pp. 789–802). Springer International Publishing., DOI: 10.1007/978-3-319-14343-9_57

70. *of Travelers Would Skip the Front Desk, Mews Survey Reveals: The Rise of Self-Check-In Hotels*. (n.d.). Retrieved October 7, 2025, from https://hoteltechreport.com/news/70-of-travelers-would-skip-the-front-desk-mews-survey-reveals-the-rise-of-self-check-in-hotels

Oracle's top hospitality industry predictions for 2023. (n.d.). Retrieved October 7, 2025, from https://www.duettocloud.com/library/oracles-top-hospitality-industry-predictions-for-2023

Ottenbacher, M., & Harrington, R. J. (2007). The innovation development process of Michelin-starred chefs. *International Journal of Contemporary Hospitality Management, 19*(6), 444–460. DOI: 10.1108/09596110710775110

Parasuraman, A. (2000). Technology Readiness Index (Tri): A Multiple-Item Scale to Measure Readiness to Embrace New Technologies. *Journal of Service Research, 2*(4), 307–320. DOI: 10.1177/109467050024001

Parasuraman, A., & Colby, C. L. (2015). An Updated and Streamlined Technology Readiness Index: TRI 2.0. *Journal of Service Research, 18*(1), 59–74. DOI: 10.1177/1094670514539730

Post, K. (2019). The neuroscience of trust. *Science, 363*(6432), 1187.16-1189. DOI: 10.1126/science.363.6432.1187-p

Sam, S. J. I., & Jasim, K. M. (2025). Diving into the technology: A systematic literature review on strategic use of chatbots in hospitality service encounters. *Management Review Quarterly, 75*(1), 527–555. DOI: 10.1007/s11301-023-00388-8

Sofi, M. R., Bashir, I., Alshiha, A., Alnasser, E., & Alkhozaim, S. (2025). Creating exceptional guest experiences: The role of engagement and relationship building in hospitality. *Journal of Hospitality and Tourism Insights, 8*(3), 891–914. DOI: 10.1108/JHTI-04-2024-0318

Tussyadiah, I. (2020a). A review of research into automation in tourism: Launching the Annals of Tourism Research Curated Collection on Artificial Intelligence and Robotics in Tourism. *Annals of Tourism Research, 81*, 102883. DOI: 10.1016/j.annals.2020.102883

Tussyadiah, I. (2020b). A review of research into automation in tourism: Launching the Annals of Tourism Research Curated Collection on Artificial Intelligence and Robotics in Tourism. *Annals of Tourism Research, 81*, 102883. DOI: 10.1016/j.annals.2020.102883

Tussyadiah, I. P., & Park, S. (2018). Consumer Evaluation of Hotel Service Robots. In Stangl, B., & Pesonen, J. (Eds.), *Information and Communication Technologies in Tourism 2018* (pp. 308–320). Springer International Publishing., DOI: 10.1007/978-3-319-72923-7_24

Vargo, S. L., & Lusch, R. F. (2004). Evolving to a New Dominant Logic for Marketing. *Journal of Marketing, 68*(1), 1–17. DOI: 10.1509/jmkg.68.1.1.24036

Vargo, S. L., & Lusch, R. F. (2008). Service-dominant logic: Continuing the evolution. *Journal of the Academy of Marketing Science, 36*(1), 1–10. DOI: 10.1007/s11747-007-0069-6

Venkatesh, V., & Davis, F. D. (2000). A Theoretical Extension of the Technology Acceptance Model: Four Longitudinal Field Studies. *Management Science, 46*(2), 186–204. DOI: 10.1287/mnsc.46.2.186.11926

Venkatesh, V., Morris, M. G., Davis, G. B., & Davis, F. D. (2003). User Acceptance of Information Technology: Toward A Unified View1. *Management Information Systems Quarterly, 27*(3), 425–478. DOI: 10.2307/30036540

Venkatesh, V., Thong, J. Y. L., & Xu, X. (2012). Consumer Acceptance and Use of Information Technology: Extending the Unified Theory of Acceptance and Use of Technology1. *Management Information Systems Quarterly, 36*(1), 157–178. DOI: 10.2307/41410412

Wade, M., & Hulland, J. (2004). *Review:* The Resource-Based View and Information Systems Research: Review, Extension, and Suggestions For Future Research1. *Management Information Systems Quarterly, 28*(1), 107–142. DOI: 10.2307/25148626

Wirtz, J., Patterson, P. G., Kunz, W. H., Gruber, T., Lu, V. N., Paluch, S., & Martins, A. (2018). Brave new world: Service robots in the frontline. *Journal of Service Management, 29*(5), 907–931. DOI: 10.1108/JOSM-04-2018-0119

KEY TERMS AND DEFINITIONS

Anthropomorphism (in HRI): Attribution of human traits to non-human agents, affecting user trust and acceptance.

CRM (Customer Relationship Management): Systems that aggregate guest data to enable personalization and lifecycle marketing.

Guest Experience: The total cognitive, emotional, and behavioural responses a guest has across the pre-stay, stay, and post-stay phases.

High-Tech, High-Touch Model: A hybrid service design where automated systems handle routine tasks, and humans deliver discretionary, empathetic service.

Human–Robot Interaction (HRI): Study of social, cognitive, and ergonomic factors in human interaction with robots/AI agents.

Personalization: Tailoring service elements to individual guest preferences, either via data-driven automation or human intervention.

Reception Automation: Technology solutions (kiosks, mobile check-in, AI agents) used to automate front-office transactions.

Robotic Process Automation (RPA): Software bots that automate rule-based back-office tasks.

Service Recovery: Processes and human actions following service failure to restore guest satisfaction.

Service-Dominant Logic (SDL): A view that value is co-created with customers through service interactions, not embedded in products.

Technology Readiness: A consumer's propensity to embrace new technologies (Parasuraman, 2000).

Value Co-destruction: When service design or implementation inadvertently reduces value for the customer.

Chapter 7
AI-Driven Smart Room Assistants:
Enhancing Personalized Guest Experiences in Hospitality

Vikas Sharma
https://orcid.org/0000-0001-8173-4548
Department of Computer Applications, SRM Institute of Science and Technology, India

Sachin Chaudhary
https://orcid.org/0000-0002-8415-0043
School of Computer Science and Applications, IIMT University, Meerut, India

Sanskriti Singh
https://orcid.org/0009-0002-7974-9817
Department of Hotel Management, SRM University, India

Pushpendra Kumar Verma
https://orcid.org/0000-0003-2777-5626
School of Computer Science and Applications, IIMT University, Meerut, India

Tarun Kumar Vashishth
https://orcid.org/0000-0001-9916-9575
Department of Computer Application, Vidya University, Meerut, India

Shahanawaj Ahamad
College of Computer Science and Engineering, University of Hail, Saudi Arabia

ABSTRACT

The hospitality industry is experiencing a transformation with the integration of AI and robotic technology. One key element of this change will be AI-enabled concierge and guest services that are changing the way in which customers are engaged in conversation, serviced and supported in the overall guest experience. This chapter highlights the evolving role of robot concierges, especially the ability of a robot concierge's ability to delivering personalized recommendations, engaging in conversation in multiple languages, providing real time responses and integrated

DOI: 10.4018/979-8-3373-6541-1.ch007

hotel management systems. This chapter reviews current applications, case studies from leading hotel organizations, and the technology of robot concierges to discuss the potential and challenges of AI concierge systems. The chapter discusses ethics around awareness of customers, data consent, privacy, and ethics, as well as how future intelligent concierge robots can provide satisfaction, operational efficiency, and some competitive advantage for hospitality businesses.

1. INTRODUCTION

The hospitality industry is changing at an accelerated pace through the introduction of Artificial Intelligence (AI) and robotics, which is leading to a new reality of personalized and technology-based guest experiences (Fernandes et al., 2024). One of the most prominent innovations is AI-powered concierge and robot guest agents that are redefining the ways hotels and resorts engage with their patrons. Traditionally, concierge services have utilized human experience to provide curated recommendations, make reservations, and solve guest problems, while the demand for efficiency, personalized service, and multilingual capabilities has created opportunities for AI robot concierges to complement and at times replace the traditional human guest experience. These AI robots can "learn" a guest's preferences, deliver solutions at the time of need, and easily interface with hotel management software. For example, robot concierges offer any time, multilingual communication, and the use of data to personalize the guest experience, all of which contribute to improved guest satisfaction and expedited operations at the hotel. Leading hospitality brands have already experimented with robotic agents to welcome guests; suggest dining or entertainment options; and service general requests. Adopting robotic guest agents can further enhance efficacy while establishing a competitive advantage over other establishments in a rapidly growing digital economy. While robot concierge systems have some positive attributes, they raise a number of issues deserving of consideration before implementation can be achieved. Issues such as acceptance of robots by guests and the community, the potential to protect sensitive personal information from hackers, and the social implications of robots taking functions away from humans or replacing human roles is topic of business discussion. Additionally, the equilibrium between the use of automation and the "human touch" in hospitality continues as an area of interest. This chapter examines the existing state, uses, and case studies of AI-powered concierge services and discusses the technological, operational, and ethical aspects of their implementation. The chapter also emphasizes how these technological advancements will shape the future of hospitality by enhancing customer satisfaction, operational efficiency, and long-term competitiveness (Avula and Sithole, 2024).

1.1 Overview of AI and Robotics in the Hospitality Sector

The hospitality industry has always been a leader in incorporating innovative technologies into guest satisfaction and operational efficiency. In recent years, the field of artificial intelligence (AI) and robotics has made strides and represents game-changing advancements for hospitality organizations, such as hotels, resorts, and travel service providers, to provide smarter, quicker, and completely personalized services. Technologies that such as AI, machine learning, natural language processing, predictive analyses and personalization, are not visions of the future, rather they are changing service delivery across the world. AI, and the many applications to engagement, capabilities, and proactive service, allows organizations in the hospitality sector to better interpret guest behaviour and requests to lead to personalized recommendations. For example, AI-powered recommendation systems, can evaluate customer reservation trends, preferences, etc., and then use to analyse for recommended customized dining, leisure, and travel options. Hotels have already embraced the use of virtual assistants and chatbots to coordinate routine guest inquiries, manage reservations, and provide ongoing service support in real time, which helps ease staff workloads while delivering a consistent standard of service. Meanwhile, robotics brings automation and a physical presence to the guest experience. Whether it's a robotic concierge greeting guests in the lobby of a hotel, or automated luggage handling and drones delivering room service, robotics provides speed, accuracy, and excitement in guest interactions. Robot concierges, specifically, combine both functionality and entertainment value, providing multilingual communication, interactive conversations with guests, and integration with property management features to accommodate guests' varying needs. Both AI and robotics create a hybrid service model where technology can assist and support human service staff to enable employees to focus on complex, personalized guest interactions while the machines take on the repetitive, mundane, and routine tasks as shown in figure 1. This has an impact on operational efficiencies but ultimately improves guests' experience overall in terms of access to service 24/7, eliminating wait times. The use of AI and robotics does not come without challenges, however, as issues regarding guest acceptance, host or staff cultural awareness, data privacy, cybersecurity, and ethical use of personal data must always be addressed. While robots and AI systems have a natural ability for accuracy and speed, the hospitality field must find a balance to nurture the "human touch" as the essence of a classic hospitality experience (Sharma et al., 2024).

Figure 1. Integration of AI and Robotics in Medical Healthcare

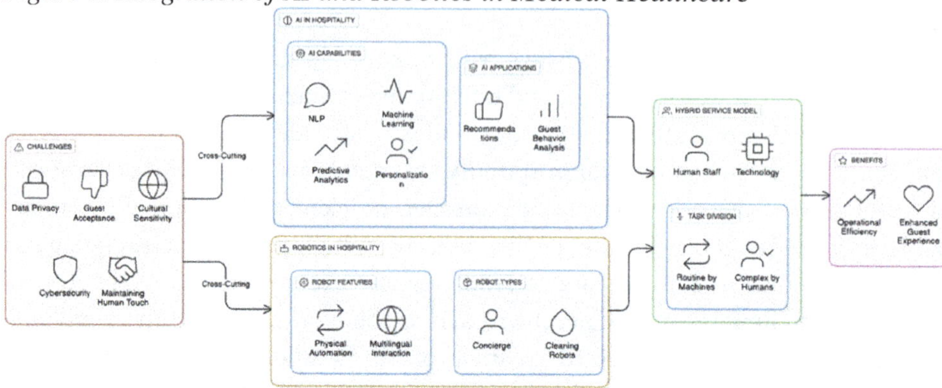

1.2 Importance of Guest Experience as a Differentiating Factor

In the modern, highly competitive hospitality industry, guest experience has become the most vital competitive advantage, influencing consumer satisfaction, brand loyalty, positioning in the marketplace, and sustainable success. Unlike other industries, where pricing practices, product attributes, or location may define the basis of competition, the hospitality sector is uniquely focused on how guests are treated and the degree to which the overall experience is memorable. Today's traveller's no longer see hotels and resorts simply as places to "stay" but rather these places must offer experiences tied into the individual's personal preferences, emotional needs, and cultural expectations. Among modern guest experiences are well-organized check-in and check-out processes, multilingual communication, dining/leisure recommendations tailored to the guest's style, and resolving specific inquiries virtually immediately. Given the rapid growth of digital technologies, especially in artificial intelligence (AI) and robotics, hospitality providers are considering and implementing new technologies to meet guest expectations. For instance, providing AI concierge services and robotic guest services, which offer real-time responses to requests, proactively predict needs based on a guest's language and existing data insights, and deliver consistent service that is not affected by human limitations such as human fatigue or unavailability. Thus, guest experience is not simply a dimension of service, but a strategic business instrument, enabling hospitality businesses to create emotional ties with guests, enhance its marketplace differentiation, and garner a competitively sustainable advantage in a more globalized and customer-focused market place (Banerjee and Bhat 2025).

1.3 Evolution from Traditional Concierge Services to AI-Powered and Robotic Solutions

Concierge services have always been a key element of the hospitality experience, with human concierges representing the human intermediary to guests who provide information and assistance, offer recommendations, and respond to guest-wants and needs. From booking a local tour or making dinner reservations to providing travel recommendations and fixing surprises, the traditional concierge entity offered the personal touch, privacy, and professionalism associated with luxury and quality products in hospitality. However, as expectations from the guest (user) experience have become more complex and demand for unique, immediate, personalized service has increased, a reliance on solely human concierges has become stratified. Public factors include access, availability, variability, language, and scalable limitations for hospitality properties responding to large guests at one time. The rapid growth of artificial intelligence, natural language processing, and robotics has led to a new generation of concierge services—AI-powered with robotic capability. These services merge AI intelligence, which is data-driven, with the efficiency and innovation of robotics and other technologies that allow hotels to provide multilingual, 24/7, and personalized service. Compared to standard concierge services, AI-powered concierges can analyze guest histories, preferences, and behaviours to predict guest needs and provide relevant suggestions in real-time, while robotic concierges can provide physical interactions like assisting with check-in, delivering room service, or managing luggage. Leading hospitality brands are increasingly trailing this technology around the world, combining a human touch with the immediacy and accuracy of machines to create an improved service approach that will change the guest experience. Rather than serve as a replacement for human experiences, the adoption of AI concierges represents an enhancement to the situational shift from traditional concierges; a technology-based solution to the repetitive operational aspects of the role that allows human concierges the flexibility to dedicate their time to more in-depth and meaningful guest interactions. (Vashishth et al., 2025). Since the mid-2010's, the evolution of concierges using new technologies has been on a continuous path toward automated and hybrid systems, moving away from the traditional models where only humans were the concierges. Strategies that were employed by early digital concierge services in the early and mid-2010's primarily focused on information delivery via web platforms and chatbots.

2. THE EMERGENCE OF AI-POWERED CONCIERGE SERVICES

AI concierge services represent a game-changer in the hospitality market and are changing the way hotels and resorts assist guests to foster guest satisfaction while managing their operations and creating value-added experiences. In contrast to a concierge based on human capability and experience, AI-concierge services are distinct systems powered by machine learning algorithms, natural language processing, and data analysis that understand the subtleties of guest preferences, allow for anticipating guest needs, and deliver real-time personalized results. They can interact with guests in multi-language conversations, offer personalized recommendations for dining, entertainment, or travel, and easily connect with hotel management and management platforms and support seamless service operations such as reservation confirmations for a room, spa, or transport arrangements. The distinct advantage of AI systems is that they do not work on time restrictions or possibility of fatigue under continuous operation and can support urgent guest requests at any hour, ultimately improve the experience through more seamless satisfaction and convenience (Jogarao, 2024).

2.1 Definition and Concept of AI Concierges

An AI concierge is defined as an intelligent, tech-based guest service model that applies artificial intelligence, natural language processing, and machine learning to deliver customized, on-demand service that is traditionally performed by human concierges at hotels and other hospitality organizations. Rather than relying on the individual knowledge/expertise and availability of a human concierge, AI concierges are designed to provide automated or robotic assistance when guests engage in conversation by recognizing guest preferences and recommending personalized solutions for dining, entertainment, transportation, local experiences, and travel options. At a conceptual level, AI concierges represent more than basic automation, they are a combination of data analytics, predictive modeling, and contextual awareness that allows hotels to assess and predict guest needs and provide solutions proactively, and in anticipation of guest demand, rather than reactively. AI concierges can also take many forms including mobile apps, in-room voice assistants, or humanoid robots operating in the shotgun spaces of hotel lobbies, and are designed to seamlessly blend natural guest interaction with property management systems and third-party service applications.

1. **Evolution of AI and Robotics in Hospitality:** In the last 20 years, automation, chatbots, the Internet of Things (IoT), and robotic solutions have become increasingly common in the hospitality sector as they have been adopted to

improve service delivery. The use of conversational AI in concierge roles in early technology trials, such as The Henn Hotel in Japan (2015) and Hilton's "Connie" (2016), exemplified early adoption of this type of solution with improved levels of efficiency and service quality. From 2017 to 2022, smart-room technologies and mobile concierge applications rapidly evolved within the hospitality sector, as hotels and other forms of hospitality worked to broaden their technology spectrum with the inclusion of "autonomous service robots" for the purposes of streamlining processes for Luggage handling, Room Service, and Check-In delivery. Following COVID-19, the shift to contactless service has strengthened and quickened the adoption of robotic solutions in the hospitality sector as robotic solutions have now transitioned from being futuristic concepts, to operational requirements. These chronological events provide the background for understanding the current evolution of-consumer robotic concierge solutions as rooted in decades of evolving technology development.

2. **Guest Acceptance and Human**: Robot Interaction Research Through their studies, researchers have identified perceived usefulness, ease of use, trust, and cultural views as primary factors influencing guest willingness to engage with AI-based solutions. Technology Acceptance Model (TAM) and UTAUT-type studies show that while guests are appreciative of the increased efficiency and speed of the service provided by automated robotic solutions, there is an expectation for the continued presence of human emotional intelligence during complex interactions. Research involving Human-Robot Interaction (HRI) has demonstrated that design elements (i.e., voice tone, gesture tracking, appearance, and cultural sensitivity) play a vital role in determining guest comfort levels. These results show that, although portions of a robotic solution may provide a logical and/or rational basis for usage, the recognition of how a person perceives and interacts with the technology is equally essential.
3. **Market Dynamics & Adoption:** The global service robot market is anticipated to experience continuous growth and has already begun to see rapid growth due to increased demand for services provided by robots, particularly in the hospitality industry. Through multiple independent studies, both globally and nationally, evidence can be gathered to illustrate an improved operational performance due to the utilization of AI concierges, such as reductions in the number of days required to check into a hotel, faster delivery of room service, increased levels of customer satisfaction and lower levels of labour required by staff at hotel front desks. Based on research studies that have been completed by hotels using AI concierge technologies, 30–40% less staff have been provided to complete similar service-related tasks, as well as much quicker response time for most common customer service requests.

4. **Ethical and Workforce:** Much of the current ethical literature concerning the use of AI and robotic technologies focuses on issues of data privacy and ethical issues such as algorithmic bias, behavioural nudging, and long-term effects of the displacement of human labour. For example, it has been suggested that guests may be "nudged" to make purchase decisions based on personalization algorithms and data analytics. The workforce literature, on the other hand, indicates that while AI and robotic technologies may allow for the elimination of manual tasks from an employee's workload, the potential anxiety associated with job security caused by the introduction of AI technology into the workplace may require employees to be retrained or their roles to be redesigned within an organization. Incorporating these factors into the body of research that exists concerning the use of AI concierge technology allows for a more balanced evaluation of the potential negative and positive implications of this emerging service.

2.2 Technological enablers: AI, Natural Language Processing (NLP), Machine Learning, IoT

The development and effectiveness of AI concierge systems have stemmed from a combination of advanced technological enablers that incorporate Artificial Intelligence (AI), Natural Language Processing (NLP), Machine Learning (ML), and Internet of Things (IoT).

5. **Artificial Intelligence (AI):** AI is the technological base infrastructure that supports AI powered concierge systems in offering the cognitive intelligence that is required to develop decisions, resolve problems, and personalize specific guest services. AI systems can read a guest profile, anticipate a guest's likes and dislikes, and make contextual, relevant, and timely recommendations for the guest. AI concierge systems are distinct in that they can replicate a human's reasoning ability to ponder, select, and implement ideas based on your most complex requests, including trip planning, food selection for a guest, catering to the multiple requests of guests at the same time, to result in overall better hotel operation and efficiency.
6. **Natural Language Processing (NLP):** Natural Language Processing (NLP) allows AI concierges to understand and communicate with user fluent in human languages via text and voice. NLP means that AI concierge systems can process requests in different languages, can understand side languages, and can respond accordingly in a conversation. This makes the interaction more natural for a user of course. NLP capabilities are especially critical in hospitality where guests or users come from a wide variety of linguistic and cultural backgrounds.

NLP capabilities support the delivery of clear communication criteria to guest services and measurably strengthen personalization through the delivery of understandings derived from language-based communication.

7. **Machine Learning (ML):** The distinct benefits of using AI concierge to consistently improve with each use is due to a technique called machine learning (ML). ML allows AI concierge systems to ingest historical data, guest interaction data, and subsequent guest feedback to improve performance. Machine Learning uses artificial intelligence (AI) technology to discover patterns of guest behavior, predict preferences in future interactions, and improve personalized recommendations over time. Using such technology ensures that each interaction is more personalized and becomes more successful, decreases the chance of error or a generic response (Ravichandra et al., 2025).
8. **Internet of Things (IoT):** AI concierges are connected to a network of smart devices in the hotel to provide services (the internet of things). This could include smart room control devices (temp, lights, TV), digital locks for room access, staircase, or elevator directions, and also service management platforms (AI, other smart devices, etc.) to maintain connections with housekeeping staff to guest room amenities. IoT allows for real-time monitoring, automation of service, and coordination between the many hotel services. For example, the concierge system could connect to the smart devices in the room catalog when a guest checked in and adjust to their preferences (ex. temperature). Also, it could coordinate the housekeeping service based on information retrieved from monitoring occupancy side data (Banu et al., 2025).

2.3 Advantages over Traditional Services

The transition from traditional concierge services to AI-enabled concierge systems offers a variety of advantages that are changing how hospitality businesses engage with and cater to their guests. One of the main benefits of AI-driven services is that they are available 24/7. AI systems provide service without the restrictions of working hours, breaks, or fatigue. Hence, guests arriving late at night or leaving early in the morning or those who require help at odd hours can be confident that someone will always be there to support them, which is not always possible with a human-only workforce. Multilingual capabilities are just as important, as AI-enabled concierge services can communicate seamlessly with guests from various cultural and linguistic backgrounds. While a human representative may need to reach for a translator or rely solely on limited language skills, AI service-enabled concierges equipped with natural language processing can swiftly communicate and respond in multiple languages, breaking down communication barriers and enabling a more inclusive experience. Personalization, at scale, is another significant benefit, enabled by data analytics and machine learning. AI-powered concierges can analyze large datasets, including previous bookings, dining habits, and other personal categories, to proactively provide tailored recommendations and experiences. For example, when a guest returns, they may receive personalized recommendations based on restaurants they previously enjoyed or activities that were aligned with their previous interests. Human concierges, who also offer personalization, may offer loyalty-based suggestions, which are informed by limited localized knowledge or memory. AI concierges will provide personalization to the same degree to all guests in a hotel, without limits of memory or prior knowledge. In operational efficiency, AI concierges will allow hotel staff to lose the low-value, repeatable operational tasks of confirming reservations, providing directions, or explaining hotel services, enabling hotel staff to redirect their energies to the high-value tasks where empathy, creativity, or judgment are useful, like handling emotional guest complaints or developing and arranging personalized experiences for guests (Manoharan et al., 2024).

3. KEY FUNCTIONALITIES OF ROBOTIC CONCIERGE SYSTEMS

Robotic concierge systems are developed to extend the conventional role of human concierges through the incorporation of automation, artificial intelligence, and smart connectivity, and to deliver a wide spectrum of guest services that are effective, consistent, and personal.

3.1 Personalized Guest Recommendations

One of the most formidable characteristics of robotic concierge services is the ability to offer personalized recommendations with the utilization of aspects of artificial intelligence, machine learning, and data analysis. Traditional concierges underpinned by their expertise and experience and other heuristics like memory, knowledge from a brand or communities of fellow concierges; limitations which a chatbot does not have. Robotic concierge is able to access and engage with a variety of data sources; booking history, dining preferences, past travel behaviour, guest feedback forms, and even real-time c-contextual data - time of day or location of the hotel. As a result, robotic concierges can suggest restaurants, cultural opportunities, attractions, wellness activities, or entertainment based on the specific interests of each guest. For example, a robotic concierge may recommend a vegan option for health-conscious travellers, suggest attractions that would be oriented to guests travelling with children, or suggest late-night options for solo business guests that may want to unwind after work. Personalization goes beyond simple recommendations by providing a timely and proactive service as an AI-enhanced approach, whereby AI-enabled systems can anticipate the needs of guests before they are even explicitly or overtly communicated. Connected to hotel management systems, IoT-enabled systems improve service delivery when they suggest hotel in-room food and beverage selections when a guest misses a meal, suggest spa services after a health-check in, or proactively adjust room climates without guest input. Proactive personalization can enhance a guests' overall experience, and create an environment of exclusivity and care that will encourage loyalty and return visitation. In addition to enhancing guest experiences, personalized recommendations can also facilitate operational efficiencies for hotels, by deliberately promoting in-house services and local partnerships that create new revenue while increasing guest satisfaction. In summary, robotic concierge systems will reshape hospitality experiences by creating personal interactions that are timely and customized to a guests' expectations and preferences, rather than generic service (Gupta and Pareek, 2024).

3.2 Multilingual Conversational Abilities and Global Guest Inclusivity

Though traditional concierges often haven't been able to fulfill the same level of personalization because of language limitations, AI-powered robots using natural language processing (NLP) and translation algorithms can speak in dozens of languages fluently and consistently. As a result, guests engage with a robot concierge in the language of their choice, ask questions comfortably, and receive responses without obstacles of miscommunication. To illustrate, a guest from Japan staying in a European hospitality establishment can easily inquire for information in Japanese and receive culturally appropriate recommendations, while an English-speaking guest in an Asian destination has the ability to converse comfortably without human interpreters. By introducing multilingual capabilities, robotic concierge services not only increase guest comfort and confidence, they further enhance the hotel's reputation as a global and guest-centric organization. By allowing for particular language experience, robotic concierges reduce stress for international guests and improve clarity in service, while decreasing misunderstandings which have a negative effect on guest satisfaction. These systems also adjust tone of voice and conversational styles as a function of cultural context, personalizing interactions in any way that guests feel recognized and valued.

3.3 Real-Time Solutions for Guest Inquiries and Service Requests

In comparison to traditional concierges, which may be restricted by workload, availability, or human error, AI-enabled concierge robot's function at a steady pace 24/7 and provide immediate answers to questions ranging from simple inquiries regarding hotel-based amenities or local attractions, to more complex requests regarding transport bookings, spa appointments, or troubleshooting issues within the guest room. These systems are integrated with the hotel property management software (PMS), customer relationship management (CRM), and devices on the internet of things (IoT) devices to carry out immediate tasks, in addition to supplying information. For example, when a guest requests extra towels, the robotic concierge can notify housekeeping immediately, or if guests wish to alter the temperature in their room, the concierge can directly interface with smart room systems to make changes in real time shown in figure 2.

Figure 2. Real-Time Solution Architecture Model for Guest Requests

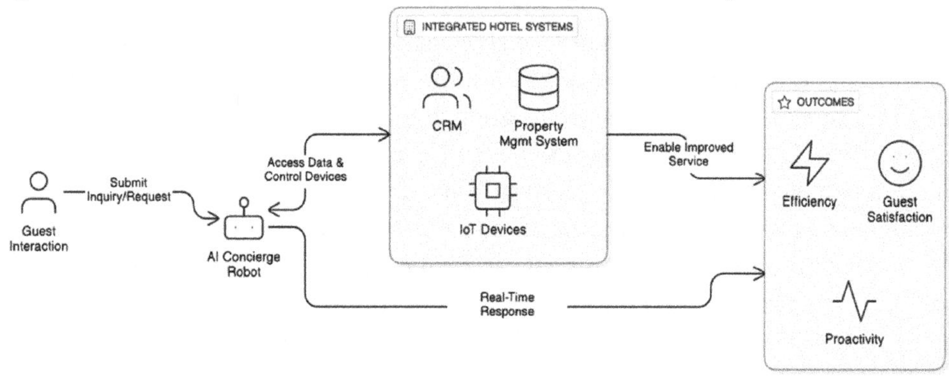

3.4 Integration with Hotel Management Systems

An important feature of robotic concierge systems is their complete connection and integration with hotel management systems (HMS) that allows them to operate seamlessly, automate hospitality service, and be very efficient. Robotic concierges integrate and connect to property management systems (PMS), customer-relationship management (CRM) software, and IoT-enabled in-room technologies that provide them access to real-time updates of the guest particulars. As a result, the hotel can provide a fully contextualized and personalized engagement at each interaction or visit. For example, if a guest requests to check-out late, the robotic concierge accesses the PMS to check if late check-out is available, and the guest's check-out time will be extended all without any interaction with hotel staff. If a guest schedules a spa appointment or a restaurant reservation, the robotic concierge syncs with the respective departments to complete the requested service where the potential for human error is reduced and response time is quicker. This centralized information also allows for the robotic concierge to be intuitive as it can anticipate guest needs based upon information from previous hotel stays, loyalty program information, and current and up-to-date logs on activity. For example, a frequent business traveler can automatically receive an opportunity to book a conference room while leisure visitors returning to the hotel will be provided with meal discounts or guided tours by a third party. Moreover, smart rooms equipped with IoT devices provide the robotic concierge with the capability to react immediately to the guest's request to change the lighting, temperature, or broadcasting system, which results in a convenient in-room experience. In operations, connected robotic concierges are a workflow management platform that liberates hotel employees from monotonous, manual, and operational activities, ensures the optimal staffing levels, and connects and collaborates with workgroups in the hotel such as housekeeping, food services, and engineering or maintenance. In addition to providing a way to be more efficient, integration also supports data-driven decision-making for hotel management by providing intelligence about service usage patterns, guest behavior, and satisfaction with their hotel stay. This information supports personalization, but if hotel management uses such information correctly, it also provides hotel managers the opportunity to identify additional opportunities for the hotel to improve either operational performance or revenue generation in the guest experience. In total, the integration of hotel management operations and robotic concierges shifts the service delivery mechanism to a connected, integrated ecosystem focused on the hotel experience where guest satisfaction, nimbleness, efficiency, and innovative capacity can together revolutionize hospitality amidst the dynamic economy.

3.5 Contactless Service and Safety in Post-Pandemic Hospitality

The COVID-19 pandemic has accelerated the acceptance of contactless technologies within the hospitality industry where robotic concierge systems have become a critical resource and solutions to ensure guest safety while still being able to provide service. With robotic concierge's ability to conduct self-service in the location of guest service, hotel human contact is minimized check-in/check-out, digital-key distribution, luggage delivery, in-room dining orders, information requests, and many other guest services. With the addition of voice recognition, mobile integration and, or touchless interfaces allow for a guest to access their ability to move throughout the service protocol, while at the same time minimizing the opportunity of viral contamination via shared surface and human contact with hotel staff. For example, a guest may request certain guest amenities or items through a robot or mobile app that a robotic concierge will order for delivery directly to the guest room to solve the issue efficiently and safely. In addition to immediate health concerns, contactless concierge services help to establish guest confidence and trust, which have emerged as new differentiators in hospitality after COVID-19. During all aspects of travel, guests will prioritize hygiene, safety, and minimal exposure as part of their overall experience. Robotic concierge services are developing the perfect service experience while incorporating guest expectations during COVID-19. As an added feature, robotic concierge service provides automation of safety considerations through application of contactless technology to take proactive ambient temperature checks; constantly monitoring air quality; and layering with sanitation/routine protocols to maintain service performance and be in compliance within health safety measures.

4. CURRENT APPLICATIONS IN THE HOSPITALITY INDUSTRY

AI powered concierge services and robotic service will no longer be a consideration for future implementation for hospitality practice, but embodiment of the reality to be established within hospitality practice as a true application reality.

4.1 Deployment in Luxury Hotels and Resorts

AI-powered concierge systems and guest assistant robots are being embraced in more elite hotels and resorts because of their perception as a means of exclusivity, personalization, and an upscale guest experience. In these high-end establishments, a robot concierge might be a new technology that meets a combination of informational service and novelty that meet the desires of affluent guests that are looking for seamless, innovative, and memorable guest experience. For example, Hilton captured headlines when it launched a robotic concierge named Connie in the mascot for every hotel brand, powered by IBM Watson. Connie is able to have a conversation with guests, make suggestions for dining and entertainment experiences, and provide travel tips for getting to nearby destinations, all in a friendly and engaging manner. Similar luxury resorts in Asia and Europe incorporate humanoid robots that greet guests, assist with check-in, and provide multilingual service to enhance efficiency while still giving a high-tech experience and decreasing guest line/wait times. Robotic concierges in luxury properties also offer personal service in guest rooms by engaging them while mapping the technology environment to manage smart-room functions, suggest wellness options, and schedule spa appointments. In some resorts, staff may operate delivery robots, which can engage guests for privacy reasons or while delivering amenities, meals, or snack items. Sometimes, guests can operate and/or interact with robots that work like tour guides around large properties, as well as navigate and explain all the property features, amenities, or activities in and around the resort. Within these luxurious environments, the successful implementation comes at a positive cost with regards to guest satisfaction, and affirms the luxury brand status of uniqueness as innovators of quality product and service, and therefore the future quality of hospitality. Luxury hotels are additionally employing robotic concierge systems in support of exclusive curated experiences as high-end travel itineraries, personalized shopping recommendations, or access to exclusive local events. By developing sophisticated capabilities through AI, and a laser-like focus on guest exclusivity, luxury hotels and resorts have elevated robotic concierges from service workers to value-addition companions.

4.2 Self-Service Kiosks and Robotic Front-Desk Assistance

In addition to luxury hotels, self-service kiosks, and robotic front desk assistants are increasingly common across different hospitality contexts, including business and budget hotels. These technologies are designed to improve the check-in and check-out time of guests, and associated wait times while also improving the speed of service delivery at the front desk, where guest traffic has historically served as a bottleneck during busy service hours. Self-service kiosks - through their AI-enabled interfaces - empower guests to complete activities such as confirming an existing reservation, selecting a room, paying for the accommodation, and printing a physical or digital room key, to name a few, that minimizes interaction with reception staff. Not only does this speed up service delivery, but it also allows independent and sophisticated travellers who favour speedy, independent service to coexist together in a way that gives the guest the feeling of convenience and being in control. Robotic front-desk assistants take that one step further by engaging in an interactive and personal experience with guests. Robotic front desk assistants are an extension of front desk automation designed to provide personal and interactive experiences to guests. These robots can greet guests, respond to common questions, assist with booking requests, accept payments, and make real-time recommendations related to dining, sightseeing, or transport through the application of natural language processing and machine learning. A few of the more robust systems come with a component to use facial recognition technology and enable secure identification of a guest for seamless check-in. Business hotels can leverage automation for their guests, providing busy professionals with the ability to save time and enhance productivity to better utilize their busy professional day, while still ensuring a consistent overall guest experience. In addition to delivering efficiency, robotic front-desk assistants can also provide brand experiences that are engaging and add an element of newness to the guest journey. For example, Marriott properties have included robots as interactive greeters and assistants, providing an unforgettable first impression that helps distinguish them from other firms in the market, particularly in Japan. Self-service kiosks and robotic receptionists further reduce costs and provide similar levels and service, ensuring that staffing shortages do not adversely impact service levels. Together, these features are representative of an important shift to a digitally mediated hospitality experience, focused on convenience, speed, and personalization, that front-desk functions represent, for contemporary travellers (Dasgupta and Jamader, 2024).

4.3 Smart Room Controls and Guest Service Robots

Integrating smart room controls and guest service robots is undoubtedly one of the most practical and guest-centric utilizations of robotics within hospitality today. More hotels are equipping rooms with Internet of Things (IoT)-enabled devices that allow for a robotic concierge to effectively control the in-room environment for their guests, allowing guests to control their hotel room lighting, temperature controls, entertainment systems, and even automated curtains or blinds through the robotic concierge. Therefore, guests can ask the AI-powered robotic assistant voice commands or use a mobile app interface to customize their in-room experience, allowing hotel guests to curate an entirely personalized and comfortable lounge and sleep experience. For example, a guest could ask the room to update to a preset temperature before returning to the room after dinner or request specific mood lighting for relaxation and music to harmony—all of which is executed seamlessly in real-time through the robotic concierge system, from start to finish. In tandem to providing environmental automation, guest service robots are also critical in automating repetitive services and improving convenience within the hotel space for guests. Robots can also facilitate meal delivery service to guest rooms at any chosen time, provide extra linen, toiletries, or other guest-requested amenities directly to the guest's door without human contact. Equipped with sensors, mobility robots can silently and discreetly navigate to bring items straight to the door, with secure compartment storage to enhance privacy, reliability, and timeliness of service. Information published in the last year from various reports highlighting Global Technology for Hospitality experiences indicates that hotels utilizing AI-enabled Concierge services or robotic Assistants & Service Robots to provide customer assistance & service delivery is creating significant operational efficiencies. Specifically, it was found that AI Concierge systems reduced response times for Guest inquiries by nearly 35% and increased the number of guests checking themselves into hotels through Self-Service Check-In by approximately 25%. Conversely, Service Robots were able to reduce Room Service delivery times from 20%-40%, while at the same time increasing future guest satisfaction ratings.

5. CASE STUDIES OF LEADING HOTEL ESTABLISHMENTS

The practical applications of AI-based concierge systems and robotic guest services can be further understood through the demonstration of actual use cases within the hotel industry (Gursoy, 2025). The case studies demonstrate how global hospitality leaders, including Hilton Hotels, YOTEL Hotels, and Marriott International have embraced robotics to enhance guest satisfaction, improve operational processes, and innovate the hotel industry.

5.1 Hilton Hotels

Hilton Hotels was among the first international chains to test cutting-edge robotics concierge by announcing "Connie", a humanoid concierge robot powered by IBM's Watson AI and WayBlazer's travel platform. Connie engages with guests using natural language to make recommendations on local restaurants, entertainment, and attractions, while also answering frequently asked questions related to the hotel. This example illustrates how AI can serve to provide supplemental staffing for hospitality operations given its 24/7 availability, multilingual capabilities, and ability to provide consistent information. Hilton saw Connie as a complimentary robot assistant or as it defined it: "not as a substitute but a valuable tool that can bring personalization and more efficiency to the guest experience." Guest responses indicated that having a robotic concierge was new, fun and something they would remember, regardless of the functional and operational advantages of providing personal and human-like service via technology.

5.2 Yotel Hotels

YOTEL's hotel chain was designed to appeal to travellers with its future-looking and technology-friendly orientation, and robotics is a significant part of their guest experience plan. All YOTEL hotels have robotic delivery systems such as "YOBOT" which takes care of luggage in the hotel, safely and securely delivering and retrieving luggage from the robot's safe storage space. Some YOTEL locations, YOTEL Chicago for example, have a room service robotic system that is responsible for the fast, safe, and contactless delivery of room service amenities to guest rooms. Furthermore, YOTEL uses smart kiosks instead of staffing personnel, creating ease for self-service check-in and check-out, an operational example of the chain's vision of transforming value and luxury by directing its hotels operation using technology to appeal to younger, tech-savvy consumers who prefer speed, autonomy, and innovation in hospitality.

6. TECHNOLOGICAL ASPECTS OF ROBOTIC CONCIERGE SYSTEMS

Robotic concierge systems within ice industry are effective, because they utilize accessible technology and empowers them to provide seamless, intelligent, personalized service. The deployment of AI-based concierge systems has become an increasingly popular trend across numerous hotel brands worldwide. However, the benefits of using AI-based concierge systems are not as extensive as previously thought; there are numerous challenges organizations encounter in the deployment and upkeep of this type of technology, which includes long implementation timeframes and significant investment, technology downtime (due to a number of different reasons), robotic mobility restrictions (i.e., only operating within a designated area), differing degrees of guest acceptance of robots as an authority etc. Some hotel chains have reported that, historically, older guests were less willing to interact with humanoid robots for various reasons that involve emotional sensitivity in relation to specific tasks, while certain other segments experienced overwhelming enthusiasm for AI-based concierge systems.

6.1 AI Algorithms and Personalization Engines

Robotic concierge systems are based on the AI algorithms and personalization engines, which enables intelligent, adaptive, and context aware guest service. The AI algorithms analyze and process substantial amounts of structured and un-structured data. In order to build a complete profile of their guest. The AI algorithms employ techniques such as: predictive analytics, clustering, and recommendation modeling to anticipate the needs of the guest, while also pre-emptively recommending customized service—including dining options, recreation activities, or tailored spa packages. While formal service models tended to be rather static in nature, with an AI-driven personalization engine, learning is occurring continually as each guest interacts with the technology, ultimately allowing service recommendations to be refined in real time resulting in even better service the next time.

6.2 Natural Language Processing for Conversational Interfaces

Natural Language Processing (NLP) is a basis for conversational interfaces in robotic concierge systems that offers human-like, interactive communication between guests and robots. NLP provides the robotic concierge's ability to understand requests, answer questions, and engage in contextual, meaningful conversation by interpreting spoken or written language that is ultimately, a more personalized guest experience that emphasizes accessibility and intuitiveness for guests. Advanced NLP systems that evaluate syntax, semantics context, and intent allows implementation of continuum of complexity in the robotic concierge's ability to interpret and process expressive queries. Expressive queries will pivot around local experience recommendations, setting adjustments within the guest room(s), as well as multi-step requests for housing requests and servicing. NLP also supports the robot's ability to demonstrate a multi-language interface empowering global guests to engage comfortably in their desired language through levels of cultural and language usage include inclusivity, equity, and hospitality for differing cultural and linguistic populations worldwide. NLP grows beyond simply question-answer dialogue engages in sentiment or how the user feels regarding the question. This may ultimately reflect urgency or tone and report those feelings back to the robot concierge while demonstrating acknowledgment and mood. The greater significance exists for robot concierges to prioritize urgent requests, prompting personalized and inclusive communication experience for guests. In addition, if the robot concierge can escalate to human convergence if needed, it supports both service and guest experience satisfaction within the conversation. NLP and machine learning in conjunction is crucial for constant improvement of each occasion with NLP for application and advancement. In summary, Natural Language Processing (NLP) turned robotic concierge systems into engaging, compassionate, and responsive partners to augment human-centric guest engagement (Vashishth et al., 2025).

6.3 Integration with IoT and Smart Hotel Ecosystems

The integration of robotic concierge technology with IoT systems or smart hotel ecosystems is critical for providing a seamless, personalized, and automated guest experience. For example, if a guest would like the temperature adjusted in their room, lights dimmed, or blackout curtains drawn, the robot concierge will attend to the request immediately without requiring any physical guest interaction. Furthermore, not only is the IoT connectivity utilized to adjust room conditions, but it also fully integrates with hotel operations to automatically connect the robot to hotel systems more generally, such as for housekeeping, maintenance, or dining services to provide exceptional service. IoT enables robots to service guests using proactively and predictively based on other systems. The beneficial interaction with occupancy sensors or smart scheduling systems becomes part of a dependency approach for improving resource schedules and reducing energy consumption and operational efficiencies. Data collected in IoT networks could also provide insights advanced AI algorithms in context of personalization, predictive maintenance, and intelligent service quality improvement. The integration of real-time data with automation and intelligent assisted decisions such as robot concierge systems enables the quickly shifts hotel operations toward a responsive, adaptable, guest-focused ecosystem that integrates technology, operational efficiencies, and guest service (Parthipan et al., 2025).

6.4 Robotics Hardware and Human-Robot Interaction (HRI)

The effective operation of robotic concierge systems does not just include sophisticated software; robotics hardware and the design of awareness of human-robot interaction (HRI) are also significant components of enhanced functionality. Robotics hardware, such as mobility-based platforms, sensors, cameras, actuators, and touch interfaces enable robots to traverse the challenging aspects of hotel environments, interact with guests, and successfully deliver items, all while safely co-existing with humans. A mobility platform relies on high-level reasoning for its movement, whether it is based on wheels or legs, and accomplishes this through obstacle detection and path planning algorithms to autonomously navigate through hotel lobbies, paths, and guest rooms. Sensors such as lidar, ultrasonic devices, and cameras can determine obstacles, identify people, and establish "safety" distances to facilitate serving in service areas where crowds can be present or were guest may be active in the space. Second, an important consideration for a comfortable and natural experience is the design aspect of Human-Robot Interaction (HRI). Robotic concierges have been designed with the ability to effectively communicate to the guests through visual cues, speech, gestures, and some level of touchscreen functionality for guests to enter and exit the information for the services requested. HRI design considers an additional body of factors, including cultural, psychological, and ergonomic, including personal space, friendly speech, and multilingual communication. Ultimately, guests engage meaningfully through tactile and auditory feedback with the robot when requesting services, asking questions, or requesting direction. Together, unethical design of robotics and HRI improve both efficiency and user satisfaction. A well-designed robot can deliver luggage, food, and room service, as well as guide guests, whilst allowing staff to focus on the guest experience, the staff, with compassion for their fellow beings, and the HRI enrollment mentally, communicating rationally and clearly, and confidence. A robotic concierge system as a social object in terms of technology and design moves from merely a tool/service to a social object taking into consideration technical reliability and social empathy, as a supplement to human team members, a richer guest experience, and part of the hotel's philosophy of innovative and technology hospitality (Yan et al., 2025).

6.5 Cloud Computing and Real-Time Analytics

Cloud computing and real-time analytics are critical infrastructure for robotic concierge systems, in that they provide the infrastructure for scalability of operations, data processing and seamless integration of the hotel's operations. Robotic concierge systems can perform a plethora of tasks, such as store, process, and analyze the chances of guests through a cloud-based platform, including guest preferences, booking histories, in room events, and guest feedback is shown in figure 3. In the cloud computing environment, several robots in several hotel locations can simultaneously identify new guest interactions based on the latest information available that is shared across the robots, resulting in standardized experiences in hospitality, thereby doing even less to assist guests with guest quality standards. The infrastructure of a cloud computing platform supports software upgrades and system functionality, system upgrades, and modifications to AI models to occur without interruption to existing hotel process functional operations. This improves operational reliability and efficiency of operations. Real-time analytics improves response time and 'IQ' of robotic concierge systems. A robotic concierge system is capable of processing information from IoT devices, guest interactions, and hotel management system applications by measuring historical engagement data in order to generate timely decisions. This would include executing automated functions from IoT devices for in-room service components, prioritizing service requests systematically to a greater extent than human concierge, or generating recommendations directed at a guest occupancy.

Figure 3. Real-Time Cloud Computing Analytics of Robotic Concierge System

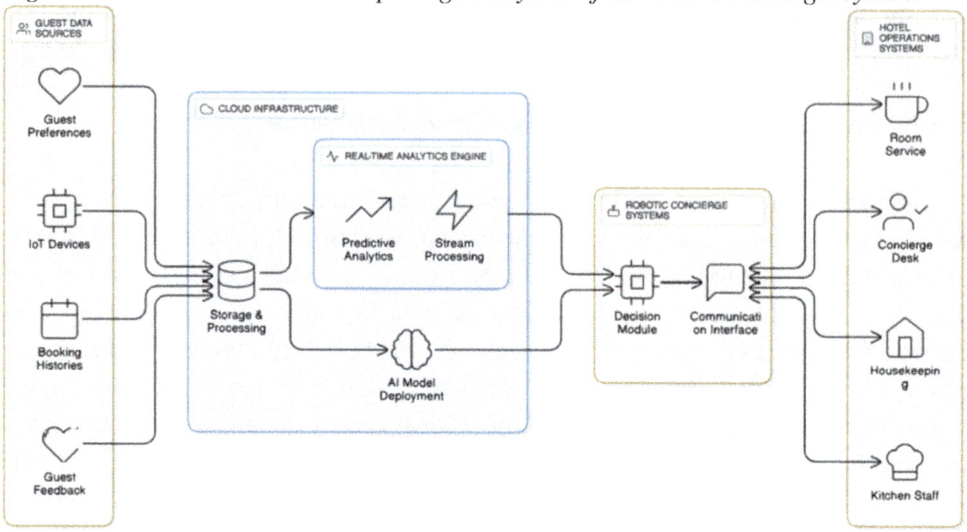

7. CHALLENGES AND ISSUES IN AI CONCIERGE DEPLOYMENT

Although AI-powered and robotic concierge systems have many advantages, they also present several challenges in the deployment of AI in hospitality, from technical, operational, and ethical standpoints.

1. **Data Privacy and Security:** An important issue of deploying AI concierge systems, or robotic concierge, is protecting sensitive guest data in the application. Robotic concierges are constructed to capture personally identifiable information, payment information, behaviour tracking and guest preferences that are vulnerable to safe storage. Any data breach or misuse of such data may result in harming guest trust and hotel enterprise reputational issues.
2. **Guest Purchase and Use Acceptance:** While some guests may appreciate AI-based services, for others there is generally a discomfort with interacting with a robot when requested assistance may be complicated and/or prompting sensitivities with and types of interactions cultural and or other. A guest-centered experience might still have human touch with embedded service protocols to offer guest-centered interaction with robots and intercultural communication by allowing use of multilingual tools and the presence of humanoid present while enjoying the drink by the guest. The service of robots is truly rewarding,

enhances the stay experience at hotels and does not detract from the guest experience of overall with their emotional connection.

3. **Support Equipment and Maintenance Issue:** To keep a robotic concierge in operation, maintenance systems and software updates and ongoing technical support become important for operation. Any glitches, failures, or downtime might potentially change guests experience or affect their guest experience. If either occurs it might lead to a negative guest experience or maintenance resulting in changes to operational efficiencies related to guest experience. A well-planned established protocol and trained personnel would be a must in a hotel State instruction and now produced evidence of what protocols to take in a stressed period claimed or observed.
4. **Ethical Issues and Bias:** Ethical issues may include eliminating algorithmic bias, attaining fairness in automated recommendations, and transparency in the AI decision-making process, etc. Even AI decisions, even if planned or scripted, can still bias or mislead a guest's experience or spending behavior, and guidelines surrounding the periodic evaluation and auditing of continued service is needed to remain ethical. AI concierge systems present several ethical challenges, which are more than just increased convenience and efficiency for guests. Data Privacy is the primary ethical concern because robots will collect sensitive data about guests including but not limited to how they behave, what they like, and how they move. When robots do not obtain informed consent from the guests to collect their personal data, the trust that the guests place in the company is at risk. Algorithmic bias presents the second ethical challenge, where recommendation engines may be biased towards certain groups of guests or businesses. In addition to potential employment displacement due to this technology being in existence, there may be social implications of guests being afraid of interacting with a robot that acts like a human. All of these ethical issues indicate that there is an urgent need for ethical guidelines related to AI Concierge technology, as well as clear policies regarding the proper handling of data collected, as well as hybrid workflows that will allow the human aspect of hospitality to remain intact.

8. FUTURE DIRECTIONS AND OPPORTUNITIES

AI, robots, and robotics systems, in the hospitality context, have never experienced the level of impending growth that is likely to occur considering technological developments in the area of artificial intelligence, machine learning, robotics, and IoT connectivity.

1. **Hyper-Personalized Experiences for Guests:** The next generation of concierge robots will develop an ability to view a guest's needs and develop hyper-personalized experiences for guests. There will be advanced AI algorithms that enable sophisticated big data analytics and link AI/robot systems as they integrate and connect with wearables and the smart hotel marketplace, so that robots will eventually be able to tailor recommendations for restaurants, leisure activities, wellness routines, and travel itineraries. Therefore, hotels will be able to create a sense of surprise and delight by providing services that are deep and personalized, which will yield and retain loyalty and fulfillment (Talukder et al., 2025).
2. **Improving Operational Efficiency and Reducing Costs:** Robotics concierge implementations will undertake increasingly broader operational tasks such as scheduling and timing of housekeeping, managing inventory, or predicting maintenance needs. With a growing amount of automations or repetitive tasks being performed with AI and robots, it is anticipated that labour costs will be reduced and staff will be optimized, as associates and professionals in guest services can devote time and effort to high-value guest interactions that require empathy, thought, and problem solving. This combination of automation and humans' ability will provide valuable service quality and value to hotel guests, while maximizing operations efficiency, (Bathla et al., 2025).
3. **Cloud-Based Analytics and Data-Driven Insights:** The innovating of cloud computing, the size of big data, along with real-time analytics will ensure hotels have the means to advocate for active monitoring of guest activity, optimization of service strategies, and determine the potential impacts of current trends based on data analytic. Data-based analytics will lead to a robotic concierge to continue to make additional recommendations, or, dynamically change the service to suit the needs of the single guest or group of guests, to enhance the individual experience and group experience, and ensure effective systems are in place for operational purposes (Sabharwal and Jain, 2025).
4. **Sustainability and Safety:** Environmental and health factors will also have an impact on the design of future robotic concierges. For example, the ongoing need for energy-efficient robotics developed from eco-friendly materials, energy literacy, and environmental practices are included in new developments in robotic concierges. The design and deployment of concierge services requiring no or limited human guest contact may be another consideration as a safe environment. The notion of robotic hotel solutions emphasizing contactless service, real-time monitoring of guest needs, or performing automatic hygiene protocols, will all likely be characterized in future hospitality service.
5. **Collaborative Human-Robot Workflows:** The ability of robots to operate in hybrid roles in the hotel/hospitality industry will enable hotels to implement

collaborative workflows to leverage human capacity to enhance positive guest experience. Robots will be assigned to perform repetitive, time-consuming, or data-driven chores, while human employees will be involved in roles that require empathy, creativity, and/or complex problem solving in response to guest needs. Together, these strategies will result in a future driven by technology to enhance human-visible capacities in operations, in a way that produces truly intelligent, adaptable, and possibly unforgettable guest experience (Venkateswaran et al., 2024).

9. CONCLUSION

The hospitality industry is being transformed by AI alleviators and robotic concierges, which are starting to generate new personalized, efficient, and responsive guest experiences through artificial intelligence and machine learning, natural language processing, Internet of Things (IoT) technologies, robotics, and cloud computing. Their application is improving operational efficiencies as well as streamlining the check-in process, controlling in-room temperatures, offering multilingual services, and answering guest questions in real-time while augmenting staff to provide a completely different staffing interaction. Even with the associated challenges regarding guest information, privacy and security, ethical concerns, culture change, challenges to converging integration, and guest reluctance to accept change, effective implementation of these solutions guarantees that AI systems in hospitality will enhance rather than replace guest experience. Our future models in hospitality will continue to include forerunners in branding for hyper-personalized, predicative service offerings, multisensory AI interactions, sustainability, and the integration of human and robotic workflows in order to personalize service encounters in intelligent, responsive, and interactive ecosystems. In sum, the opportunities provided by robotic concierge systems and AI-based alleviators will afford hospitality settings the opportunity to innovate by bringing technology and human service together to improve guest satisfaction, operational effectiveness, and competitiveness in hospitality in this day and onward.

REFERENCES

Avula, M., & Sithole, T. (2024). Artificial Intelligence's Potential to Improve Operational Efficiency and Customer Experience in the Hospitality and Tourism Sectors. In Impact of AI and Tech-Driven Solutions in Hospitality and Tourism (pp. 351-382). IGI Global. DOI: 10.4018/979-8-3693-6755-1.ch018

Banerjee, P., & Bhat, A. (2025). Transforming Smart Hospitality: Leveraging AI and IoT for Sustainable Tourism, Enhanced Guest Experience, and Cybersecurity Resilience. *Eco-Tourism and Sustainable Development*, *1*(1), 14–26. DOI: 10.63385/etsd.v1i1.87

Banu, S., Samantaray, R. R., Ayesha, N., Fathima, A., & Tabassum, Z. (2025). Robotics and Automation in Hospitality Services. In Cutting-Edge Technologies for Business Sectors (pp. 367-400). IGI Global.

Bathla, G., Raina, A., Rana, V. S., & Bhushan, B. (2025). The Impact of Artificial Intelligence and Automation in Transforming the Hospitality Sector. In *Transforming the Service Sector with New Technology* (pp. 541–552). IGI Global Scientific Publishing. DOI: 10.4018/979-8-3693-7447-4.ch033

Dasgupta, D. S., & Jamader, A. R. (2024). Revolutionizing tourism and hospitality: Harnessing AI for sustainable transformation. Available at *SSRN* 4896697. DOI: 10.2139/ssrn.4896697

Fernandes, M. D., Wani, P. S., & Shah, G. D. (2024). The Integration of Artificial Intelligence in the Hospitality Industry. In *Hotel and Travel Management in the AI Era* (pp. 525–550). IGI Global. DOI: 10.4018/979-8-3693-7898-4.ch025

Gupta, P., & Pareek, R. (2024). AI-Enabled Guest Experience: Disrupting Hospitality through Next-Generation AI Technologies. In Technology and Luxury Hospitality (pp. 165-181). Routledge.

Gursoy, D. (2025). Artificial intelligence (AI) technology, its applications and the use of AI powered devices in hospitality service experience creation and delivery. *International Journal of Hospitality Management*, *129*, 104212. DOI: 10.1016/j.ijhm.2025.104212

Jogarao, M. (2024). AI in the hospitality industry: Revolutionizing guest experience. In *Technology and luxury hospitality* (pp. 199–210). Routledge. DOI: 10.4324/9781003488248-14

Manoharan, G., Ashtikar, S. P., & Kumar, S. (2024). Extenuating digital transformation: Revolutionizing hotel operations through AI technology. In *Hotel and Travel Management in the AI Era* (pp. 257–274). IGI Global. DOI: 10.4018/979-8-3693-7898-4.ch012

Parthipan, S., Guruprasath, R., & Maha Sriram, S. (2025, June). Optimizing Hotel Performance through AI-Powered Revenue Forecasting and IoT Data Analysis. In *2025 International Conference on Emerging Technologies in Engineering Applications (ICETEA)* (pp. 1-5). IEEE.

Ravichandra, T., Murugeswari, P., Revathi, M., Senthilkumar, S., Rathore, D. S., & Sudhakar, M. (2025). The Future of Machine Learning and Robotics in Digital Technology for Hospitality. In *Cutting-Edge Technologies for Business Sectors* (pp. 401–428). IGI Global.

Sabharwal, D., & Jain, V. (2025). Transforming Hospitality through AI, Big Data, and Automation: The New Era of Service Innovation. In Addressing Contemporary Challenges in the B2B Hospitality Sector (pp. 309-336). IGI Global Scientific Publishing.

Sharma, M., Tyagi, P. K., Semwal, R., Tyagi, P., Singh, V. J., & Kumari, N. (2024, June). Robotic concierges and smart spaces: A vision for the future of hospitality services. In *International Conference on Innovation and Regenerative Trends in Tourism and Hospitality Industry (IRTTHI 2024)* (pp. 511-525). Atlantis Press. DOI: 10.2991/978-94-6463-437-2_34

Talukder, M. B., Rahman, M. Z., & Khan, M. R. (2025). Enhancing Guest Experience through Smart Hotel Technologies. In *Adapting to Evolving Consumer Experiences in Hospitality and Tourism* (pp. 301–324). IGI Global.

Vashishth, T. K., Sharma, V., Sharma, K. K., Chaudhary, S., Kumar, B., & Kumar, P. (2025). Graphs, language models, and NLP: The future of search engines. In *Applied Graph Data Science* (pp. 91–104). Morgan Kaufmann. DOI: 10.1016/B978-0-443-29654-3.00011-9

Vashishth, T. K., Sharma, V., Sharma, M. K., & Sharma, R. (2025). Enhancing Hotel Customer Service with AI-Powered Chatbots. In Marketing Technology-Infused Hospitality: Upskilling Frontline Employees for Competitiveness (pp. 83-114). IGI Global Scientific Publishing.

Venkateswaran, P. S., Sriramkumar, M., Vaddy, R. K., Kotagiri, A., Rajest, S. S., & Regin, R. (2024). *Applications of artificial intelligence and robots in service industries. In advancing intelligent networks through distributed optimization*. IGI Global.

Yan, Z., Konar, R., & Balasubramanian, K. (2025). The Future of Hotels Robotics, AI, and Service Automation in Practice. In Sustainability, Innovation, and Consumer Preference (pp. 249-278). IGI Global Scientific Publishing.

Chapter 8

Robotics in Luxury Hospitality:
Transforming Service Experience in Five-Star Hotel Restaurants

Gurucharan Singh
https://orcid.org/0009-0006-3432-6270
MMICT&BM (Hotel Management), Maharishi Markandeshwar University, Mullana, India

Rakesh Sharma
https://orcid.org/0000-0002-3762-1500
MMICT&BM (Hotel Management), Maharishi Markandeshwar University, Mullana, India

ABSTRACT

Robotics is transforming five-star hotel restaurants by integrating artificial intelligence (AI) and automation into guest welcoming, ordering, service, and billing. These technologies improve efficiency, accuracy, and consistency while addressing labor shortages, rising costs, and sustainability goals. Rather than replacing staff, robots complement human service by managing repetitive tasks, allowing employees to focus on personalized engagement and emotional connection. Guest responses are mixed: many value speed, hygiene, and novelty, while others express concerns about warmth, cultural sensitivity, and privacy. This highlights the importance of balanced human–robot collaboration and ethical adoption strategies. Future trends point toward emotionally intelligent robots, data-driven personalization, and integration with smart city ecosystems. Ultimately, robotics enhances competitiveness and sustainability in luxury dining, provided it preserves the human essence of hospitality.

DOI: 10.4018/979-8-3373-6541-1.ch008

I. INTRODUCTION

The restaurant business cannot be left out in change since it has been adapting to the shifting consumer expectations, behaviour in the society and economic realities. The priorities to visit a restaurant and eat became the quality of food and human hospitality. They were not visiting the restaurants to eat as a single person, but also seeking individual service, atmosphere and human experience (Liu et al., 2024). But in the twenty first century service excellence is being re-defined with the advancement of technology that is becoming extremely common place almost in all dimensions of dining experience. The customers of the day demand safer, faster, and more customized services, which drives restaurants to introduce robotics and AI, among other intelligent services to their front of house (FOH) service. And this shift was intensified due to the COVID-19 pandemic, which turned contactless service and hygiene into the key demands of the guest satisfaction (Zeng et al., 2020). What seemed to be the emergency actions that would be used in the event of a health crisis yesterday become the normalcy today. The food service experience provides the emerging trends that involve the delivery of digital food menus to the guests, self-service lanes, robots, and no-friction payments as a fresh representation of the contemporary restaurant experience. Simultaneously, the global issue of the lack of workforce within the hospitality industry has only added to the pace of the necessity to automate. Restaurants are thus not just embracing robotics and any other digital technology as a fad and new thing, but as the means of running the smooth and steady way of doing things to stop any long-term sustainability (Ivanov et al., 2022).

In Vo (2019), the design and implementation of a restaurant on a line-following approach based on a robot is highlighted. This system makes use of the infrared sensors to identify and trace the preset routes so that food can be accurately delivered in a restaurant environment. Use of such robots will minimize human work, enhance efficiency in operations as well as satisfaction of customers. In addition to that, the plan renders it economical and dependable in structured service environments. Such innovations form a very strong step towards automation within the hospitality sector (Sun and Zhang, 2023).

According to the recent scholarship, the role of robotics in hospitality has evolved beyond offering operational support but it is now a key element of the guest experience design. New studies have pointed out how the interaction between humans and robots in luxury settings has been changing with the advent of emotional AI, affective computing and adaptive service algorithms. Research also highlights emerging issues of ethical AI regulation, data privacy, and open use of customer analytics. In addition, the post-pandemic informational sources demonstrate that sophisticated automation of the fine dining sector is now linked to the sustainability agenda, such as the waste management, energy-saving robots, and the digital food management systems. The combination of these revised concepts would help to enhance the scholarly basis of the chapter and reflect its compatibility with the latest developments in the intelligent automation of services (Zhao et al., 2024).

Evolution of Technology in Restaurants

The application of technology in restaurants is not a novel concept. The streamlining of restaurant management had started early with electronic cash registers and point of sale (POS) systems in the 1970s. Another breakthrough came in the 1990s, with the development of online booking systems and online payment systems. Nevertheless, robotics and advanced AI-based solutions have only become a part of the mainstream restaurant operations in the past two decades. Kiosks, AI drive-thru and mobile apps became a common feature in fast-service and casual restaurants. As an example, McDonalds, Dominoes and Starbucks have effectively used digital ordering to enhance speed and minimize human error. Robotics in the form of robotic hosts, robotic waiters and automated service carts have now enabled the meaning of hospitality to be redefined in the fine-dining and casual form of the restaurant (Lu et al., 2019). The development of robots such as Pepper, BellaBot, and PuduBot emphasizes the way robots can be used to carry out interactive functions, including welcoming visitors and carrying dishes on the tables. Such developments are not at all an outcome of technological innovation, but a demand-driven development. In particular, younger generations and modern visitors are digital natives that are not afraid of automation. To them, a restaurant providing intelligent and seamless services is more in accordance with their lifestyle expectations (Liu et al., 2024).

Changing Guest Expectations and Industry Drivers

Guest expectations are at the core of why technology is transforming restaurant services. Five primary drivers can be identified:

- **Speed and Efficiency:** The modern customers demand short waiting times, fast order processing, and instant payment options. The technologies like AI-powered reservations, self-order kiosk, and contactless billing address this need of instant gratification.
- **Personalization**: Guests will no longer desire generic experiences. Artificial intelligence-based systems are capable of retaining customer preferences, dietary limitations, or favourite meals. Digital tools and robotics allow restaurants to provide customized services, thus producing the feeling of exclusivity.

- **Hygiene and Contactless Service**: Dining post-pandemic will require less physical contact. Food delivery robots, QR code menus, and digital wallets decrease the number of unnecessary human-to-human interactions; meanwhile, they remain efficient.
- **Novelty and Experience**: Robotics in the service is seen as a novelty by at least some of the younger diners, and something that makes the restaurant experience better.
- **Labor Shortages and Cost Pressures**: Hospitality industry is grappling with staff shortages and increasing wages all over the world. Robots and AI systems are perceived as the machinery that can replace the monotonous duties, and human employees can be spent on doing jobs that demand either creative skills, compassion, or problem-solving (Liu et al., 2024).

Focus on Front-of-House (FOH) Services

Although robotics has been widely discussed in the context of kitchen automation (robotic chefs, food preparation, etc.), this chapter specifically focuses on front-of-house services the guest-facing processes that span from arrival to billing. The FOH is the most visible aspect of restaurant operations, directly influencing guest perception and satisfaction.

- **Guest Arrival and Reservation Systems:** The first impression a restaurant creates, now enhanced through AI-driven reservations, robotic greeters, and digital check-in kiosks.
- **Ordering Experience:** The critical interaction point, transformed by digital menus, self-ordering kiosks, mobile apps, and AI-driven recommendations.
- **Table Service:** Robots delivering food, clearing tables, and interacting with guests to ensure efficient and hygienic service.
- **Billing and Payment:** The final stage of service, where smart billing systems, mobile wallets, and invisible payment solutions streamline the checkout process.

Global Trends and Future Demands

One of the broader trends across the globe in the hospitality and retail sector is the introduction of robotics in the restaurant services (World Economic Forum, 2023). Other countries like Japan, China and South Korea have been on the forefront in the area of interaction with their guests by the use of robots with robotic waitress becoming a common phenomenon in themed restaurants. The chains in Europe and North America have been swayed towards digital kiosks and contactless ordering and payment innovations. Robotic theme restaurants such as Robot Restaurant in Chennai have also sprung up in India with robot servers which signifies the trend of welcoming the trend in the new markets (Sun and Zhang, 2023).

Looking ahead, future demands point towards:

- **Emotionally Intelligent Robotics:** Robots capable of interpreting guest emotions and adapting service tone accordingly.
- **Augmented Reality (AR) and Virtual Reality (VR) in Ordering:** Menus that allow guests to visualize dishes in 3D before ordering.
- **Invisible Payments:** Systems where billing occurs automatically as guests leave, without requiring physical transactions.
- **IoT Integration:** Smart tables, chairs, and environments that interact with guest devices for a seamless dining experience.
- **Sustainable Technologies:** Energy-efficient robots and paperless systems aligning with global sustainability goals.

The useful uses of robotic service can be practiced in some luxurious hospitality environment. As an example, Henn-na Hotel in Japan employs humanoid robots as concierge, check-in, dining, and other assistance, and it shows how, by providing automation, one can redesign FOH experience without losing brand novelty. Equally, robotic concierges and automated service carts have been tested by Marriott and Hilton hotels in Asia and Europe to improve productivity and decrease the reliance on labor. Premium dining areas in India that implement BellaBot and PuduBot represent the increased favorability of robotics among high-end customers who are looking to have an innovative and convenient experience. These examples can be used to place theoretical discussions in a context of how the integration of robotics occurs in the real operational environment (Ivanov et al., 2022).

Opportunities and Challenges

The opportunities for robotics and technologies in restaurants are vast. They can enhance guest satisfaction, reduce operational costs, and help restaurants differentiate themselves in competitive markets. However, several challenges must be considered:

- **High Costs:** Advanced robotics require significant investment, which may be prohibitive for small and medium restaurants (World Economic Forum, 2023).
- **Cultural Acceptance:** Some guests value human interaction more than automation and may find robots impersonal.

- **Data Privacy:** AI-driven personalization requires extensive guest data, raising privacy and security concerns.
- **The Hospitality Paradox:** The essence of hospitality lies in warmth and empathy, which robots cannot fully replicate.

Financially, the payback of robotic systems related to financial returns on investment (ROI) is subject to the business volume of services, cost savings on labor, and efficiency benefits in the long-term. Fast-paced restaurants or those that have limited labor in the market usually record faster ROI since robots promote less staffing and fewer errors in taking orders. Over time, automation also reduces training expenses, operational errors, and employee turnover. Including even a brief economic comparison between human-driven and robot-supported service models would provide readers with clearer insight into the financial feasibility of robotics adoption (World Economic Forum, 2023).

II. TECHNOLOGICAL TRANSFORMATIONS ACROSS THE RESTAURANT SERVICE JOURNEY

Welcoming and Greeting the Guest

First few minutes of the guest interaction will be important in determining the experience of the dining experience. Historically, human hosts or hostesses have been tasked with this role where they offer glimpses, take care of queues and escort guests to their tables. But this point of entry has been redefined by the recent use of robotics and digital tools in the hospitality sector. The introduction of robotic greeters at the entrance of restaurants is one of the most obvious innovations. Such robots will help to give friendly and regular welcomes, respond to simple questions, and even recommend signature meals or offers. As an example, humanoid robots such as Pepper made by SoftBank Robotics are deployed in restaurants in Japan and Europe where they do not just welcome the customers but also participate in minor dialogues. Combining natural language processing (NLP), these robots provide a feeling of individualism, which appeals to technologically oriented guests. Besides robotized greeters, AI-controlled check-in desks are also becoming widespread (Lee et al., 2021). These online platforms enable the guest to make reservations, wait listings or real time bookings when they arrive. The kiosks that usually have multilingual features are very helpful especially in multicultural restaurants. This eliminates the obstacles that exist due to language differences and helps in reducing the time that is taken in waiting periods to allow a smoother flow of the arrival of guests. Another technology in guest welcoming is the introduction of facial recognition. Facial recognition is employed in the smart restaurants in China to recognize the customer when they walk in. The system recalls their past preferences of dining and allows the staff or service robots to give a personalized welcome message. This will not only help in the development of loyalty but it will also simplify the personalization of the service without necessitating the guest to retrace previous selections. Moreover, the gap between convenience and hospitality has been filled using mobile-based check-ins. Customers are now able to make their reservations through restaurant applications and do not have to go to the queues to announce their arrival (Lee et al., 2021). Push notifications are sent to the staff or robotic hosts to prepare the table beforehand. This is an activity that was initially popularized by the ongoing COVID-19 pandemic because of contactless needs but nowadays, it has become a standard service expectation among tech-savvy diners. These technologies have more than one field of benefits in the welcoming stage. The guests enjoy minimized waiting periods, culturally flexible interaction, and a smooth flow between entrance and seating. In the case of restaurants, it entails better control of the inflows of guests and the establishment of a futuristic brand image. However, challenges persist. Humanoid robots or complex recognition systems are expensive to install and this limits their usage to high end or testy dining areas. In addition, other diners would feel that robotic greetings are impersonal and they would rather receive human touch. A strike between automation and hospitality will then become critical (Sun & Zhang, 2023).

Reservation and Seating Management

The second key touchpoint that follows the first one, reservation confirmation and seat selection, is the second one. This stage directly affects the performance of operations where inefficient seating arrangement could cause congestion, lengthy queues and dissatisfied customers (Zhao et al., 2024). The technological advances have transformed this profession through the AI-powered reservation systems and robot seating-helps. The current reservation systems are not just a simple reservation rates platform such as OpenTable or EazyDiner in India. Predictive analytics allow them to optimise the allocation of tables according to the length of time the diners are expected to spend at the restaurant, the tastes of their customers, and the busiest hours. Using a computer as an example, it is possible to predict how long a couple may wait sitting at a table; using the historical data, the restaurant may ensure that the maximum number of people is served without making them rushed. The other novelty is the smart seating management software that has digital floor maps. The hosts (human or robotic) will be able to see the availability of the table in real time and be able to assign guests to it. Autonomous robots in high-tech restaurants follow their guests to their respective tables with ease in overcrowded restaurants (Ivanov et al., 2022). These robots have sensors and obstacle-avoidance technology and this means that seating is efficiently controlled with minimal dependence on the floor staff. Contactless seating confirmation has been popular too. Digital table numbers or QR codes are provided to the guests on the arrival, allowing a guest to scan it and find their place of seating without the help of human staff. This will reduce the number of people who jam the entrance points and give power to customers to move into their tables. Seating customisation is also improved by facial recognition and guest profiling. Within the luxury dining environment, recurring customers are automatically recognized, and they do reserve a table or a seat, in advance (Zhao et al., 2024). As an example, a couple who often takes a table on the window-side can have this option respected without having to request it every time (Park, 2020). Loyalty and the general dining experience are promoted through such personalization. On the part of the restaurant, the technology-driven seating management saves the aspect of human error, avoids double-booking, and enables to forecast the guest flows better. Employees no longer need to be assigned seats manually, but can instead devote their time and energy to more valuable activities, such as giving recommendations or making sure their customers are happy (Sun & Zhang, 2023). Still, there are obstacles to be taken into consideration. Algorithms may at times be over-relied upon to make any form of rigid decisions when it comes to seating arrangements, unable to adapt to any change of heart such as a table change request. Moreover, such advanced systems can be too expensive to small or middle-sized restaurants. Another issue is that guests are not comfortable with face recognition, particularly in areas where the role of privacy is a heavily discussed topic. Altogether, introducing robotics and electronic platforms to the reservation and seating management can be taken as a massive step towards operational excellence. The capacity to blend speed, accuracy and personalization does not only allow the guest to be satisfied but also aids restaurants in managing the changing demand as the marketplace is becoming highly competitive.

Ordering and Menu Interaction

Ordering food has always been among the most interactive parts of the restaurant service and involves consistent communication between the staff and the guests. It is as well the place where brand identity, efficiency, and personalizing experiences of the restaurant are most manifested. Due to the active development of robotics and the digital world, ordering and menu interaction faced a major shift, leaving the monotonous paper menus and the orders accompanied by verbal communication in favor of the interactive, smart, and contactless system.

Evolution from Physical to Digital Menus

In the past the printed menus were the first thing that the guest would see upon entering the restaurant. Although the design and descriptive text were considered significant, the static menus had no flexibility in regards to real-time updating, customization and activity. With the advent of digital menu boards and tablet-based menus, this experience was transformed as restaurants are now able to update dishes, pricing and promotions immediately. Indicatively, tablet menus such as Presto and Ziosk have been prevalent in the United States, and they allow the guests to view the pictures of dishes and get information about the ingredients and customize their dish with ease. Operational inefficiencies are also solved by means of digital menus. The inconvenience of a seasonal shift in the ingredients or the lack of the certain dishes or the fluctuation of the price depending on the demand may be conveyed instantly, and the guests will not be disappointed. Another way that restaurants save money on printing is by becoming paperless to meet sustainability objectives.

Role of QR Codes and Mobile Ordering

The QR code menus have gained popularity, and they are still used in the post-pandemic period because the COVID-19 pandemic increased their usage. Restaurants are also integrating technology that allows customers to get the digital menu of the restaurant on their smartphones by scanning a code written on the dining table. This product is affordable, hygienic and conforms to the increasing consumer demands of contactless interactions. Mobile ordering applications go a step further to integrate with loyalty programs, dietary preference profiles and payment gateways. The guests will be able to recalculate the order of their favorite dishes, divide bills online, or even order meals in advance. The mobile ordering system adopted by Starbucks is considered as one of the most well-known systems, however, other fine-dining restaurants have implemented the same to facilitate service at their busiest times of the day (Sun & Zhang, 2023).

Augmented Reality (AR) and Virtual Reality (VR) Menus

One of the recent trends is the implementation of AR and VR as the means of menu interaction. AR menus enable the guests to view a 3D representation of the dishes prior to ordering. This way of scanning a menu image or operating on a tablet will give diners a realistic preview of portion sizes, the way the food is plated, and even how it is cooked. This will decrease uncertainty and improve decision making especially in multicultural dining whereby the guests might not be conversant with the cuisine. VR menus are not very prevalent, but luxury restaurants are testing them out where customers get to be immersed in an imaginary world surrounding the recipe of the dish.

Robotics in Order-Taking

Robots have begun to be actively involved in processing and accepting orders. Voice recognition humanoid robots can communicate with the guests using natural language, memorize their preferences, and even transmit the orders to the kitchen system. Such robots remove the possibility of miscommunication due to noise, accents, or fatigue of the staff. In China and South Korea, restaurants such as Robot. He and Foodom employ robotic waiters, which in addition to serving food also take orders autonomously. The robots will be combined with AI and allow them to suggest dishes, depending on the time of the day, demographics of the customers, or previous purchasing patterns. In a busy or large-scale restaurant, the chatbots or voice assistants (AI-powered) running specifically on Hospitality with Amazon Alexa or Google Assistant can also offer this option to place an order. All guests need to do is speak out their orders and the system would translate the order into actionable orders to the service staff or robotic servers (Tussyadiah et al., 2020).

Personalization through AI and Data Analytics

The modern ordering systems are being characterized by personalization. Artificially intelligent applications use previous history of dining, dietary constraints, and behavior to suggest custom menu items. As an example, a vegetarian diner could be offered trending plant-based meals or specials of the chef depending on their profile. Predictive analytics are also used to upsell by restaurants. In certain cases where a guest customarily orders main course but not dessert, the system can give an exclusive discount on desserts shortly before confirmation of the order. On the same note, suggestions on wine pairing can be automatically produced to match specific dishes enhancing the guest experience and increasing revenue. Emerging tech-based platforms such as Dineout are now introducing AI-driven recommendation engines to their partner restaurants in India, which suggests that the trend of personalization is spreading internationally.

Contactless Ordering with Wearables and IoT

Other than the mobile devices, the Internet of Things (IoT) has increased the number of channels used to order to the wearables including the use of smartwatches. Here, the guests are able to make quick re-orders (re-order another drink) using a wearable interface connected to the restaurant. This ease of access lowers waiting times at the services and facilitates multitasking to diners who do not want to be interrupted a lot when conversing (Tussyadiah et al., 2020). The other innovation in the future is IoT-enabled tables. Such tables are fitted with inbuilt touch screens which enables customers to read menus, make orders, call service and even play games during waiting (Zhao et al., 2024).

Language and Accessibility Enhancements

The technological ordering systems also deal with the issue of inclusivity. Digital menus that are in more than one language can overcome the language barrier among foreigners. Services such as text to speech menu support the blind customer and the use of simple icons and images is to support the children or the elderly customer who are not well acquainted with digital technology. These innovations not only increase the number of customers visiting the restaurant, but it is also in line with the global standards of accessibility, which is socially responsible and increases satisfaction.

Advantages of Technology-Driven Ordering

The integration of robotics and digital tools into ordering offers several tangible benefits:

- **Accuracy:** Orders are transmitted digitally, reducing miscommunication.
- **Speed:** Instant updates and automation reduce service delays.
- **Upselling:** Personalized recommendations increase average spend.
- **Satisfaction:** Enhanced engagement through AR/VR creates memorable experiences.
- **Efficiency:** Staff workload is minimized, allowing more focus on hospitality.

Robotic Servers and Waitstaff

One of the most visible applications of robotics in the restaurant service space is robotic waitstaff. These robots are designed to deliver dishes, beverages, and cutlery from the kitchen pass or bar station to guest tables.

- **Humanoid Robots:** Robots like *Pepper* (Japan) or *Nadine* (Singapore) not only carry food but also engage with customers through small conversations, greetings, or entertainment. Their humanoid design allows them to mimic aspects of traditional waitstaff, providing an innovative, futuristic dining feel.
- **Tray-Carrying Robots:** Restaurants in China, such as Haidilao Hot Pot, use robots that follow pre-programmed paths to deliver dishes. Equipped

with sensors and LiDAR technology, they can avoid obstacles and navigate through crowded dining rooms.

- **Collaborative Service Bots:** Instead of replacing waitstaff entirely, some restaurants adopt service bots like *BellaBot* (Pudu Robotics), which carry trays to the table, while human servers handle guest interaction and plating. This hybrid approach maintains human warmth while maximizing efficiency.

Automated Beverage Service

Beverage service ranging from water refills to elaborate cocktails is another area transformed by automation.

- **Robotic Bartenders:** Restaurants and bars increasingly use robots for mixing and serving drinks. For example, *Makr Shakr* (Italy) features robotic arms that can craft cocktails with precision, consistency, and flair. Guests place orders via apps or kiosks, and the robots prepare the drinks in real-time.
- **Self-Serve Beer Taps:** Smart beer taps linked to RFID wristbands or prepaid cards allow guests to pour their own drinks while monitoring consumption. This reduces dependency on bartenders during peak hours and enhances interactivity for guests.
- **Automated Coffee Machines:** In cafés and casual dining spaces, robotic baristas like *Cafe X* (San Francisco) prepare personalized coffee orders in under a minute. Guests place orders via kiosks or apps, and robotic arms deliver customized beverages.

Drones in Restaurant Service

While drones are commonly associated with external food delivery, some futuristic restaurants have begun using drones inside dining halls for on-table service.

- **Indoor Drones:** Controlled by AI navigation systems, drones can carry lightweight items like drinks, snacks, or condiments to guest tables. Restaurants in Singapore and China have experimented with such systems, especially in open dining spaces.
- **Outdoor Dining Service:** In venues with outdoor seating or rooftop arrangements, drones offer a practical solution for overcoming long walking distances for staff.

Smart Tables and Automated Assistance

In addition to robotic servers, some restaurants are experimenting with smart dining tables that blend service technology with entertainment.

- **Embedded Service Functions:** Guests can call for service, request refills, or order additional items using table touchscreens. Some systems automatically notify servers or robots, ensuring swift response times.
- **Automated Refill Systems:** Smart dispensers for water, soda, or wine installed at the table reduce wait times. Guests can refill drinks as needed without staff intervention.
- **Gamified Service Features:** Interactive table systems allow guests to play games, browse restaurant stories, or watch preparation videos while waiting for food. This not only improves satisfaction but also reduces perceived waiting times.

Human–Robot Collaboration in Service

A growing trend is not complete automation but collaborative service models, where humans and robots work together.

- Robots handle repetitive, labour-intensive tasks (delivering plates, carrying heavy trays).
- Human servers focus on personalized tasks (explaining dishes, upselling, handling complaints, maintaining guest rapport).

This cooperation enables restaurants to solve shortage issues in staffing and also preserve the spirit of hospitality. As an example, certain Japanese fine-dining establishments have physical wine delivery with robots, but still, human sommeliers advise on wine. Integration of robots in the operations of FOH also requires new skills among the employees in the hospitality industry. The employees should be trained on how to supervise robots, fundamental troubleshooting, and how to serve the guests using data. Instead of displacing staff, robots change job specifications to more valuable interpersonal services like engaging the guest and resolving conflicts and offering personalized services. It would be relevant to mention the staff reskilling as an indication of how the restaurant can cope with the workforce shift and help the employees adjust to the hybrid service model.

Enhancing Hygiene and Safety

Post-pandemic dining emphasized the importance of hygiene and contactless service. Robots provide an advantage here:

- They minimize direct contact between guests and staff.
- UV sterilization features in some robots ensure sanitized trays and cutlery delivery.
- Automated dispensers for condiments and sauces reduce shared touchpoints.

This not only reassures health-conscious diners but also reduces the chances of cross-contamination in crowded dining environments.

III. ORDERING EXPERIENCE: THE SHIFT TOWARDS AUTOMATION

The past years changed the concept of food ordering in restaurants radically as the most sophisticated digital technologies, artificial intelligence, and user interfaces appeared. Previously, ordering could only be mediated by servers whose functions were to create a path between the customers and the kitchen. This system was highly personal but had the propensity to cause inefficiencies, errors or delays during the rush periods. The current state of the dining industry is however taking a direction towards automation-linked ordering systems which have been ensured to be quick, precise, convenient, and customized (El-Said, 2020). This is not a technology hype, but a customer shifting demand trend in the age where efficiency and digital integration have been regarded highly as much as taste and ambience. The ordering experience has now become the locus of restaurant innovation, with four key trends that are being pursued: introduction of interactive digital menu, use of AR/VR to display menu, introduction of self-ordering interfaces and app, introduction of voice-based AI-based ordering systems, and the greater emphasis on highly personalized recommendations (Çeltek, 2023). All these developments show that a new era of dining has started where automation is in effect that determines the interaction between customers, convenience and satisfaction.

Interactive Digital Menus and AR/VR-Based Menu Displays

One of the significant changes in the service in the restaurant has been the replacement of the paper menu with electronic or digital menu in the tablets, smart phones or interactive screens attached to the tables. These platforms offer multiple advantages: they can give information on the availability of the item in real-time, give more detailed information about the ingredients, the nutritional values, allergy warnings, and, even, dynamic pricing based on the demand or the promotion. The digital menus are also able to use high-resolution images and videos of food items unlike the non-dynamic menus that are used to help the guests make better choices. Going one step further, other restaurants are already thinking about integrating Augmented Reality (AR) and Virtual Reality (VR) menu, which will transform the whole procedure of decision taking. Using AR devices, visitors can see on their table what they want to order and they have a chance to see the portion sizes, garnishing information, and the way of plating in 3D. VR applications go even further and provide thematic experiences, e.g., a seafood restaurant could provide the consumer with the opportunity to investigate a virtual ocean environment when dining out and browsing the seafood menu (Lee et al., 2018). Such sensory presentation, in addition to making the customer more excited, will decrease the decision fatigue because the guest will have greater confidence in what he/she is ordering. Interactive menu also offers operational advantages to the restaurants. They help to reduce the risk of miscommunication between employees and customers, lower the cost of printing, and facilitate upselling in real-time by suggesting add-ons or combos. It has been found that customers who encounter interactive menus are more prone to spend more due to the alluring visualization of the food and directed ads to have drinks or desserts. In the immediate future, AR menus connected with AI-based dietary analysis may propose personalised choices based on health objectives, allergies, or even wearable health statistics of a diner.

Self-Ordering Kiosks and Mobile Ordering Apps

The increased using of self-ordering kiosks and mobile applications is another significant force behind automation in restaurant ordering. Self-ordering kiosks (which can be located at the entrance or inside the dining area) enable the customers to browse the menu, order dishes, and pay without using a server (Çeltek, 2023). This system is common in fast as well as casual restaurants where rapidity and convenience are the main priorities (Hwang et al., 2020). Kiosks minimize queues, dependency of staff, and offer accuracy of orders. Mobile ordering apps go even further in providing this convenience by allowing the guests to make their orders remotely and later arrive at the restaurant. In the case of dine-in services, it is possible to browse menus, divide bills between parties, and even plan meal time to lessen the wait options using apps. Having in-built loyalty programs, accrual of reward points and cashless payment embedded features, these platforms reinforce customer engagement and loyalty towards brand. Not only that, mobile apps provide the restaurants with a treasure trove of customer data. Monitoring user behaviour, including most ordered meals, favourite times of the day to eat, and average expenditure can assist restaurants to create specific promotions and enhance menu engineering. Mobile ordering decreases the work load of service staff during the peak hours, therefore making the operations easier and quicker to turnover. The need of self-ordering systems increased worldwide after the pandemic because of health and safety issues. People became used to contactless communication, and even restaurants with fine cuisine started experimenting with online ordering. The integration of kiosks and mobile platforms with AI-powered chatbots, in the future, will further automate the work of the guests, offering them conversations that feel human to guide them through the ordering process.

Voice-Enabled Ordering Through AI Assistants

The use of voice technology in the hospitality sector is a new trend quickly becoming widespread due to the popularity of artificial intelligence (AI)-based assistants that include Amazon Alexa, Google Assistant, and Siri. In the context of a restaurant, voice-enabled ordering provides a non-hands-on, streamlined experience, which makes it easier to use by a variety of customer categories as well as older and visually impaired people. Restaurants are taking voice assistants that are powered by AI into their apps and even on the table. The guests can just say, I want a vegetarian pasta and more cheese, and the AI system enters the order, validates the information and orders it, without human involvement (Tussyadiah et al., 2020). The sophisticated natural language processing (NLP) enables such systems to take into consideration complicated orders, accents, and even propose complementary products. A example is asking a diner to order a burger and then the AI assistant could reply, "Would you like fries and a soft drink with that?a computerized version of up-selling which is traditionally performed by waiters. Voice assistants can also answer questions on ingredients, portion size or time of preparation and this will increase customer confidence on their choices. As the voice recognition technology advances to a mature stage, then it will find application in smart wearables and in-car systems in such a way that the customers will be able to place orders with the food in restaurants when they are on the road. This can happen when you arrive at home and tell your car voice assistant to order you dinner at your preferred restaurant, which is prepared when you arrive. The meaning of convenience is not the only aspect of these innovations which alters, and it also establishes a larger ecosystem of restaurant services beyond the physical place (Çeltek, 2023).

Future Demand: Hyper-Personalized Recommendations

The hyper-personalization is the new technology in automated ordering where the technology predicts the needs and preferences of the guests with incredible precision. Using customer data (past orders, dietary limitations, seasonal preferences, and even social media presence, among others), AI-driven systems may suggest highly personalized recommendations that seem personally curated to each diner. As one of the examples, a frequent client who regularly orders vegan meal items can be welcomed to the business with a note: Welcome back! Try our new vegan quinoa salad? In the same way, predictive algorithms would recommend wine pairing, offer a suggestions list of dishes trending, or provide a list of the chef specials based on the history of a guest (Lu et al., 2019). In addition to upselling, the tactic establishes a feeling of personal rapport thus leaving diners to feel important and heard. The hyper-personalization can be developed to real-time contextual recommendations in the future. As an example, weather information may affect recommendations, such as the sale of cold drinks on a hot summer day or warm soups on a cold day. Social networking with health applications and wearables may further customize the personalization by suggesting low-calorie food items to exercise-lovers or recommending foods rich in iron to those with nutritional deficiency. It is an area of data-driven personalization that is increasing at a very fast pace, particularly in younger generations, including Millennials and Gen Z, who treasure the convenience that technologies make possible and require brands to understand their individual preferences and adjust to them. Hotels that implement hyper-personalized ordering systems are likely to have increased customer loyalty, satisfaction, and competitive edge.

Table 1. Summary of Technology-Driven Ordering Innovations

Technology / Innovation	Key Features	Guest Benefits	Restaurant Benefits
Interactive Digital Menus	Real-time updates, images/videos, ingredient transparency	Better decision-making, clearer expectations	Fewer errors, reduced printing costs, faster updates
AR/VR Menu Displays	3D dish visualization, immersive previews	More confidence in choices, enhanced experience	Higher upselling, reduced ordering uncertainty
Self-Ordering Kiosks	Touch screens, payment integration	Reduced wait time, control over ordering	Lower staff workload, improved accuracy
Mobile Ordering Apps	Menu browsing, loyalty integration, pre-ordering	Convenience, personalization, contactless service	Valuable customer data, improved flow management

continued on following page

Table 1. Continued

Technology / Innovation	Key Features	Guest Benefits	Restaurant Benefits
Voice-Enabled Ordering (AI Assistants)	Natural language processing, hands-free interaction	Accessibility for elderly/ visually impaired guests	Faster order processing, reduced staff involvement
Hyper-Personalized AI Recommendations	Data-driven suggestions, predictive analytics	Tailored dining experience	Increased average spend, targeted upselling
IoT & Wearable-Based Ordering	Smartwatches, smart tables	Instant reordering, minimal disruption	Faster service cycles, seamless integration

IV. ENHANCING TABLE SERVICE WITH ROBOTICS

While the previous sections addressed automated ordering and menu interaction, the discussion now shifts to table-side service innovations, avoiding overlap by focusing specifically on physical delivery, interaction, and human–robot collaboration. Dining table has been the most prominent feature of the restaurant experience. Other than food, it is at the table where hospitality is mainly experienced in terms of keenness, warmth and smooth service. Conventionally, this has been achieved by servers who are efficient and have emotional intelligence, such that the guests feel welcome, taken care of, and satisfied. Nonetheless, emerging with fast-paced technological growth and increasing demands of customers, robotics has become one of the game changers, in terms of reinventing the table service. Robotization of restaurants is not an issue of substituting human warmth with mechanical accuracy but rather making the services more efficient, less erroneous, and more satisfying to the guests. The delivery of food and beverages using robot servants, robotic tray carriers and table helpers, and human-robot cooperation patterns are becoming increasingly popular all over the world. In the future, the long-term goal is to come up with emotionally intelligent robots that can have adaptive service styles that are comparable to human empathy (Sun & Zhang, 2023).

Service Robots for Food and Beverage Delivery

One of the most obvious types of robotics in restaurants nowadays is the food and beverage delivery robot. The designs of these robots, which tend to have smooth, contemporary touch, can independently move around food areas to bring food at the kitchen to the table of a customer. They are fitted with sensors, cameras and navigation software, which help them to navigate through obstacles, engage the guests, and deliver on time. PuduBot and BellaBot are widely used examples which are famous in Asia, Europe, and North America. They have multi-level trays in these robots and can carry a number of orders simultaneously, relying on the LIDAR and SLAM (Simultaneous Localization and Mapping) technologies. Visitors are informed on arrival by the visual displays or sound alerts and they can receive their meals by collecting them by the robot tray (Jang et al., 2020).

The integration of such robots offers several benefits:

- **Operational Efficiency**: Robots reduce wait times during peak hours, ensuring food arrives fresh and on time.
- **Error Reduction**: Minimizing human handling reduces mix-ups or spillage during table delivery.
- **Labor Optimization**: Robots handle repetitive delivery tasks, allowing human staff to focus on high-value interactions such as upselling, resolving guest concerns, or creating memorable experiences.
- **Novelty and Branding**: Guests often perceive robotic service as futuristic and entertaining, creating a unique restaurant identity that attracts tech-savvy diners.

Robotic Tray Carriers and Automated Table-Side Assistance

In addition to delivery, robotics is also being utilized in table-side service with machines becoming an automated tray carrier, a server, or a helper system to supplement the efforts of human employees. Some of the tasks that these robotic assistants are capable of executing include filling water glasses, carrying extra cutlery, empty plates, or serving pre-requested beverages without the need of being prompted (Liu et al., 2024). Robotic tray carriers are especially helpful in a large restaurant or multiple levels where employees are used to walking long distances, which leads to the exhaustion of the staff and slow service. These robots tend to be built to have autonomous navigation systems, weight-balancing systems, and have the ability to transport heavy trays over long distances (Tuomi et al., 2020). There are also sophisticated models which have arms that can be extended to put the items on the table instead of the guests picking them. Interactive robots are also being programmed to be used at the table. Individually, an example of a robot that could be helpful in fulfilling the goals and objectives of the hotel business is one that is able to introduce himself, offer specials, or even sing Happy Birthday to guests celebrating a special occasion. Such gestures provide an element of entertainment besides reducing the repetitive work of the human staff. The application of AI-based humanoid robots that can imitate gestures and simple social actions is one of the new solutions. The robots will be able to stand around the tables, provide feedback after taking meals, and instruct guests on the digital payment options (Qiu et al.,

2020). Although in the initial phases, it is clear that there is increased expansion of interactive automation in the dining table due to such innovations.

Human-Robot Collaboration for Personalized Guest Engagement

Though the efficiency of robots is excellent, now they do not possess the richness of emotional understanding and cultural sensitivity that human waitresses add to the dining experience. Therefore, the idea of the human-robot collaboration is not a full automation but the most promising model (Choi et al., 2020). Under this model, robots are used to deal with repetitive activities that are more labour-intensive, whereas humans are engaged in personalised interactions. This partnership is also economically favorable. High turnover and increasing labor costs represent a significant problem in many restaurants across the globe, leading to shortages in the staff. Restaurants can overcome these issues by implementing robots as assisting employees and retain the same level of service at the same time. Guest perceptions also tend to favour HRC models. Surveys indicate that while customers enjoy novelty and speed from robots, they still value the human touch for creating warmth and emotional connection. A well-balanced model where robots complement, not compete with, human servers create an optimal service experience that blends efficiency with empathy.

Figure 1. Human-Robot Collaboration (HRC) Model for Restaurants

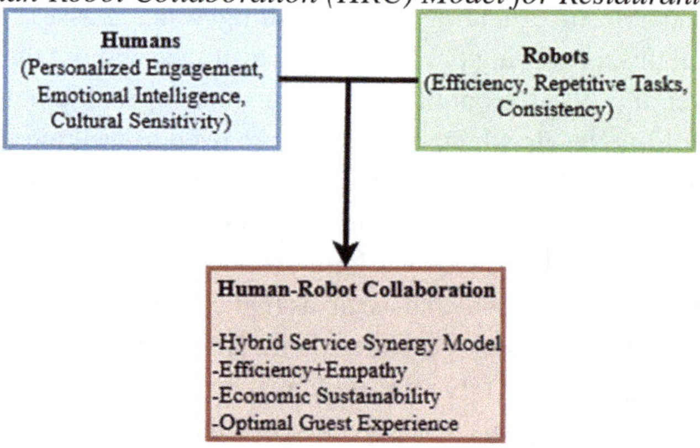

See Fig 1, Although robots have speed, consistency, and efficiency, they do not have emotional intelligence and cultural awareness that human waiters provide at the table. That is why the future of restaurant service is the collaboration of humans and robots (HRC), not complete automation. Robots play the role of efficiency backbone in this model, since they will handle repetitive and labour intensive jobs like the delivery of trays, transfer of orders and clearing tables. In the meantime, people are the emotional heart, as they are devoted to the engagement of guests, upselling, and the elimination of individual service cases. The separation of functions guarantees the precision of operations and coziness at the same time. HRC assists the restaurants economically by enabling them to endure shortage of staff, high turnover and the increasing labor costs (Hou et al., 2021). Robots also alleviate the workload of repetitive tasks, and human employees are focused on value-adding interactions. This increases productivity, and also, staff satisfaction, eliminating repetitive tasks. As a guest, the surveys indicate that customers are pleased with the newness and effectiveness of robots yet they still appreciate the human touch in making unique dining experiences. A balanced hybrid model where robots do not replace people servers, they go hand in hand to produce the best experience is the most desirable resulting in efficiency and empathy (Çeltek, 2023).

The proposed *Hybrid Service Synergy Model (HSSM)* highlights this balance:

- Humans → emotional intelligence, personalization, flexibility.
- Robots → efficiency, accuracy, speed.
- Collaboration → efficiency + empathy, economic sustainability, and optimal guest experience.

Future Demand: Emotionally Intelligent Robots Capable of Adaptive Service Styles

The future of robotics in table service is closely tied to the development of emotionally intelligent robots. Unlike current machines, which operate based on pre-programmed commands and basic interactivity, emotionally intelligent robots would be capable of recognizing guest emotions, adapting communication styles, and providing empathetic responses.

This development requires advancements in three areas:

- **Emotion Recognition Technologies**: Robots equipped with cameras, voice analysis, and biometric sensors could interpret a guest's facial expressions, tone of voice, or body language.
- **Adaptive Communication**: Emotionally intelligent robots would tailor their language, tone, and behavior to match the guest's mood. A cheerful robot might crack a light joke for relaxed diners, while adopting a more formal demeanor for business meetings.
- **Learning from Interaction History**: By storing and analyzing past interactions, robots could build guest profiles that inform future service styles. A

frequent diner who prefers privacy might receive minimal interaction, while another who enjoys small talk could engage in longer conversations.

The need of such robots can be explained by the fact that Millennials and Generation Z have increased expectations and expect to receive personalized and meaningful dining experiences with the help of technology. These consumers tend to be receptive to robotic service but would want it to be more of a human approach than a robot one. Moreover, robot servers that have high levels of emotional intelligence would help in bridging the gap when human servers are not available. As an example, 24-hour-dining establishments may have late-night patrons engage with robots that can exhibit empathy and thus the service standards will be ensured at every time of the day (Wirtz et al., 2018). The path to such progressive systems, however, seems to be problematic. The cultural disparity that exists in the perception of the people towards the robots, the ethical issues that are raised by privacy and data collection and the expensive nature of the emotionally intelligent robotics are all challenges. Nevertheless, with the development of AI, machine learning, and affective computing, there is a high likelihood that emotionally adaptive robots will come into existence.

Table 2. Summary of Robotic and Hybrid Table-Service Innovations

Type of Technology / Robot	Primary Functions	Guest Benefits	Restaurant Benefits
Humanoid Service Robots	Greeting, light conversation, basic service tasks	Novelty, entertainment, futuristic experience	Brand differentiation, reduced FOH workload
Tray-Carrying Robots (e.g., BellaBot, PuduBot)	Deliver dishes, avoid obstacles, follow mapped paths	Faster food delivery, hygienic handling	Reduced staff fatigue, better efficiency in peak hours
Collaborative Service Bots	Carry trays while humans handle interaction	Balanced automation + human warmth	Streamlined operations without losing hospitality
Robotic Bartenders / Beverage Systems	Mix cocktails, prepare drinks with precision	Consistency, speed, interactive experience	Reduced wastage, precision dosing, faster service
Smart Tables / IoT-Enabled Tables	Ordering, calling service, refills, entertainment	Convenience, engagement, minimal waiting	Optimized table turnover, reduced staff intervention
Automated Refill Systems	Water/soft drink refills, self-serve beverage stations	No waiting for staff, autonomy	Time-saving for staff, smoother service flow
Indoor Service Drones	Lightweight delivery of drinks/condiments	Unique experience, quick access	Ideal for large/outdoor layouts, reduced staff movement

continued on following page

Table 2. Continued

Type of Technology / Robot	Primary Functions	Guest Benefits	Restaurant Benefits
Human–Robot Collaboration Models (HRC)	Robots handle repetitive tasks; humans handle emotional tasks	Efficient service with interpersonal warmth	Lower labor pressure, better guest satisfaction
Emotionally Intelligent Robots (Emerging)	Detect emotions, adapt communication style	Personalization, improved comfort	Stronger engagement, future-proof service model

V. SMART BILLING AND PAYMENT SYSTEMS

The final stage of a restaurant dining journey, billing and payment, has undergone a major transformation in recent years. Once a manual, time-consuming process, it is now one of the most technologically advanced aspects of service. Smart billing systems not only ensure faster checkouts but also integrate personalization, security, and customer engagement, making the payment process an extension of the overall dining experience (Kattara et al., 2013).

Contactless Payments: QR Codes, NFC, and Mobile Wallets

These innovations in billing are one of the most visible, the use of contactless payments. Visitors do not have to wait until the servers come with a hardcopy of the bill or swipe card (Lee, 2011). Rather, there are several non-frictional alternatives (Liu et al., 2024):

QR Codes: QR codes on tables or bills, which are used by diners to scan and see their bill in digital and pay immediately, have become very popular in the times of the pandemic. This saves time and limits the involvement of staff.

NFC Technology: Near Field Communication allows making transactions like paying with a credit/debit card or performing transactions with smart devices such as phones and smartwatches in a Tap-to-Pay way. This is a fast and hygienic method; therefore, it is especially attractive in a large restaurant business.

Mobile Wallets: Mobile payments have been incorporated into restaurant payment systems, like Paytm, Google Pay, Apple Pay, and Samsung Pay, which means that customers can pay using their devices. The younger demographic of customers who are more tech-savvy and like cashless experiences is adopting mobile wallets.

AI-Enabled Smart Billing Systems and Automated Receipt Generation

Artificial Intelligence (AI) is playing an increasingly critical role in billing by ensuring accuracy, personalization, and operational efficiency.

Smart Billing Automation: POS (Point-of-Sale) systems powered by Artificial Intelligence systems will automatically generate error-free bills by synchronizing order from digital menus or self-ordering Kiosks. This reduces human error and also makes the process faster.
Dynamic Advanced Receipt: Automated receipts are able to be customized with guest names, order information, discount, or recommendations tailored to their next visit. Some advanced systems even make recommendations about which items are complementary or offer promotions during the checkout process to make up selling opportunities available.
Predictive Analytics: AI can be used to analyse data on guests, such as how often they visit and what they prefer to eat, to tailor suggestions for payment or to emphasize loyalty benefits. For example, a guest who has ordered recommended new additions might receive an on-the-spot discount for trying a new dessert product upon billing time.

Table 3. Summary of Smart Billing and Payment Innovations

Billing Technology	Key Features	Guest Benefits	Restaurant Benefits
QR Code Payments	Scan-and-pay, digital bill access	Fast checkout, no contact, easy splitting of bills	Reduced staff involvement, faster table turnover
NFC Tap-to-Pay Systems	Card/phone "tap" transactions	Convenience, hygiene, secure transactions	Quick payments, improved service speed
Mobile Wallets (e.g., Paytm, Google Pay, Apple Pay)	Seamless smartphone payments	Familiar, user-friendly, loyalty integration	Reduced cash handling, efficient POS operations
AI-Driven Smart Billing Systems	Automatic bill generation, syncing with orders	Error-free bills, personalized offers during checkout	Fewer billing errors, real-time financial tracking
Dynamic Receipt Generation	Customized receipts with recommendations	Clear details, personalized suggestions	Upselling opportunities, guest engagement
POS–Loyalty Program Integration	Auto-crediting/redeeming points	Rewards without extra steps, smoother experience	Stronger customer loyalty, targeted marketing

continued on following page

Table 3. Continued

Billing Technology	Key Features	Guest Benefits	Restaurant Benefits
Customer Data Platforms (CDPs)	Analyze spending patterns, preferences	Tailored benefits on future visits	Enhanced menu design, improved CRM decisions
Invisible Payments (Emerging)	Auto-charging as guest exits	Zero-wait checkout, seamless experience	Maximum efficiency, optimized guest flow

Integration with Loyalty Programs and Customer Data Platforms

Modern billing systems are no longer isolated functions; they are strategically connected to loyalty programs and customer data management platforms.

- **Seamless Loyalty Integration:** Guests can automatically earn or redeem points during payment, eliminating the need for separate processes. This integration strengthens brand loyalty and encourages repeat visits.
- **Customer Data Platforms (CDPs):** Smart billing systems capture guest data, such as spending habits, preferred items, and visit frequency. This data is then used to personalize future offers, design targeted marketing campaigns, and enhance overall customer relationship management.
- **Omnichannel Consistency:** Whether a guest dines in, orders online, or uses takeaway services, integrated billing ensures that loyalty points and payment histories remain consistent across all platforms.

VI. CHALLENGES AND CONSIDERATIONS

The adoption of robotics and automation in restaurants is a promising step toward future-ready service models. However, this transformation is accompanied by several challenges that restaurants must address before large-scale integration can succeed. These challenges extend across financial, cultural, ethical, and operational dimensions.

Cost of adoption and ROI is one of the priorities (Liu et al., 2024). To buy robots, install AI-based systems, and install smart payment technologies, it can cost a lot of money. In addition to the upfront expenses, training employees to operate machines will also be required by restaurants, as well as maintenance and software updates. Large-scale hospitality brands might be able to handle such investments, but obstacles to such investments arise in smaller and independent restaurants since it is difficult to gauge the long-term benefits against the financial allocation (Kim et al., 2021). The other important factor is the cultural acceptance and customer comfort levels. Eating is not an act of efficiency alone, rather it is a social and emotional act. Human touch like customized greetings, small talk or cultural nuances are part of the dining experience to many customers (Zeng et al., 2020). Younger, technologically inclined guests will find robotic service to be enjoyable, but older or more traditional diners will find it impersonal. In this way, restaurants have to think over their customers and cultural background thoroughly and only then implement high-tech solutions. There is also the balance between automation and the human touch which is also a challenge. Robots are fast and accurate at repetitive and labour-intensive work and may have a lack of empathy, flexibility, and emotional intelligence. Full automation may end up becoming a completely mechanical experience. A hybrid solution, in which robots care about efficiency and humans build relationships, is thus proving to be the most efficient solution (Qiu et al., 2020).

In addition, the increasing dependence on digital systems raise the concerns of data privacy, cybersecurity and ethics (Wang, 2021). Smart billing platforms, artificial intelligence (AI) based recommendations and loyalty programs often rely on the collection of sensitive guest data. A breach or misuse of this information has the potential to hurt the trust and the reputation of guests and the reputation of the brand (Park, 2020). Ethical debates also exist surrounding the replacement of human jobs with machines, such that questions have been posed regarding the issue of workforce displacement as well as the social responsibility of restaurants (Wirtz et al., 2018).

To summarize, the major challenges can be outlined as:

- **Cost and ROI**: High initial investment, ongoing maintenance, and unclear profitability for smaller restaurants.
- **Cultural acceptance**: Variations in customer comfort levels depending on demographic and cultural background.
- **Human touch vs. automation**: Risks of depersonalized dining experiences if balance is not maintained.
- **Data security and ethics**: Risks of data breaches, privacy violations, and workforce displacement.

Since robotics and AI systems are very dependent on guest data, they need to be ruled ethically. When processing, storing, or collecting customer information, luxury restaurants should make sure to stick to the global data protection frameworks like GDPR. Easy consent, data transparency, and effective anonymization are in place to foster the confidence of guests in AI-based personalization. The ethical considerations are also applied to algorithmic fairness, inclusiveness in automated service construction, and responsible communication on how guest data are used to make service recommendations. By solving these issues, there is greater confidence in smart service environments by the guests (Liu et al., 2024). The views of guests on robotic service are also very different in cultural changes. Robotic servers have become highly acceptable and even unforgettable in terms of their newness and efficiency in the East Asian markets, especially Japan, China, and South Korea. European and Middle Eastern diners, on the contrary, tend to appreciate warmth, tradition and human contact and hence are more sensitive to entirely automated service. Knowing such cultural peculiarities, luxury restaurants that intend to implement robotics in various markets should consider these peculiarities very carefully because the acceptance, comfort, and perceived quality of services are largely dependent on cultural expectations.

VII. FUTURE TRENDS AND DEMANDS IN RESTAURANT SERVICE

The restaurant business is on the edge of the revolution when the service is no longer about food and welcoming as the service is becoming more and more about the combination of the latest technologies and information-based systems. Since consumers are seeking faster, safer, customized, and greener experiences, restaurants need to look further than conventional models of service and adjust to the new trends (Liu et al., 2024). The future will see dining redefined as it is completely automated, hyper-personalized, integrates into the ecosystem, and is sustainable. One of the trends that will play a critical role in the future is the introduction of fully automated service environments commonly known as robotic dining concept. These areas are imagined to be restaurants where all processes of service, such as receiving a guest to serving him with food and processing the payment, are handled by smart machines. Robotic dining is already being experimented in various regions in Asia like in Japan, South Korea, and China, using conveyor systems, robot servers, and automated delivery units. The beauty of these ideas is not just in their novelty but in the fact that it deals with the actual issues such as lack of labor, increased salaries and need to have faster service. To the younger generation who espouse efficiency and technological advancement, these settings are a futuristic and involving dining experience. But restaurateurs will have to fight to keep it warm and emotionally connected in areas where machines reign (Hwang et al., 2019). Another significant force of change is the need to make solutions personalized using big data and AI analytics. Visitors are demanding more and more customized food services, which will meet their individual requirements, preferences and their eating habits. The AI systems can use the information of purchasing history, rewards programs, and even browsing history to recommend dishes to a guest even before the guest looks at the menu (Zeng et al., 2020). In more sophisticated situations, biometric data or bracelets, can notify the restaurant about any dietary limitation or health objectives and serve the chefs and waiters to provide the most specific meals. An example is a personalised digital menu in which a repeat customer will be greeted with their favourite dishes together with AI-suggested wine. This kind of hyper-personalization can lead to the satisfaction, and increase customer loyalty since the guest will feel appreciated and acknowledged. Another important dimension is the integration with smart cities as well as IoT enabled dining ecosystems. As cities use smart infrastructure, restaurants will no longer work in an isolated way but will actually be the service nodes of an interconnected network. Imagine a situation in which a city's transportation system, retail centers, and restaurants have data that is shared seamlessly (Tussyadiah et al., 2020; Park, 2020). A commuter finishing work could be given individualized restaurant recommendations depending upon location, available time, dietary preferences, and even the current traffic pattern. IoT devices could be used to connect kitchen operations, front of house system and guest devices to facilitate a seamless flow of service from booking to billing. Delivery services could benefit, too, with drones and autonomous vehicles incorporated into city-wide logistics services. This level of connectivity will redefine the guest journey of dining not only as an event, but to be a part of a larger urban lifestyle ecosystem. Equally as important is the introduction of sustainable and energy-efficient technologies for the environmentally conscious diners. Increasingly, restaurants are being held accountable for their impact on the environment, particularly by guests who are younger demographics. The future is likely to see a wide range of energy-efficient robotics, AI responsible systems for reducing food-waste, and digital solutions that will minimize papers and single-use plastics. Restaurants have the option to incorporate blockchain technology and ensure transparency of sourcing, therefore guests can trace ingredients from farm to plate. Automated waste tracking systems could notify management about inefficiencies, while smart energy management could use lighting, heating and cooling systems optimally in real time. In return, guests will reward such initiatives with their loyalty, as sustainability will not remain a choice, but a demand (Jang et al., 2020). Beyond these dominant trends, several emerging demands are also likely to shape the future of restaurant service:

- **Augmented Reality (AR) and Virtual Reality (VR) experiences**: Guests may preview dishes in 3D before ordering or engage in immersive dining environments.
- **Voice-activated dining interfaces**: AI-powered voice assistants could handle reservations, order modifications, and billing seamlessly.

- **Health-conscious dining technologies**: Integration with wearable devices may allow restaurants to offer meals aligned with real-time nutritional needs.
- **Flexible service models**: Hybrid dining experiences combining in-restaurant dining with tech-enabled takeout and delivery will become standard.

Collectively, these trends suggest a future where restaurant service is faster, smarter, more sustainable, and deeply personalized. Guests will expect seamless integration of robotics and AI with human interaction, where efficiency is complemented by empathy. For restaurants, the challenge will not simply be adopting the latest technology, but ensuring that these innovations enhance the overall dining experience rather than replacing its human essence. Although emerging trends such as emotionally intelligent robots, hyper-personalization, and smart-city integration offer significant value, they also introduce new complexities. Concerns related to data governance, guest autonomy, workforce displacement, and cultural sensitivity require ongoing evaluation. A conceptual framework illustrating the transition from traditional human-led service to hybrid robotic models would help clarify these interrelated issues. Such a model could map guest touchpoints, identify risks, and highlight opportunities for balanced technological adoption.

Ethical Hospitality Technology Framework

A responsible technology adoption framework is essential for ensuring that innovations enhance, rather than diminish, the hospitality experience. Restaurants should follow ethical guidelines that address transparency, user consent, inclusivity, data minimization, and guest autonomy in AI-driven decisions. Establishing clear protocols for human override, error escalation, and cybersecurity monitoring is also crucial. By adopting these frameworks, hospitality organizations can balance efficiency with empathy and uphold ethical standards while integrating advanced robotic systems (Çeltek, 2023).

VIII. CONCLUSION

History of restaurant service has never been a vacuum of changes, it is always a mirror of the overall societal, technological and cultural changes. Since the hospitality industry has always been about humanity and addressing every customer on a personal level, the modern adoption of digital tools and automation still makes restaurants reinvent the service concept. This chapter has discussed the increasing importance to robotics and advanced technologies in determining the restaurant service experience of welcoming guests all the way to processing billing services and addressing the opportunities and challenges they pose. The necessity to implement a seamless, personalized, and sustainable dining experience in the future will only accelerate the use of new technologies, but it should be said, the human element of hospitality cannot be disregarded.

As stated in the first discussion, first impressions in restaurants are evolving as a result of using technologies in welcoming guests. The traditional methods of reception, which demanded the generosity of a host, are being complemented or even replaced with reception technology based on AI, online check-in and robot hosts. These innovations make it more efficient, reduce waiting time and introduce something new in the experience of guests. However, they also bring into question the emotional cordiality and cultural suitability, and underscore the need to achieve a compromise between robotization and human touch in the early phases of the interaction with a guest. One of the stages in which the shift of technology becomes more evident to the customers is arguably the ordering process. The communication between the restaurant and the guest has been altered by the enhancement of computerized menus, AR/VR visuals and the self-ordering kiosk. These technologies put more power in the hands of the diners since they can have more control, more efficiency, and accuracy in their ordering process and the restaurants can stream-line their operations and gather some beneficial data. Large Data-driven and AI-enhanced hyper-personalized recommendations will allow personalization of menus to the specific preferences of the guest in the future to create the experience that will be unique to each diner. But, as in other stages of service, the trouble is, not to make the process over-mechanized, and not to lose the touch of the personal.

The biggest transformation, maybe, is the advancement of the table service with the help of robotics. Nowadays, robots can deliver food, move trays and even provide some table service. They are incapable of substituting the empathy, cultural sensitivity and emotional intelligence that human servers add to the dining experience, but their effectiveness and novelty could be quite attractive. This has given rise to Human-Robot Collaboration (HRC) model whereby robots take part in tedious work which is intensive in labour, but human servers take part in relationship building and development of enjoyable experiences. This model is not merely a technological model but rather an economical alternative, as it does not ruin the spirit of hospitality as it resolves the issues of increasing labor expenses and staff shortage. The other area that technology has improved is the payment and billing procedure. Contactless payments, QR, NFC-enabled gadgets, and mobile wallets have become a ubiquitous feature, particularly at the time when the world did not stop feeling the necessity to be safe and hygienic after the global pandemic. Besides convenience, AI-driven billing systems can be cross-connected to the loyalty program, automatically produce the receipt and provide useful data insights to restaurants. It leaves the customers with a quicker, smoother and a more personalized checkout process. In the restaurant case it is an opportunity to have a better relationship between the customer that optimizes operations. Despite this kind of development, there are several challenges and problems that are considered critical to the use of robotics and automation in the restaurant service. Small establishments might not be encouraged to use such technologies because of the costliness of investing and low returns. Happiness in the culture is also critical as robotic service may be cold or even culturally incorrect to certain diners. In addition, automation versus the human factor is a sensitive one since excessive reliance on machines might cut the emotional element of the hospitality business. The ethical and security concerns, particularly, the data privacy, cybersecurity, and workforce displacement concerns, are added to it.

The restaurants must be not only worried about the problem of innovation, but ethical, culturally aware, and financially viable in order to succeed in the integration of these technologies. In the future, the chapter has established some of the trends and requirements that characterize the restaurant industry in future. The automated concepts of robotic dining can also be implemented in the immediate future, particularly in large cities where labor and cost are a problem. Individualization is an artificial intelligence phenomenon that will continue to grow with restaurants offering more personalized menus, dietary adjustments and change-overs of real-time services. Smart cities and IoT systems integration that will combine reservation, transportation, and eating will turn dining into a larger aspect of an urban lifestyle. In addition, the idea of sustainability will become one of the attributes of the restaurants of tomorrow when the technology will be able to produce energy in an efficient way and reduce waste and provide transparency in the supply chain. All those tendencies, cumulatively, work towards the rise of demands of diners to see restaurants serve convenience and novelty, in addition to being consistent with values of personalization, connectivity, and environmental responsibility. The future of the restaurant service is, however, not one of full automation or all the way back to tradition, but in the compromise between technology and people.

Robotics, AI and smart systems will be able to drive efficiency, personalization, and sustainability, but humans may still be necessary to provide empathy, cultural awareness and genuine connection. The restaurants that will be successful will be those that acknowledge the unique capabilities of any machine and a human individual and create models of service that exploit each of them respectively. In conclusion, the shift in the service of restaurants can be considered the part of a bigger social trend of becoming more sustainable, automating, and digitalizing. As the guests evolve over time, it is high time restaurants become innovative not only to be on par with their competitors but also to keep in touch with the changing world, where dining is becoming more of an experience and less of a transaction. It is not without its challenges but the road to take is fruitful. Careful implementation of robotics and progressive technologies where efficiency is equally considered alongside empathy, automation and authenticity restaurants would be able to establish the service model of the future that will not only tolerate the timelessness of the hospitality industry but also meet the needs of the future client base.

REFERENCES

Çeltek, E. (2023). Analysis of smart technologies used in smart hotels. *Journal of Business Research – Turk, 15*(4), 2880–2892. DOI: 10.20491/isarder.2023.1754

Choi, Y., Choi, M., Oh, M., & Kim, S. (2020). Service robots in hotels: Understanding the service quality perceptions of human-robot interaction. *Journal of Hospitality Marketing & Management, 29*(6), 613–635. DOI: 10.1080/19368623.2020.1703871

El-Said, O. A., & Al Tall, T. (2020). Studying the factors influencing customers' intention to use self-service kiosks in fast food restaurants. In *Information and Communication Technologies in Tourism 2020* (pp. 206–217). Springer., DOI: 10.1007/978-3-030-36737-4_17

Hou, Y., Zhang, K., & Li, G. (2021). Service robots or human staff: How social crowding shapes tourist preferences. *Tourism Management, 83*, 104242. DOI: 10.1016/j.tourman.2020.104242

Hwang, J., Kim, H., & Kim, W. (2019). Investigating motivated consumer innovativeness in the context of drone food delivery services. *Journal of Hospitality and Tourism Management, 38*, 102–110. DOI: 10.1016/j.jhtm.2019.01.004

Hwang, J., Park, S., & Kim, I. (2020). Understanding motivated consumer innovativeness in the context of a robotic restaurant: The moderating role of product knowledge. *Journal of Hospitality and Tourism Management, 44*, 272–282. DOI: 10.1016/j.jhtm.2020.06.003

Ivanov, S., Webster, C., & Garanina, A. (2022). *Robots, Artificial Intelligence, and Service Automation in Travel, Tourism and Hospitality 2.0.* Emerald Publishing.

Jang, H. W., & Lee, S. B. (2020). Serving robots: Management and applications for restaurant business sustainability. *Sustainability (Basel), 12*(10), 3998. DOI: 10.3390/su12103998

Kattara, H. S., & El-Said, O. A. (2013). Customers' preferences for new technology-based self-services versus human interaction services in hotels. *Tourism and Hospitality Research, 13*(2), 67–82. DOI: 10.1177/1467358413519261

Kim, S., Kim, J., Badu-Baiden, F., Giroux, M., & Choi, Y. (2021). Preference for robot service or human service in hotels? Impacts of the COVID-19 pandemic. *International Journal of Hospitality Management, 93*, 102795. DOI: 10.1016/j.ijhm.2020.102795 PMID: 36919174

Lee, C. J. (2011). Understanding bank service quality in customers' terms: An exploratory analysis of top-of-mind definition. *International Journal of Business and Social Science*, *2*(21), 1–7.

Lee, W. H., Lin, C. W., & Shih, K. H. (2018). A technology acceptance model for the perception of restaurant service robots for trust, interactivity, and output quality. *International Journal of Mobile Communications*, *16*(4), 361–376. DOI: 10.1504/IJMC.2018.092666

Lee, Y., Lee, S., & Kim, D. Y. (2021). Exploring hotel guests' perceptions of using robot assistants. *Tourism Management Perspectives*, *37*, 100781. DOI: 10.1016/j.tmp.2020.100781

Liu, X., Zhang, L., Chen, H., & Li, Y. (2024). The evolution of smart hotels: Current trends and future directions. *Tourism Management Perspectives*, *51*, 101081. DOI: 10.1016/j.tmp.2024.101081

Lu, L., Cai, R., & Gursoy, D. (2019). Developing and validating a service robot integration willingness scale. *International Journal of Hospitality Management*, *80*, 36–51. DOI: 10.1016/j.ijhm.2019.01.005

Park, S. (2020). Multifaceted trust in tourism service robots. *Annals of Tourism Research*, *81*, 102888. DOI: 10.1016/j.annals.2020.102888

Qiu, H., Li, M., Shu, B., & Bai, B. (2020). Enhancing hospitality experience with service robots: The mediating role of rapport building. *Journal of Hospitality Marketing & Management*, *29*(3), 247–268. DOI: 10.1080/19368623.2019.1645073

Sun, J., & Zhang, R. (2023). Ethical AI in hospitality: Data privacy and customer trust. *Journal of Hospitality Technology*, *19*(3), 145–163.

Tuomi, A., Tussyadiah, I., & Stienmetz, J. (2020). Service robots and the changing roles of employees in restaurants: A cross-cultural study. *Ereview of Tourism Research*, *17*(5), 836–841.

Tussyadiah, I. P., Zach, F. J., & Wang, J. (2020). Do travelers trust intelligent service robots? *Annals of Tourism Research*, *81*, 102886. DOI: 10.1016/j.annals.2020.102886

Vo, N. T. (2019). Development of restaurant serving robot using line following approach. *Journal of Science and Technology Issue on Information and Communications Technology*, *17*(12.1), 1. DOI: 10.31130/JST-UD2019-087

Wang, X. V., & Wang, L. (2021). A literature survey of the robotic technologies during the COVID-19 pandemic. *Journal of Manufacturing Systems*, *60*, 823–836. Advance online publication. DOI: 10.1016/j.jmsy.2021.02.005 PMID: 33612914

Wirtz, J., Patterson, P. G., Kunz, W. H., Gruber, T., Lu, V. N., Paluch, S., & Martins, A. (2018). Brave new world: Service robots in the frontline. *Journal of Service Management*, *29*(5), 907–931. DOI: 10.1108/JOSM-04-2018-0119

World Economic Forum. (2023). *Travel & tourism development index 2023: Rebuilding for a sustainable and resilient future*. https://www.weforum.org/reports/travel-tourism-development-index-2023

Zeng, Z., Chen, P., & Lew, A. (2020). From high-touch to high-tech: COVID-19 drives robotics adoption. *Tourism Geographies*, *22*(3), 724–734. DOI: 10.1080/14616688.2020.1762118

Zhao, Y.. (2024). Emotional intelligence in service robots: Adaptive human–robot interaction models. *IEEE Transactions on Affective Computing*.

Chapter 9
From Room Service to Green Service:
How Hospitality Robots Can Redefine Luxury Through Sustainability

Arzoo Rai
https://orcid.org/0009-0000-8103-5819
Lovely Professional University, India

Amrik Singh
https://orcid.org/0000-0003-3598-8787
Lovely Professional University, India

ABSTRACT

Robotics, artificial intelligence (AI), and sustainability are changing the face of luxury hospitality. Technological invention and environmental awareness are redefining luxurious after having been set in the identity of personalized human service. Service robots provide new opportunities to hotels to achieve operational efficiencies, enhance customer experiences and integrate environmental responsibility initiatives into their brand value. This chapter provides an argument that hospitality robots are not just practical tools, but also symbolic reminders of a responsible luxury ethos, by reviewing the related recent literature in the field of hospitality management, smart tourism, and responsible AI. However, even their introduction suggests certain dubious issues of ethics, openness to visitors, employee resilience, and greenwashing threats. The chapter also forms a system of incorporating robotics in infused sustainable luxury service, which parallels paradigm shift of replacement of room service, to green service system.

DOI: 10.4018/979-8-3373-6541-1.ch009

INTRODUCTION

Historically, luxury in the hospitality sector has been connected with material abundance, exclusivity, and human-focused and personalized services. The initial theories of the luxury hotel industry focused on value-rich decadence with extravagance of the interior, cuisine, and high-class privacy as indicators of luxury and social status. But with the changes in the values of the consumers and the priorities of the society, luxury has started becoming more of an experience and a sense of responsibility rather than a waste of material possession. Visitors are now requiring individualized, authentic, and ecological services that meet their own value and social sustainability interests (Ali et al., 2018). Such redefinition echoes more broadly a market shift within the service sector, which has seen the introduction of personalization, innovation, and sustainability into defining quality become ubiquitous (Jin et al., 2016; Ansari et al., 2024). Global environmental compulsions influence this development of luxuries. Hospitality firms, especially luxury hotels, have been compelled by climate change and the urgent need to cut down on their carbon footprints, to rethink their operating models. Hotels are historically significant users of water and energy and significant sources of waste generation and greenhouse gas emissions (Chan & Hsu, 2016; Ansari & Singh, 2023; Berry & Singh, 2024). According to the UNWTO (2023), the global hotel industry contributes nearly 1% of total global emissions, with luxury hotels accounting for a disproportionately high share due to their higher per-room energy intensity. Integrating robotics aligned with SDG 12 (Responsible Consumption and Production) and SDG 13 (Climate Action) presents a strategic opportunity for luxury hospitality to transition from high-impact operations toward measurable sustainability outcomes. Robotics enable precision energy management, automated waste reduction, and data-based tracking of carbon savings, thereby linking innovation directly with global sustainability goals. This practice places the sector at odds with global sustainability goals. Gössling & Higham (2021) argue that a low-carbon imperative forces urgent transformational change in tourism. This position is supported by the United Nations World Tourism Organization (UNWTO, 2023) in its official pronouncement, the Glasgow Declaration, which urges the industry to speed up practical action on climate. Since most luxury hotels are over 5 stars with a visible and influential image, they should be at the forefront of this change as it entails integrating sustainability at the core of the luxury promise.In the same direction as environmental pressures, the hospitality industry is exposed to radical technological disruption (Berry & Singh, 2024; Bhalla et al., 2023). The development of robotics, artificial intelligence (AI), augmented reality (AR), and virtual reality (VR) is transforming the nature of service provision and customization led by hotels. Smart hospitality systems are linked and interoperable and allow lodging facilities to continuously engage their guests in the experience of the hotel staff and improve performance (Buhalis & Leung, 2018). Smart technologies also allow hotels to predict the needs of guests, manage the use of energy resources, and offer individual services based on personal interests (Neuhofer et al., 2015). Another opportunity connected to AR and VR applications is that prospective visitors can learn more about rooms and amenities through virtual objects, which can help influence the purchase and generate more satisfaction (Ansari & Singh, 2023). These solutions demonstrate that technology has already become not a peripheral resource, but a central part of luxury branding and offering services that are environmentally aware. Luxury is developing along symbolic and ethical lines as well. What guests are also taking is finding experiential goods that support their own sense of being a responsible consumer, a concept Delmas & Burbano (2011) term as conspicuous conservation. Hotels that incorporate the utilization of green technologies (e.g., automated waste collection systems or artificial intelligence-based energy controls) as a significant element of their design can be seen as a signifier of environmental responsibility to this group of travelers. Simultaneously, the COVID-19 pandemic extended the process of adopting technology, as the hygiene, safety, and other resilience innovations advanced hotels to adopting robots and automation (Bhalla et al., 2023; Chourasia et al., 2025; Francis et al., 2024). Not only did robots involved in contactless check-in, cleaning, and delivery reassure guests during uncertainty times, but they are also reinventing exclusivity as protection and trust (Baum et al., 2020; Shin & Kang, 2020).However, applying technology in the context of luxury service provokes some fundamental concerns regarding the nature of authenticity, human interaction, and morality. Though AI and robots can quite effectively do the routine job, they will never replace the care, compassion, and interpersonal understanding human employees have on duties (Tung & Au, 2018). It is presumed that a luxury life destiny is likewise findable in hybrid service patterns in which man and robot engage each other since robots have the capability to carry out convention internal services, and man have the capability to carry out high-preferred services, which are emotional engagements (Han et al., 2021; Francis et al., 2024; Maheshwari et al., 2024; Nadda et al., 2024). The balance here will necessitate organizational creativity and adaptation, i.e., the redesign of facilities to suit robotic devices (Ivanov et al., 2017), rethinking strategies toward the workforce (Singh & Kumar, 2025), and inculcating sustainability concepts into HR (Singh & Sihmar, 2025).Simultaneously, there is an upsurge of complexities within the ethical environment of robotics and AI in hospitality. We may in addition execute emotion algorithms based technology, e.g., in the chance of more inching individualization, and also in the chance of alerting privacy concerns and ethical fears, too (Arora & Singh, 2026). Equally, AI-powered decision making tools would tend to be biased or opaque, which is the basis of the luxury experiences (Kapoor et al., 2023). Indeed, regulation is a crucial feature of the Artificial Intelligence Act (EU, 2021), which states that AI use in sensitive industries, including tourism, must be more accountable, transparent, and regulated by human beings. Such luxury hospitality firms must be able to juggle all the opportunities of technology whilst complying with restrictive regulations and ethical requirements.These challenges and opportunities are also starting to be identified in the academic community. A systematic research agenda has been widely proposed to comprehend better the role of robotics and AI in hospitality and tourism (Cain et al., 2019; Ivanov & Webster, 2019, 2020; Tussyadiah et al., 2020). Their results indicate that technology has a potential but must also be implemented with a critical eye, considering the implications of technology to the workforce, the acceptance of this technology by the guests, and its sustainability in the long run. Research additionally highlights that it is essential to train and enable future personnel in the hospitality industry to work in a technology-mediated space (Murphy et al., 2017; Sharma & Singh, 2024). Meanwhile, the emergence of digital healthcare and wellness tourism (Sharma & Singh, 2025) reveals the way robotics and AI can transform hotels, healthcare and wellness industries, and the service ecosystem as a whole. This chapter argues that hospitality robots should not be viewed merely as efficiency-enhancing technologies, but as operational and symbolic mechanisms capable of redefining luxury through responsible consumption, ethical innovation, and measurable sustainability outcomes. Given their high energy intensity and symbolic brand visibility, luxury hotels occupy a critical position in the global sustainability discourse. By integrating robotics into energy management, waste reduction, and service delivery systems, luxury hospitality can move beyond symbolic sustainability toward quantifiable environmental performance aligned with global climate commitments.

The Rise of Hospitality Robots

Implementation of robots in the hospitality industry is one of the most revolutionary technological innovations in services. Robots are no longer far-fetched; they are now part of the daily working environment of most hotels. Today they are used in fulfilling features like concierge, food and drink delivery services, cleaning, luggage handling, as well as entertainment services offered to guests. It has evolved with the development of artificial intelligence (AI), machine learning, sensor technologies, and automation of services and with the requirements of the industry to achieve efficiency, guest experience, and mitigate adverse environmental effects (Cain et al., 2019; Ansari et al., 2024; Ansari & Singh, 2023; Berry & Singh, 2024). Turning out to be anything but gimmicks, today's hospitality robots have emerged as a cornerstone reinvention of service provision, especially in high-end markets where innovation and sustainability are most sought after.Initial applications of robotics in hotels were typically highly promoted and focused on demonstrating innovation and capturing the media interest. One popular case study is the Henn-na Hotel in Japan, which has positioned itself as the first robot hotel on the globe by installing humanoid robots at the reception desk and robotic arms in changing rooms. Despite inherent issues involved in these deployments, such as breakdowns and dissatisfaction with the service offered by a robot, they demonstrated the promise of robotics to transform hospitality operations, and prompted researchers to explore the economic, social, and operational implications of robotics in the tourism sector (Ivanov & Webster, 2019). Later studies focused on hospitality as one of the best experimental environments where robotics in service could be tested, since the area is labor intensive and most of the work done in a hotel is repeatable (Murphy et al., 2017). In recent industry initiatives, Marriott International integrated autonomous delivery and waste-handling robots across several properties, reporting a 15% reduction in energy use and improved waste segregation (WTTC, 2024). Similarly, Accor's 'Smart Sustainable Concierge' pilot combined robotics and AI to automate guest communication while cutting paper use by 40%. These examples demonstrate that robotics, when framed within sustainability programs, offer tangible benefits beyond novelty enhancing both efficiency and environmental performance (Francis et al., 2024; Maheshwari et al., 2024; Nadda et al., 2024; Prakash et al., 2024). There are different categories of robots used in the hospitality industry that have different roles. Concierge robots primarily work as front desk and assist with hotel check-in procedures, directions, and suggestions (Tung & Au, 2018). Food, drinks, and newspapers are delivered to guest rooms by delivery robots, which shortens waiting time, minimizes the use of human resources, and, additionally, comes in handy especially during the COVID-19 outbreak (Shin & Kang, 2020). Robot cleaning and maintenance can do monotonous tasks like vacuum cleaning, surface disinfection, laundry delivery, etc., which employees can no longer be physically exhausted because of such jobs (Kuo et al., 2017). Most commonly anthropomorph, playful in nature, and built to entertain, robotics can interact with children and families in the services of a lobby or restaurant to create those memorable moments, which solidify brand distinctiveness (Tussyadiah et al., 2020). Each of these categories demonstrates how robotics have been applied in so many different ideas and ways and how it can be used to advance not only efficiency but also guest communication and brand image.Academic research on robots has focused heavily on guest acceptance of robots, since consumer perceptions are essential factors influencing the adoption of robots (Maheshwari et al., 2024; Nadda et al., 2024; Prakash et al., 2024). Research indicates that cultural background and generational differences, as well as the level of anthropomorphism in the design of a robot, all influence the level of its acceptance (Lu et al., 2019; Han et al., 2021). Using the technology acceptance and behavioral intention models, Kuo et al. (2017) investigated the impact of the views on usefulness, ease of use, and trust of the security staff on the willingness to use the robots as a tourist guest, revealing that these readings can largely influence the intention to be served by a robot. Better late than never, a follow-up study by Kuo et al. (2022) confirmed the aforementioned finding, citing the fact that acceptance goes up when robots are seamlessly incorporated as parts of valuable service ecosystems instead of add-ons, which are novel and introduced as a feature. The insights provided above mean that the most valuable use of robots would be as supplementary components that do not end the interactions with humanity (they complement each other, rather than replace). The COVID-19 crisis increased the adoption of robots within the hotel industry, particularly in high-end hotels that had to reassure their clientele that health and safety would not lead to a drop in service quality. Robots were used to introduce non-contact check-in, room service delivery, and sanitizing of the public areas to reduce human-to-human interaction (Baum et al., 2020; Shin & Kang, 2020). Notably, the pandemic commoditized robotic services moving them beyond a novelty to a category of commonplace practices. There were also the hotels that implemented this type of innovation, not only calmed down the guests, but also redeemed exclusivity related to safety, privacy and resilience.In addition to efficiency and safety, organizational and workforce considerations are also critical with hospitality robots. Introducing robots means that hotels must re-consider its internal layout facilities, service protocols, and other elements, as Ivanov & Webster (2020) assert. As an example, the corridors might have to be expanded to allow delivery robots to pass, and IT would also have to be updated to ensure robot interoperability. According to Ivanov et al. (2017), the frontier of robot-friendly hospitality facilities is explained as a concept whose scale cannot be implemented alongside current facilities but rather exists at a level of planned investment and conscious design. On the same note, Melián-González & Bulchand-Gidumal (2016) point out that technology uptake not only influences the satisfaction of the guests but also that of the employees who are forced to learn to work with robot co-workers. This way, provided that the robotics are well managed, the workload stress would decrease, and the employees would concentrate on the emotional aspects of communication with the guests. But when not incorporated properly they can amount to job-insecurity and employee bad-blood (Sharma & Singh, 2024a; Sharma & Singh, 2024b).

LITERATURE REVIEW

There are threats and restrictions associated with the usage of robots, notwithstanding their potential. Osawa et al. (2017) warn that the presence of a robot poses both real and perceived risks to both workers and guests. Technical breakdowns, information safety, and privacy risks can undermine trust, and the luxury setting is the place where discretion and reliability are valued most. Secondly, this underperformance can detrimentally affect the face-to-face interaction associated with luxury hospitality, leading to a technological performance paradox where technology adds to efficiency rather than authenticity (Tung & Au, 2018). This highlights why design, implementation, and control should be conducted with much care, because robotics should actually enhance the guest experience and not be a direct detriment to core principle concepts of luxury and service.What is emerging is that robots in hospitality are not an attempt to substitute human labor with machines. Rather, they are a fresh paradigm of efficiency, safety, sustainability, and branding overlap (Sharma et al., 2024b; Singh & Hassan, 2024; Singh & Kumar, 2025a). Luxury hotels are uniquely placed to use robots as a functional tool and symbol of innovation and environmental friendliness (Ali et al., 2018). A robotic system will enable these hotels not only to strengthen their positive environment responsibility and forward service, but also to stand up against an increasingly competitive market. The following sub-topic will discuss how this shift transcends operations effectiveness to the level of integrating sustainability as a characteristic attribute of luxury hospitality. Collectively, the literature suggests that hospitality robotics research has primarily focused on operational efficiency, guest acceptance, and workforce implications. While these studies offer valuable insights, they remain fragmented and largely disconnected from sustainability metrics particularly within the luxury hospitality context. Limited research integrates robotics adoption with measurable environmental outcomes, stakeholder perspectives, and long-term brand positioning. This gap highlights the need for a holistic framework that situates robotics as a strategic sustainability instrument rather than a technological novelty.

From Efficiency to Sustainability

A majority of the early references made towards the implementation of robotics in hospitality was in the context of efficiency. The robots were believed to provide low-cost labour, since the frequency of human errors and heterogeneity in service will be reduced. Activities like room cleaning, food delivery, or check-in, among others might be automated to minimise wait time and improve consistency. Although these helped hoteliers in the short term, they just helped to cover a small section of the problems which the luxury set up was experiencing. Gradually it became clear that efficiency was not all that would make the future of hospitality robotics. Rather, sustainability has become a philosophy that is transforming the role of robots in the luxury hospitality industry (Sharma et al., 2024a; Sharma & Singh, 2024c; Sharma et al., 2024b; Singh & Hassan, 2024; Singh & Kumar, 2025a). That trend correlates with the social demands of low-carbon development and sustainable consumption, which overlap with the perception of urgency that exists in the writings of Gössling & Higham (2021), who argue that players in the tourism industry are supposed to adopt a low-carbon imperative, lest they be pushed out by the climate age.Luxury hospitality can be sustainable more than merely making sure that waste is minimized or lessened or that the energy used is as low as possible. It also includes the re-inventive thinking on the hotel activities, supply chains, and service experiences. As noted in the UNWTO declaration on tourism in Glasgow (2023), the tourism businesses should be encouraged to ensure that they do not only minimize emission productions, but also establish clear and quantifiable channels toward net-zero. Hotels are supposed to be the first of their kind in the costly segment where brand image holds a symbolic significance. Robotic systems add to this agenda through facilitating operational practices which will reduce ecological footprints. Empirical data now supports this transformation: hotels using robotic housekeeping and AI-based resource optimization systems report 10–20% reductions in electricity consumption and up to 12% reductions in water usage (Singh & Kumar, 2025). Although initial installation costs are high, lifecycle cost assessments suggest a payback period of three to five years, making robotics both sustainable and financially feasible (Singh & Hassan, 2024; Singh & Kumar, 2025a; Singh & Kumar, 2025b). However, hotels must communicate these achievements transparently to avoid the perception of greenwashing.An example could be that delivery robots will reduce human journeys made by multiple employees, reducing the energy used. The efficiency of recycling can be improved with the help of automated waste sorting robots, and the heating, lighting, and water consumption can be optimized with the help of AI-controlled resources management systems. Hotels can communicate their sustainability effort by incorporating these practices into visible guest-facing services, which can also help lower efficiency.The observable aspect of sustainability is also important in the perception of the guests. Tourists and consumers expect more authenticity and responsibility in their travel experiences, and too many of them want to identify with a brand that shares their values. This has led to so-called "conspicuous conservation" in which consumers demonstrate their awareness of environmental awareness publicly by the decisions that they make (Delmas & Burbano, 2011). With luxury hospitality, robot technology can be codified as a symbolic instrument of conspicuous conservation. To explain this, a robot to supply the facilities and prioritise on the carbon saving engaged by the hotel can not only serve the target purpose, but it can communicate the brand meaning. This kind of prominent inclusion in sustainability reports improves brand performance as well as satisfaction among guests.Robotization in sustainability strategies also has an impact on the workforce and organizational culture of luxury hotels. As Melián-González & Bulchand-Gidumal (2016) note, the connection between the hotel and employees is transformed by the existing information technologies. Using robots to run administrative and manual procedures decreases the labour, workload of employees and allows them to concentrate in tasks that need creativity, understanding, and eliciting feelings (Singh & Singh, 2024; Singh, 2024; Singh & Bathla, 2023). Carrying along with this redistribution of labor in addition to the rise in job satisfaction are the prevailing tendencies of green human resource management today, whereby both longevity of the employee and organization are in equal assortment. Moreover, according to Singh & Kumar (2025), the issue of adaptation and resilience of organizations is conditional with the possibility to incorporate innovation into the fundamental strategies of the organization: thereby, hospitality companies should reposition robotics within an expanded sustainability and resilience program.The sustainability issue on the educational side also needs to be looked at. Sharma & Singh (2024) outlined that digital technologies may be applied towards improving education and supplying the ability to develop sustainability understanding and skills in the personnel. Within the framework of luxury hospitality, the idea of promoting employees to collaborate perfectly with robots through training programs is not only more effective in terms of operational performance, but it also puts the concept of sustainability into the professional culture of this industry. This capacity building focus would help in making sure the adoption of robotics is not seen as a challenge to jobs, but rather a way to transform the workforce. Additionally, with regards to digital innovations in wellness and medical tourism, Sharma & Singh (2025) demonstrate that technology can be utilised to generate sustainability value in a wider service network, which is valuable for luxury hotels to go beyond operational efficiency in sustainability initiatives.Meanwhile, one of the areas that should be anxiously handled is guest acceptance of sustainability-oriented robotics. Kuo et al. (2022) champion that innovative services provided by hospitality robots are more acceptable when they are continuously incorporated into the service delivery experience as opposed to when they stand as singular and separate features. To demonstrate that, when combined with a larger eco-acceptable brand narrative such as using renewable energy and prominent recycling efforts, the robot room service will be well-received by a guest. This shows the need to ensure that robotic implementations are matched with the sustainability that lies over and above the routing of the hotel. Maheshwari et al. (2024) take this a step further by applying this thought to the implementation of AI where success relies significantly on its framing as a strategic component of sustainability and responsible innovation, instead of a cost-cutting action.However, there are continued challenges of matching robotics to sustainability agendas. The first one is the threat of the so-called greenwashing when hotels want to be assured by some content, but the critical changes in the technological solutions implemented to position a hotel as environmentally responsible should not be hidden (Delmas & Burbano, 2011). The undermining of credibility is one such aspect whereby one could use a single robot as an eco-marketing tool without necessarily having to confront the higher structural inefficiency. This threat points to the importance of an act of authentic integration, i.e., robotics, rather than an ad hoc tool but a part of a pronounced sustainability policy and approach encompassing not just chain upgrades and innovation in the supply chain but regulations at the region. After all, the shift towards sustainability in favor of efficiency defines a paradigm shift in the role of robotics in the field of luxury hospitality (Nadda et al., 2024; Prakash et al., 2024; Sharma & Singh, 2024a; Sharma & Singh, 2024b; Sharma et al., 2024a). Robots are no longer considered as a cost-cutting and service standardization tool. Rather they are becoming visible, symbolic, and practical assets that help hotels to commit to a sustainable position. Robotics support luxury hospitality in response to global climate needs by assisting in the reduction of carbon footprints, improving recycling, rationalizing resource utilization, and transforming the functions of the workforce. More to the point, they re-spin the very definition of luxury not as a luxurious experience via excessive consumption, but rather a luxurious experience via responsible, innovative, and long-term ecological care.

Robotics and Measurable Sustainability Outcomes

One of the key limitations of sustainability initiatives in hospitality is the absence of clearly defined and measurable outcomes. For robotics to contribute meaningfully to sustainability in luxury hotels, their deployment must be evaluated through tangible environmental performance indicators rather than symbolic adoption. Hospitality robots contribute to sustainability by reducing internal transport, optimizing energy and water usage, and improving waste management efficiency. Autonomous delivery robots minimize staff movement and energy use, AI-enabled housekeeping systems reduce water and chemical consumption, and robotic waste-sorting systems enhance recycling accuracy. Lifecycle cost assessments indicate that despite high initial investment, such systems often achieve financial payback within three to five years, reinforcing both environmental and economic feasibility.

Stakeholder Perspectives

Hotel Owners & Investors - From the perspective of hotel owners and investors, robotics adoption is evaluated through return on investment, operational resilience, and brand differentiation. Sustainability-driven robotics strengthen premium brand positioning while reducing long-term environmental and regulatory risks.

Employees - For employees, robotics represent both opportunity and challenge. When supported by reskilling initiatives and green human resource management, robots reduce physical workload and enable staff to focus on emotionally intensive and creative service roles.

Guests - Guests particularly younger and sustainability-conscious travelers—perceive robotics positively when they enhance convenience while visibly supporting environmental responsibility. Acceptance declines when robots are perceived to replace human warmth rather than complement it.

Regulators - Regulators increasingly influence robotics adoption through data protection laws and AI governance frameworks, requiring luxury hotels to ensure transparency, accountability, and ethical compliance.

Guest Experience and Acceptance

Guest acceptance is one of the most important factors that will influence the success of robotics in luxury hospitality. Not to be confused with other back-end efficiency measures (e.g., energy optimization or various waste management processes), but being the most visible to guests, robots directly influence their ideas of a luxury offering and service condition. Guest acceptance at luxury hotels, where exclusivity and customization are two fundamental brand promises, must rely on the functionality of robots upon which it depends, as well as its symbolism as a signifier of innovation, sustainability, and care. Studies indicate that the reception of guests towards robots is complex and depends on cultural, psychological, and situational factors. Service Robot Integration Willingness (SRIW) is a scale of acceptance drivers, such as perceived usefulness, ease of use, and trust, created and validated by Lu et al. (2019). A recent survey (Han et al., 2023) revealed that 68% of luxury travelers under 40 perceive robotic services positively when linked to sustainability initiatives, compared to only 37% of guests above 55. This demographic gap implies that robotic acceptance depends not only on usability but also on generational attitudes toward technology and ecological responsibility. Thus, hybrid service models where robots enhance efficiency while human staff deliver emotional warmth remain essential for luxury positioning (Singh & Singh, 2024; Singh, 2024; Singh & Bathla, 2023). According to their findings, guests are more likely to accept the use of robots when they see the robots as improving convenience and minimizing effort. In a similar way, Han et al. (2021) note that positive impressions are reinforced when guests hold the belief that robots do not substitute but complement human employees. This implies that acceptance does not merely concern the novelty of robotics but rather, their existence as seamless service ecosystems contributing to the luxury experience.The definition of luxury, however, causes confusion in the adoption of robots. Luxury hospitality is predicated on human interaction, where employees offer a highly personalized approach, considering cultural nuances and applying emotional intelligence. Tung & Au (2018) believe that these aspects of empathy, creativity, and spontaneity cannot completely be imitated by robots. Visitors who prioritize the human touch might feel that robots diminish the authenticity of the luxury experience. This turns out to be a paradox: the acceptance of robots by some guests due to efficiency and novelty, and the opposite: some visitors oppose them because they fear the loss of a human touch that has become a characteristic feature of hospitality.Design is mainly involved in the mediation of guest acceptance. According to Tussyadiah & Park (2018), anthropomorphism in designing robots as humans influences the perceptions of robots by guests of a hotel or other service establishment (Singh, 2024; Singh & Bathla, 2023). Humanoid robots are also more likely to be tolerated in front-of-house positions because they are a source of familiarity and relatability. Nevertheless, pushed to the limit, anthropomorphism can become incongruent, reaching the point where robots appear human enough yet real enough to produce an uncanny valley effect (Osawa et al., 2017). This means that the design must balance it such that the robots are not threatening and dull. It follows that hotels have to mind both the practical value and personal emotional reaction to the choice or design of robotic systems.Expectations of guests vary depending on their demographics and culture as well. Robots are usually approved of by younger and highly technology-focused tourists as the digitalization of hospitality creates an image of progress and novelty (Cain et al., 2019). However, more traditional or older customers would appreciate face-to-face interaction and be opposed to robotic communication. Ali et al. (2018) note that due to the heterogeneity of tastes related to staying in luxury hotels, luxury markets cannot be assumed to share a homogeneous structure. Rather, hotels need to profile consumers of robotics services and differentiate them according to their categories, in addition to providing choices of personalization that consumers of specific sections might be interested in. Some guests would like to have a robot serve their food because it is more private and novel, whereas others choose human staff to enjoy personal interaction (Ansari et al., 2024; Ansari & Singh, 2023; Berry & Singh, 2024; Bhalla et al., 2023; Chourasia et al., 2025). The concept of experiential value, that has been demonstrated to have a strong impact on customer loyalty, also overlaps with the integration of robotics. As Jin et al. (2016) discovered, experiential value and relationship quality have a considerable influence on repeat patronage in the hospitality environment. In the context of robotics, this implies that in addition to tasks, the robots need to add value to the experience. For example, a delivery robot can create an improved experience through efficient service delivery, as well as a novel interaction that guests will remember. The capacity of robots to bring utility as well as delight becomes crucial in luxury hotels, where this ability to provide memorable experiences often serves as a form of differentiation.Sustainability is an added-value to guest acceptance. As mentioned above, a high percentage of travellers, particularly in the high-end market, are eager to prove their environmental awareness with the decisions that they make (Delmas & Burbano, 2011). Guests are more likely to perceive robots favorably when they are deployed as a green service strategy in terms of energy consumption, waste reduction, and necessitating fewer human journeys to the hotel (i.e., reducing needless trips made by humans). This follows the pattern of conspicuous conservation, in which open displays of environmental responsibility augment the symbolic importance of consumption. Therefore, one of the most effective methods to make robotics acceptable by sustainability-related guests is to position it as not only efficient but also ecologically friendly.Despite these opportunities, certain challenges exist. When the robots fail or provide erratic service, the guest would soon lose their trust. Moreover, service disruptions, rather than enhanced service, can be frustrating because many robots have been poorly integrated. Additionally, the gradual advance in enhancement will have to be listed, along with its intuitive nature that is required to be adapted so that robotics develop steadily in regard to the needs of the guests (Singh & Kumar, 2025a; Singh & Kumar, 2025b). As Han et al. (2021) stated, it simply must communicate to the guest at all times about the specific functions of the robot within the hotel. And this involves investing not only in technology but also in staff training because employees become the most important channels to mediate between guests and robots, as well as to troubleshoot challenging situations as they come.

Workforce and Organizational Adaptation

Not just does the introduction of robotics in the hospitality industry change the experience of guests perceiving them; it is also changing the dynamics through which luxury hospitality has operated: large teams of employees will now transition to smaller teams focused on experienced, semi-personalized service. The age of robotics is changing the composition of traditional hotel workforce. Likewise, an increasing number of repetitive, physically demanding, or low-complexity jobs are performed by robots and employees are increasingly able to concentrate on higher-order work, including emotional engagement, creativity, and individualized service. But along with this change, workforce adaptation, organizational culture, and job displacement threats emerge as significant additional challenges.According to Baum et al. (2020), the COVID-19 pandemic further exacerbated existing workforce vulnerabilities, and increased the stressors of not just job security, but precarious employment in hospitality. Promotion of robotics in this time under the pretext of health and safety, elicited concerns among employees of eventual replacement of man power in the long term (Ansari et al., 2024; Ansari & Singh, 2023; Berry & Singh, 2024; Bhalla et al., 2023; Singh & Singh, 2024; Singh, 2024; Singh & Bathla, 2023). Though automation minimized possible risks of spreading a virus due to the decrease in human-to-human interactions, it contributed to feelings of fear over potential displacement of staff by robots. To manage all these issues, the hotels are to introduce robots as something that goes along with the work, not to replace it and make their employees think that technology is a coworker and is not stranger, threatening them with unemployment. Hotels can adopt Green Digital Literacy Programs that combine sustainability education with robotic operation training. This initiative fosters confidence among employees while promoting environmentally responsible service practices. Collaboration between HR and sustainability managers ensures that robotic integration supports both technological innovation and workforce well-being. Another approach of organization adaptation is the reorganization of facilities and processes to suit robotics. According to Ivanov et al. (2017), there is such concept as robot-friendly hospitality facilities that implies redesign of hotel since it needs to be redesigned to be integrated with robots. Changes in structure may be required to enable free movement of robots through corridors, elevators, and back-of-house systems. Elaborating on the same topic, Ivanov & Webster (2020) note that successful adoption will depend on the interoperability of the robotic systems with the existing systems used in the property management. This implies a substantial investment in both physical and digital infrastructure (Francis et al., 2024; Maheshwari et al., 2024; Nadda et al., 2024; Prakash et al., 2024). In the case of luxury hotels where quality, smooth services are the expected qualities, any integration failure can destroy the brand credibility.Equally important is the human resource implication of adoption of robotics. The discussion raises the point mentioned by Melián-González & Bulchand-Gidumal (2016) when the authors emphasize that the information technology not only has an impact on efficiency, but also on morale and workforce satisfactions. Robots might also alleviate manual and repetitive tasks, which might reduce stress and workload. Thus, a bad adoption management would cause resistance as the workers might experience a threat of being moved down or undermined. According to Singh & Sihmar (2025), the trick is in embracing green human resource management (GHRM), where workforce management applies the concept of sustainability. By showing robotics as a tool that can lead to greater sustainability, hotels would be able to create a culture in which employees feel proud to work in an environment that embraces innovative technologies that demonstrate the ethical and environmental philosophy of their company.One of the elements of workforce adaptation is training and capacity building. Sharma & Singh (2024) introduce the importance of digital technologies in enhancing the process of education and training and state that, technology can be used to equip employees with the necessary skills needed to develop the skills of the future. Within the context of luxury hospitality, training programs can be based on the concept of hybrid skill set, when employees should know how to cooperate with robots, how to solve technical problems, and how to improve the communication with the guests mediated by technologies. These efforts can guarantee the economic efficiency of operations but also promote an environment in which staff members regard robotics not as a potential career killer but as a career-enhancing tool.Lessons of organizational adaptation can be applied to the broader service ecosystem, as well. To mention but a few, Sharma & Singh (2025) exemplify how wellness and medical tourism applied to digital innovations could not only improve the quality of provided services but also open up sustainable growth opportunities. The same principles can be applied to luxury hotels that, instead of treating robotics as an efficiency-oriented measure, start embracing it as a part of a bigger wellness-based, environment-friendly brand story (Maheshwari et al., 2024; Singh & Kumar, 2025a; Singh & Bathla, 2023). This framing helps employees view robotics as an expanding presence, reducing resistance and promoting workforce development.Another important dimension is organizational resilience. According to Singh & Kumar (2025), innovation and adaptation play a significant role in the resilience aspect of organization in the hospitality and tourism sector. During times of disruption, such as during a pandemic, economic recession, or an environmental crisis, hotels embracing robotics and digital solutions will be better equipped to carry on with their operations and protect both visitors and their workers. Robots increase the continuity of the background services as human operators shift towards the expressive phase of crisis management and strategic realignment. In such a way, robotics, in addition to increasing operational efficiency, facilitate resilience, allowing luxury hotels to maintain brand promise in unstable environments.However, workforce consequences of robotics adoption will have to be handled with care, so as not to deepen the inequalities. Already, as Baum et al. (2020) indicate, precarious working conditions exist in the hospitality setting, so not well-thought-out automation could put vulnerable staff members at risk. Hotels need a policy which focuses on retraining and redeployment instead of redundancy (Singh, 2024; Singh & Bathla, 2023). One of the main areas that will be invested in is the well-being of the employees and encouraging the culture of collaboration between human and robot to achieve positive results. This should also incorporate candid communication relating to the goal of robotics, transparency in decision making concerning adoption, and developing new possibilities and prospects that allow the employees to contribute to the overall development with regards to the implementation of technologies in practices.

DISCUSSION AND INTERPRETATION

The findings reveal a strategic tension at the core of luxury hospitality: efficiency-driven automation versus experience-driven service. Rather than replacing human labor, robotics enable hybrid service models in which machines perform routine and resource-intensive tasks while employees focus on personalization and emotional engagement. From a sustainability perspective, robotics function as both operational tools and symbolic communicators of responsible luxury. However, their effectiveness depends on authentic integration into sustainability strategies rather than superficial adoption, as transparency and measurable outcomes are essential to avoid perceptions of greenwashing.

Ethics, Regulation, and Responsible AI

The increased use of robotics and artificial intelligence (AI) in luxury hospitality services prompts not only operational and experiential issues but also numerous issues of ethics and regulations. Where efficiency, personalization, and sustainability are certain benefits, the implementation of robots and AI systems will bring challenges concerning privacy, transparency, accountability, and trust into hospitality services (Sharma et al., 2024b; Singh & Hassan, 2024; Singh & Singh, 2024; Singh, 2024; Singh & Bathla, 2023). In the case of luxury hotels, where reputation and loyalty to the hotel matter significantly, these aspects of ethics are especially important to manage. Any lapse in dealing with ethical risks can rapidly wear away the credence that lies at the core of luxury branding.Data privacy is one of the topical issues. Sensors, cameras, and machine-learning algorithms have become an indispensable part of modern service robots: they analyze the behavior of guests, customize their services, and optimize their resources. As much as these technologies significantly improve the guest experience, they also capture sensitive data on preferences, habits, and to an extent, emotional states (Singh & Singh, 2024; Singh, 2024; Singh & Bathla, 2023). As far as it has been properly ethically conscionable in the sense of providing this proximity of personalization on numerous occasions, the so-called technology of the emotional recognition is allied, however, by numerous ethical dilemmas that have the potential to produce trespass into the emotional existence of its visitors (Arora & Singh, 2026). Mishandling or misuse of such data may be irreversibly damaging to trust in the luxury context where, as part of the guest experience, discretion and privacy form part of the experience.Privacy is not the only problem with responsible AI adoption. Responsible AI at the service interface, write Kapoor et al. (2023), needs to discuss bias, fairness, and explanability. Beyond compliance, ethical AI governance can act as a differentiator in luxury hospitality. Transparent data handling and explainable AI systems enhance guest confidence and reinforce brand credibility, transforming ethical responsibility into a competitive advantage. Robots employed at hospitals are potentially reproducing all sorts of biases within their training sets that cause a guest to be treated unequally. Speaking of which, when proposed as an example, the AI systems that allow profiling visitors based on their behavior in the past do not require any intent to stereotype or even block certain groups. Bias in AI systems will arise in the luxury hospitality sector where the imperatives of inclusiveness and equality are progressively gaining prominence, contradicting brand promises of excellence and personalized service. To ensure ethical alignment, it is hence important that AI is transparent, auditable, and accountable (Prakash et al., 2024; Sharma & Singh, 2024a; Sharma & Singh, 2024b). Policymakers also start to respond to risks posed by AI and robotics at a regulatory level. The Artificial Intelligence Act (2021) of the European Commission is one of the first widespread efforts to develop compatible rules regarding AI. The Act classifies AI systems by their levels of risk and requires those with the highest level of risk, such as systems concerning human rights, safety, and personal liberties, to be more rigorously supervised. When crafting luxury hospitality, where sophisticated AI applications may find use in the personalization of its guests and the automation of its services, compliance regulations and proactive ethical governance become especially evident. Hotels introducing robotics then have the duty not only to provide a technical functionality but also to correspond to the shifting legal mechanisms and be responsible towards regulators and assure their visitors. These regulatory developments notwithstanding, there are still ethical issues at the organizational level. According to Maheshwari et al. (2024), the use of AI in tourism services tends to swing up and down, being perceived as both a boon or bane, depending upon the manner of implementation. When expressed as a sustainability and careful innovation strategy, AI promotes guest confidence and increases competitive longevity. But taken just as a method to achieve short-term efficiency or as a marketing tool, it will be seen as shallow or exploitative. This is consistent with Delmas & Burbano's (2011) critique of greenwashing that opaque or dishonest displays of responsibility tend backfire, harming credibility. Brand authenticity challenges insincere adoption of robotics as perceived to be green or ethical in the luxury hospitality sector, where customers are also becoming highly demanding. Another area of ethical concern is associated with safety and reliability. The robots deployed publicly and privately must be non-dangerous, reliable, and user-friendly. As stressed by Osawa et al. (2017), not only technical but also perceptual risks of robotics also exist. Although robots may be statistically safer than human workers, guests will feel endangered by them because they do not know them well or read about them in the media. Hotels operating in luxury settings, where the key priorities are reassurance and peace of mind, need to handle both perceived and actual risks and make sure that robots emit confidence, but not suspicion. The input of employees to resolve ethical and regulatory issues also plays an important role (Singh & Singh, 2024; Singh, 2024; Singh & Bathla, 2023). Being the interface creators between the guest and technology, the staff should be motivated to learn the ethical aspects of robotics and AI. This would involve the understanding of how the details on guests are gathered and processed, responding to privacy enquiries and the possibility of stepping in when the technology fails. When employees are allowed to become ethical custodians, they enhance the compliance protocol, hence, the power that builds trust between the guests and the brand. Lastly, the ethical discussion has to be processed in terms of sustainability and corporate social responsibility. Technology can be coupled with elevated moral intent when luxury hotels are seen as part of larger sustainability initiatives, which include and can encompass energy conservation, waste reduction, healthy guest experiences, etc. But that will need a shift to a level of authentic integration rather than symbolic usage. Ethical adoption should thus take three dimensions, namely, (1) safeguarding the right of guests in terms of privacy and fairness, (2) adhering to changing regulatory implementation guidelines, and (3) accumulating technology to sustainability and organizational principles. Ethical AI governance should therefore be viewed not merely as regulatory compliance, but as a strategic asset in luxury hospitality. Transparent data practices and explainable AI systems enhance guest trust and reinforce brand credibility, transforming ethical responsibility into competitive advantage.

CASE STUDIES AND APPLICATIONS

Robotics and artificial intelligence (AI) in hospitality have shown tangible implications that could help luxury hotels to shift their traditional service models to the sustainability, resiliency, and innovation models. To make guests enjoy their stay, make the robotics system more effective, and help green branding, researchers continue to explore the possibilities and challenges of robotics. In these case studies, we see that robotics has ceased to be something of the distant future but has become a major part of the hospitality luxury strategy. Guest-facing services are some of the most explicit examples of the adoption of robotics. In Japan, the Henn-na Hotel was made a popular hotel all over the world, gained worldwide fame by employing humanoid concierge and robot arms to serve checking in and cloak rooms. Although the special attention was clearly paid to such features of the experiment as faults and dissatisfaction, malfunctions in guests, the latter revealed that robotics can transform the character of service interactions (Ivanov & Webster, 2019). More recent implementations, including food and amenity delivery robots, are now standards within high-end hotels worldwide in Asia, the U.S., and Europe. The robots are portable and super clean, and this point holds particular significance during the COVID-19 period (Shin & Kang, 2020; Ansari et al., 2024; Ansari & Singh, 2023; Berry & Singh, 2024; Singh & Bathla, 2023). More remarkably is that when they are presented as environmentally friendly solutions, e.g., because of their energy efficiency, they are aligned with the low-carbon impetus that Gössling & Higham (2021) give priority to. In addition to physical robotics, immersive technologies like augmented reality (AR) and virtual reality (VR) are becoming a growing part of luxury hospitality. Ansari & Singh (2023) observe that AR/VR applications enable a potential guest to virtually visit a hotel room, spa, or conference location before he makes a booking. Such immersive previews hold value in luxury settings, where perception and a sense of exclusivity matter, and where avoiding redundant travelling can benefit the sustainability agenda. Combined with robotics, AR and VR serve as an example of the way that technological innovation can be used to enhance the luxury experience without directly harming the environment. At the Parkroyal Collection Marina Bay, Singapore, AI-enabled food waste management robots have reduced organic waste by 25% since 2022 (WTTC, 2024). Similarly, Jumeirah Group's Dubai properties introduced autonomous housekeeping carts that improved operational efficiency by 12% and elevated staff satisfaction levels. These data-driven outcomes reinforce the feasibility of robotics as a sustainability instrument rather than mere technological ornamentation. Another important case insight is organizational innovation. According to Singh & Kumar (2025), hotels that adopt open innovation principles, such as robotics, respond more effectively when crisis-affected. As COVID-19 unfolded in early 2020, properties that had seriously invested in service robots managed to be operational with less human interaction, which ensured the safety of both guests and employees. It was made continuous by means of robots and the redistribution of human labor to the field where the human skills required are empathy and management of crisis. The mentioned examples reveal that robotics can reinforce organizational resilience and help luxury hotels to provide continuous service. Another critical area has to do with workforce adaptation. As Melián-González & Bulchand-Gidumal (2016) point out, information technologies can influence efficiency as well as the satisfaction of employees. Hotels that prepare employees to work with robots say it integrates easier and requires less opposition. Singh & Sihmar (2025) also advocate for green human resource management as various robotics will be linked to the essence of sustainability and empowering the people who will be working within this organization. We could provide an example, in certain European luxury hotels, hybrid training is formed: technical troubleshooting is mixed with sustainability training in such a way, that the employees could begin to think about robots as the collaborative partners in achieving the environmental goals. Sharma & Singh (2024) also work to develop this concept of the staff preparation to the environment where one technology is a major constituent, as well as the alteration of workforce is housing a direction consistent with the long-term sustainability plan, are both poised by the aid of the digital educational tools (Sharma et al., 2024b; Singh & Hassan, 2024; Singh & Kumar, 2025a). The other area of application of robotics outside of the hospitality setting is in related fields, which a luxury hotel can learn. Robot assistants and AI-enabled systems are improving service quality, as well as service safety in the most medical and wellness tourism projects. Sharma & Singh (2025) say that improvement of patient experience based on introduction of digital innovations via a medical tourism facilitates increase in sustainability by means of innovative resource use. The concept of using technology to combine exclusivity with health can be seen in wellness resorts, for example, when they train their fitness experts or offer spa techniques using robots. These cases indicate that the same can be said of the luxury hotels, i.e., they could potentially use robotics to facilitate not only their efficiencies but also the development of holistic and wellness-based guest experiences. However, issues have not disappeared. Ivanov & Webster (2020) observe that challenges relating to integration with current IT systems might restrict the performance of robotics. According to Osawa et al. (2017), the acceptability of guests can be negated by real and perceived risks (e.g., safety concerns or the feeling of uneasiness with anthropomorphic designs). Further, Delmas & Burbano (2011) warn that the use of robots on a symbolic or superficial basis in the name of sustainability could be nothing more than greenwashing. As an example, hotels simply displaying one robot to sell their own products, without making significant structural investments, may sound suspicious to ecologically aware clients. Then authenticity is one of the main conditions so that robotics can play a significant role in luxury and sustainability. The evidence presented in the case provides some lessons in luxury hospitality. The most effective way to use robotics is, of course, as part of an extended system of innovation, such as AR, VR, and AI resource management (Francis et al., 2024; Maheshwari et al., 2024; Singh & Hassan, 2024; Singh & Kumar, 2025a). Second, guests accept more in case the robots are located not as show-business elements but as genuine contributors to sustainable and experience attributes. Third, positive workforce adaptation must be provided by green HR practices and employee training. Last but not least, authenticity and transparency are paramount: robotics should be integrated into a viable sustainability model and not be implemented as a symbolic gesture.

LIMITATIONS AND FUTURE RESEARCH

Despite the revolutionary possibilities of robotics and artificial intelligence (AI), their implementation in the luxury hospitality sector is not completely free of restrictions. Similar to any disruptive innovation, the use of robots presents issues that are technological, organizational, ethical, and social in nature. The existence of such limitations can also be critically analyzed to provide not only a way to temper expectations but also point out future research directions.Technological reliability is one of the most obvious constraints. Even with developments in the field of robotics, service robots have continued to prove vulnerable to system errors, failures, and malfunctions, particularly in integrations with existing property management systems.

The old infrastructure used at these hotels generally fails to connect with the robotic system, resulting in service disruption and breakdown, as Ivanov & Webster (2020) state. Technical faults are especially difficult in the luxury setting where conformity and absolute delivery are the main manifestations of brand promise. Future research needs to focus on how interoperability standards, price-modular design, and adaptive systems can assist in making integration easier so that robots would contribute to seamless guest throughput.Limitations also continue to persist in guest acceptance. Although guests largely enjoy the use of robots as a sign of innovation, some feel that robots reduce the originality of luxury hospitality. According to Tung & Au (2018), robots cannot fully replicate the empathy and cultural sensitivity of human staff, which means their usage may be viewed as a deficit in personalization. On a somewhat similar point, Osawa et al. (2017) note that installations that evoke a sense of discomfort through anthropomorphic designs can avert acceptance through contextual encoding of risks or the danger of surveillance. These observations mean that a new study on the cultural, demographic, and psychological forces defining guest perceptions toward robotics is necessary. Inter-regional and guest-based comparative studies could help illuminate the fluctuation of acceptance in varied global locations and ways the hotel can project robotic services to varied markets. Future research should also compare cultural and regional variations in robotic adoption between Asia, Europe, and the Middle East, while lifecycle analyses assessing the embodied carbon of robot manufacturing and disposal can provide more accurate sustainability benchmarks. The labour impacts of the adoption of robotics need to be looked at as well. Baum et al. (2020) also highlight that precarious work is already a feature of hospitality, and the deployment of robotics is a threat of increasing anxiety across all professions. Although proponents claim that robots take away the need to do repetitive duties, employees might view them as threats. Halikia & Patrick (2019) compare studies by Singh & Sihmar (2025) and Sharma & Singh (2024), who find that a workforce needs to be adapted through training, education, and working on green human resource management. Future studies might examine how the hybrid models of working with robots may be developed further, including the psychological implications of work with apparatuses and the effectiveness of retraining. It also may be possible to find answers to the question of whether employees will eventually consider robotics to be their intermediary towards the idea of sustainability or as something that threatens their employment. There is additional complexity due to ethical and regulatory restrictions (Ansari et al., 2024; Chourasia et al., 2025; Francis et al., 2024; Maheshwari et al., 2024). Arora & Singh (2026) caution that the issue concerning emotion recognition technologies is consent and privacy, and Kapoor et al. (2023) highlight the risks of bias and untrustworthiness related to the choices made by AI. The Artificial Intelligence Act (2021) introduced by the European Commission offers a legal instrument; its adoption within the hospitality setting is not properly studied yet. Since luxury hotel businesses operate on an international customer base, certain jurisdiction-specific challenges are evident in their navigation through varying regulatory standards. The need to understand how regulatory compliance correlates with brand positioning and the ways in which hotels can be responsible without limiting future innovation is an area that needs to be given attention in future research.Sustainability has both opportunities as well as limitations. Even though robot invention can optimize resource utilization and minimize waste and help with low-carbon processes, the energy consumption, manufacturing, and maintenance of the units can raise concern over the sustainable environment. Gössling & Higham (2021) are making it clear to us that tourism must aim for a real deal in the decarbonization and not in the fake-decarbonism. Similar concerns are expressed by Delmas & Burbano (2011) who caution against the practice of greenwashing where signaling or window dressing of technology conceals underlying efficiency flaws. Researchers will thus need to study the entire lifecycle footprint of robotics supply chains such as emissions, energy consumption, and disposal potential. At that point, the robotics in relation to sustainability can be reviewed as precise and credible. Another constraint is the symbolic hyper-emphasis on technology at the cost of comprehensive service. Although robots are a frequent focus nowadays, luxury hospitality runs the risk of positioning robots as a gimmick as opposed to a device within the larger sustainability agenda and service approaches. Maheshwari et al. (2024) believe that the application of AI should be positioned as a long-term initiative instead of a quick fix. Further studies need to explore how hotels can strike a balance between technological innovation and the continuation of cultural heritage, authenticity, and human warmth—standards that remain synonyms of luxury to many guests.Lastly, robotics in hospitality is an academic field which has yet to be studied in detail. Though there are useful works that consider the dimensions of acceptance (Lu et al., 2019), workforce adaptation (Melián-González & Bulchand-Gidumal, 2016), and guest experiences (Tussyadiah & Park, 2018; Han et al., 2021), comprehensive approaches incorporating these dimensions are scarce. According to Murphy et al. (2017), another obstacle that robotics may create at the pedagogical level is the need to train future managers to operate in technology-intensive settings. Future studies must thus have an interdisciplinary direction using computer science, sustainability studies, psychology, and management as reference points.

CONCLUSION

Robotics and artificial intelligence (AI) applied in luxury hospitality is not a simple technological addition, but a paradigm shift in the conception, enactment, and perception of luxury. Luxury, traditionally characterized by extravagance, elitism, and personalized service which addresses human concerns, is currently being re-shaped in the context of sustainability, innovation, and responsibility. No longer viewed as a futuristic gimmick, robots are now seen as devices that do not only help boost productivity but also protect human health and assist with global sustainability. Their existence in luxury hotels is representative of a greater shift: a shift in room service, based on indulgences, to green service based on not only ecological stewardship, but also moral responsibility. One of the key messages in this chapter has been the dualistic nature of robotics as a useful resource and a symbol. Rationally, the most obvious value of robots is that they execute tasks calmly and strictly, like cleaning, food delivery, a check-in process, etc. Not only do these applications optimize performance through their ability to eliminate human error, but they also make better use of resources as well as to reduce or at least minimize waste. Symbolically, robots denote innovation, privilege, and sustainability. In the case of luxury hotels, where discretion remains at the very heart of the brand promise, it makes ethical governance not just a regulatory mandate but a strategic necessity. When applied smartly, robotics will make the guests feel responsible, thus ensuring exclusivity and global standards. Perhaps the most influential force influencing the adoption of robotics in luxury hospitality is sustainability. Collectively, these innovations will elevate an overall model of smart and sustainable luxury that goes beyond operational efficiency to include resiliency, wellbeing, and sustainable ecological stewardship in the long run.Even with such developments there are limits. The complexity in the adoption of robotics can be presented by technical reliability, acceptance among guests, employee concerns, ethics risk, and the issue of the product sustainability throughout the lifecycle. As part of future research, those gaps need to be bridged through additional interdisciplinary studies on interoperability, living with differences across cultures, mixed models of workforce, and most effective sustainability impacts. The difficulty facing practitioners is how to match robotics adoption to brand identity while having the technologies complementing, as opposed to degrading, the specific value proposition of luxury hospitality.For practitioners and scholars alike, robotics in luxury hospitality now represent a dual commitment to innovation and to planetary responsibility. Three strategic imperatives emerge for practitioners: (1) integrate robotics into measurable sustainability KPIs and public reporting; (2) adopt hybrid workforce models emphasizing reskilling and green human resource management; and (3) ensure transparent communication of sustainability outcomes to prevent greenwashing. When embedded authentically, robotics redefine luxury not as indulgence, but as a symbol of responsible innovation and sustainable elegance.

REFERENCES

Ali, F., Qiu, Z., & Gao, J. (2018). Luxury hotel guests' perceived authenticity and behavioral intentions: The role of traditional and modern luxury. *International Journal of Contemporary Hospitality Management, 30*(3), 1599–1616. DOI: 10.1108/IJCHM-12-2016-0683

Ansari, A. I., & Singh, A. (2023), "Application of Augmented Reality (AR) and Virtual Reality (VR) in Promoting Guest Room Sales: A Critical Review", Tučková, Z., Dey, S.K., Thai, H.H. and Hoang, S.D. (Ed.) Impact of Industry 4.0 on Sustainable Tourism, Emerald Publishing Limited, Leeds, pp. 95-104. DOI: 10.1108/978-1-80455-157-820231006

Ansari, A. I., Singh, A., Pattanaik, S., Patnaik, R., & Suryawanshi, U. (2024). Exploring the Role of AI-Powered Virtual Assistants in Optimizing Self-Service Experiences for Customers Within the Service Sector. In Nadda, V., Tyagi, P., Singh, A., & Singh, V. (Eds.), *Integrating AI-Driven Technologies Into Service Marketing* (pp. 225–238). IGI Global., DOI: 10.4018/979-8-3693-7122-0.ch013

Arora, A., & Singh, A. (2026). Ethical considerations for emotion recognition technology in luxury hospitality: Balancing personalization and privacy. *Journal of Travel Research, 12*(1), 1–15.

Baum, T., Mooney, S., Robinson, R. N. S., & Solnet, D. (2020). COVID-19's impact on the hospitality workforce—New concerns and new considerations. *International Journal of Contemporary Hospitality Management, 32*(9), 2893–2900. DOI: 10.1108/IJCHM-04-2020-0314

Berry, K., & Singh, A. (2024). AI-Driven Service Marketing: Transforming Customer Experience and Operational Efficiency. In Nadda, V., Tyagi, P., Singh, A., & Singh, V. (Eds.), *AI Innovations in Service and Tourism Marketing* (pp. 35–56). IGI Global., DOI: 10.4018/979-8-3693-7909-7.ch003

Bhalla, A., Singh, P., & Singh, A. (2023). Technological Advancement and Mechanization of the Hotel Industry. In Tailor, R. (Ed.), *Application and Adoption of Robotic Process Automation for Smart Cities* (pp. 57–76). IGI Global., DOI: 10.4018/978-1-6684-7193-7.ch004

Buhalis, D., & Leung, D. (2018). Smart hospitality: Interoperability and the future of hotel operations. *International Journal of Contemporary Hospitality Management, 30*(3), 1365–1383. DOI: 10.1108/IJCHM-08-2017-0471

Cain, L., Thomas, A., & Alonso, A. (2019). Robots and the tourism industry: An analysis of current applications and future considerations. *Tourism Management Perspectives*, *31*, 218–226. DOI: 10.1016/j.tmp.2019.05.004

Chan, E. S. W., & Hsu, C. H. C. (2016). Effects of carbon management on the financial performance of hotels. *International Journal of Hospitality Management*, *57*, 172–177. DOI: 10.1016/j.ijhm.2016.06.002

Chourasia, S., Kaur, S., & Singh, A. (2025). The Future of Indian Restaurants: The Role of Technology in the Success of Cloud Kitchens. In Singh, K., Kumar, S., & Talukder, M. (Eds.), *Impact of AI and the Evolution of Future Ghost Kitchens* (pp. 15–34). IGI Global Scientific Publishing., DOI: 10.4018/979-8-3373-0432-8.ch002

Delmas, M. A., & Burbano, V. C. (2011). The drivers of environmental performance: The role of "conspicuous conservation". *Journal of Environmental Economics and Management*, *61*(2), 195–211. DOI: 10.1016/j.jeem.2010.10.001

European Union (EU). (2021). *Proposal for a Regulation of the European Parliament and of the Council Laying Down Harmonised Rules on Artificial Intelligence (Artificial Intelligence Act)*. European Commission.

Francis, R. S., Anantharajah, S., Sengupta, S., & Singh, A. (2024). Leveraging ChatGPT and Digital Marketing for Enhanced Customer Engagement in the Hotel Industry. In Bansal, R., Ngah, A., Chakir, A., & Pruthi, N. (Eds.), *Leveraging ChatGPT and Artificial Intelligence for Effective Customer Engagement* (pp. 55–68). IGI Global., DOI: 10.4018/979-8-3693-0815-8.ch004

Gossling, S., & Higham, J. (2021). The low-carbon transition in tourism: A challenge for business, a necessity for humanity. *Journal of Sustainable Tourism*, *29*(6), 843–856. DOI: 10.1080/09669582.2021.1901035

Han, H., Yu, J., & Kim, M. (2021). The role of robot presence on customer satisfaction and behavioral intention in a hotel context. *International Journal of Hospitality Management*, *97*, 103004. DOI: 10.1016/j.ijhm.2021.103004

Ivanov, S. H., & Webster, C. (2019). The next wave of automation in tourism, travel, and hospitality—A new field of research. *Journal of Tourism Futures*, *5*(2), 114–120. DOI: 10.1108/JTF-05-2019-048

Ivanov, S. H., & Webster, C. (2020). *The robotisation of travel, tourism and hospitality: Research, theory and practice*. CABI.

Ivanov, S. H., Webster, C., & Berezina, K. (2017). Adoption of robots and service automation by travel, tourism and hospitality companies. *International Journal of Contemporary Hospitality Management*, *29*(10), 2901–2917. DOI: 10.1108/IJCHM-08-2016-0492

Jin, N., Line, N. D., & Goh, B. (2016). Experiential value, relationship quality, and customer loyalty in the context of upscale hotel brands. *Journal of Travel Research*, *55*(3), 356–370. DOI: 10.1177/0047287514555848

Kapoor, A., Balaji, M. S., & Kannan, P. K. (2023). Algorithmic bias in service robots: A path to ethical service delivery. *Journal of Service Research*, *26*(4), 509–529. DOI: 10.1177/10946705231154561

Kuo, M. H., Chen, C. M., & Tseng, Y. C. (2017). The effect of perceived robot usefulness, ease of use, and trust on the intention to use hotel service robots. *International Journal of Hospitality Management*, *67*, 28–39. DOI: 10.1016/j.ijhm.2017.07.001

Kuo, M. H., Chen, C. M., & Tseng, Y. C. (2022). Examining the continuous use intention of service robots in the hospitality industry: A perspective of service-dominant logic. *International Journal of Contemporary Hospitality Management*, *34*(3), 1269–1290. DOI: 10.1108/IJCHM-05-2021-0676

Lu, L., Cai, R., & Gursoy, D. (2019). Developing and validating a Service Robot Integration Willingness (SRIW) scale. *International Journal of Hospitality Management*, *80*, 23–32. DOI: 10.1016/j.ijhm.2019.01.005

Maheshwari, N., Singh, V., Singh, A., & Ansari, A. I. (2024). Understanding Artificial Intelligence Adoption in Tourism Services and Marketing: Boon or Bane? In Nadda, V., Tyagi, P., Singh, A., & Singh, V. (Eds.), *AI Innovations in Service and Tourism Marketing* (pp. 359–374). IGI Global., DOI: 10.4018/979-8-3693-7909-7.ch018

Maheshwari, P., Singh, S., Singh, A., & Ansari, H. M. (2024). Framing AI adoption in tourism: The role of sustainability and responsible innovation. *Journal of Vacation Marketing*, *30*(1), 1–19. DOI: 10.1177/13565381231189474

Melián-González, S., & Bulchand-Gidumal, J. (2016). The role of information technology in increasing job satisfaction in the hotel sector. *Tourism Management*, *55*, 220–229. DOI: 10.1016/j.tourman.2016.02.008

Murphy, J., Hofacker, C., & Gretzel, U. (2017). The robots are coming: Implications for the tourism industry. *Tourism Management Perspectives*, *21*, 5–7. DOI: 10.1016/j.tmp.2016.11.006

Nadda, V., Tyagi, P. K., Singh, A., & Singh, V. (Eds.). (2024). *Integrating AI-Driven Technologies Into Service Marketing* (Vols. 1–2). IGI Global., DOI: 10.4018/979-8-3693-7122-0

Neuhofer, B., Buhalis, D., & Ladkin, A. (2015). Smart tourism ecosystem: Frameworks for smart destination development. *International Journal of Tourism Research*, *17*(5), 458–469. DOI: 10.1002/jtr.2001

Osawa, H., Minato, T., Shimada, M., & MacDorman, K. F. (2017). The uncanny valley: A pervasive phenomenon. *Frontiers in Psychology*, *8*, 1943. DOI: 10.3389/fpsyg.2017.01943

Prakash, J., Singh, V., Singh, A., & Kumari, U. (2024). Investigating the Synergistic Impact of AI and Marketing Within Virtual Tourism Platforms for Millennials: A Review. In Nadda, V., Tyagi, P., Singh, A., & Singh, V. (Eds.), *Integrating AI-Driven Technologies Into Service Marketing* (pp. 419–438). IGI Global., DOI: 10.4018/979-8-3693-7122-0.ch020

Sharma, M., & Singh, A. (2024a). Enhancing Competitive Advantages Through Virtual Reality Technology in the Hotels of India. In Kumar, S., Talukder, M., & Pego, A. (Eds.), *Utilizing Smart Technology and AI in Hybrid Tourism and Hospitality* (pp. 243–256). IGI Global., DOI: 10.4018/979-8-3693-1978-9.ch011

Sharma, M., & Singh, A. (2024b). Embracing Technological Innovation: A Review of Hi-Tech Services in Hospitality Industry. *Evergreen*, *11*(4), 2818–2830. DOI: 10.5109/7326926

Sharma, M., Singh, A., & Rana, S. (2024a). Empowering Hospitality Brands AI Strategies for Enhanced Customer Experiences in Emerging Markets. In Nadda, V., Tyagi, P., Singh, A., & Singh, V. (Eds.), *Integrating AI-Driven Technologies Into Service Marketing* (pp. 173–190). IGI Global., DOI: 10.4018/979-8-3693-7122-0.ch010

Sharma, R., Sachdeva, T., Singh, A., & Nadda, V. (2024b). A New Era of Engagement and Satisfaction: Transforming Customer Experience With AI-Driven Technologies. In Nadda, V., Tyagi, P., Singh, A., & Singh, V. (Eds.), *AI Innovations in Service and Tourism Marketing* (pp. 1–16). IGI Global., DOI: 10.4018/979-8-3693-7909-7.ch001

Sharma, R., & Singh, A. (2024c). Blockchain Technologies and Call for an Open Financial System: Decentralised Finance. In Vardari, L., & Qabrati, I. (Eds.), *Decentralized Finance and Tokenization in FinTech* (pp. 21–32). IGI Global., DOI: 10.4018/979-8-3693-3346-4.ch002

Shin, H., & Kang, J. (2020). The effect of service robots on customer emotional experience and behavioral intention in the hotel industry during the COVID-19 pandemic. *Journal of Hospitality and Tourism Management*, *44*, 290–298. DOI: 10.1016/j.jhtm.2020.07.009

Singh, A. (2024). Virtual Research Collaboration and Technology Application: Drivers, Motivations, and Constraints. In Chakraborty, S. (Ed.), *Challenges of Globalization and Inclusivity in Academic Research* (pp. 250–258). IGI Global., DOI: 10.4018/979-8-3693-1371-8.ch016

Singh, A., & Bathla, G. (2023). Fostering Creativity and Innovation: Tourism and Hospitality Perspective. In P. Tyagi, V. Nadda, V. Bharti, & E. Kemer (Eds.), Embracing Business Sustainability through Innovation and Creativity in the Service Sector (pp. 70-83). IGI Global.

Singh, A., & Hassan, S. C. (2024). Service Innovation Through Blockchain Technology in the Tourism and Hospitality Industry: Applications, Trends, and Benefits. In Singh, S. (Ed.), *Service Innovations in Tourism: Metaverse, Immersive Technologies, and Digital Twin* (pp. 205–214). IGI Global., DOI: 10.4018/979-8-3693-1103-5.ch010

Singh, A., & Kumar, V. (2025a). Organizational Innovation and Adaptation in Hospitality and Tourism: Strategies for Resilience and Leadership. In C. Ramos, T. Costa, R. Madeira, & J. Ferreira do Rosário (Eds.), Open Innovation and Technology in Tourism and Hospitality (pp. 355-372). IGI Global Scientific Publishing. DOI: 10.4018/979-8-3693-8633-0.ch018

Singh, A., & Kumar, V. (2025b). User-Centered Hospitality: UI/UX as the New Frontier in Hotel and Restaurant Innovation. In C. Ramos, T. Costa, R. Madeira, & J. Ferreira do Rosário (Eds.), Open Innovation and Technology in Tourism and Hospitality (pp. 395-410). IGI Global Scientific Publishing. DOI: 10.4018/979-8-3693-8633-0.ch020

Singh, V., & Singh, A. (2024). Revolutionizing the Hospitality Industry: How chatGPT Empowers Future Hoteliers. In Bansal, R., Ngah, A., Chakir, A., & Pruthi, N. (Eds.), *Leveraging ChatGPT and Artificial Intelligence for Effective Customer Engagement* (pp. 192–203). IGI Global., DOI: 10.4018/979-8-3693-0815-8.ch011

Tung, V. W. S., & Au, N. (2018). Exploring customer experiences with robotics in hospitality. *International Journal of Contemporary Hospitality Management*, *30*(7), 2680–2697. DOI: 10.1108/IJCHM-06-2017-0322

Tussyadiah, I. P., & Park, S. (2018). Consumer evaluation of robotic service: The role of anthropomorphism. *International Journal of Tourism Research*, *20*(7), 804–815. DOI: 10.1002/jtr.2198

Tussyadiah, I. P., Zach, F., & Wang, J. (2020). Attitudes toward service robots in the hotel industry: Insights from an online review analysis. *Journal of Hospitality and Tourism Technology*, *11*(1), 173–193. DOI: 10.1108/JHTT-05-2018-0050

United Nations World Tourism Organization (UNWTO). (2023). The Glasgow Declaration on Climate Action in Tourism.

United Nations World Tourism Organization (UNWTO). (2024). Tourism and Climate Action: Global Commitments and Progress.

World Travel & Tourism Council (WTTC). (2024). [A new era for Travel & Tourism.]. *Sustainability Report*, •••, 2024.

Chapter 10
Empowering Guests: The Rise of Self-Service Technology in Hospitality

Amrik Singh
https://orcid.org/0000-0003-3598-8787
Lovely Professional University, India

ABSTRACT

The hospitality industry is undergoing a remarkable transformation with the integration of self-service technology (SST), which is redefining the guest experience and operational efficiency. From mobile check-in kiosks and digital room keys to AI-powered chatbots and self-ordering systems in restaurants, these innovations empower guests by giving them greater control, convenience, and personalization. Self-service technology is a great way to positively impact the hotel experience, allowing guests to take their hotel journey into their own hands, providing convenience and speed while enhancing the level of service, even during off-hours. Self-service technology (SST) is quickly transforming the hospitality sector by opening new avenues for service delivery and giving clients more choices. A workforce that can quickly adjust to change is necessary to prepare for the implementation of successful self-service technology delivery projects. This chapter delves into critical findings regarding the adoption and implications of SST in tourism and hospitality

INTRODUCTION

Self-service technology is a kind of technology based on a digital interface and/or Artificial Intelligence to enable the customers to receive assistance without any human interference being required. This is applicable to the hotel industry in that the hotels will be able to interact with their clientele via the application of digital

DOI: 10.4018/979-8-3373-6541-1.ch010

tools to make them the creators of their experiences (Al-Romeedy et al., 2025a). The rapid growth of self-service technology is affecting hospitality businesses in a significant way by offering new opportunities as well as challenging both customers and employees (Singh & Hassan, 2024; Singh & Kumar, 2025a). An article in Time magazine about the popularity of self-service cites that in the previous year (2006) U.S. customers spent 128 billion on self-service kiosks, which is a 80 percent increase over the past year and by the year 2007 it could be 1.3 trillion. Hospitality organizations have developed much of their success in the last few hurdles in self-service strategies, with the application of technology, but the human capital investment is needed to suit the personal traits needed in order to optimize the full potential of self-service benefits (Singh & Hassan, 2024; Singh, 2024). According to a recent article in The Economist, it was stated that, Self-service is now performing to the service industry what mass production did to manufacturing which has automated the processes and has cut down the costs to a considerable level. The use of self-service has become more and more common through self-service check-in desks in hotels, restaurants and airports (Singh & Bathla, 2023; Srivastava et al., 2025). The other example of the swift growth in the field of technology is the spread of online learning in business and education. The self-service concept is based on self-directed learning even when skill limits the abilities of self-service tactics as dictated by the user. This study was aimed at examining the relationship between self-efficacy and some demographic factors and the readiness of the employees in hospitality firms to engage in self-directed learning (Al-Romeedy et al., 2025b; Ansari et al., 2024; Srivastava et al., 2025). Due to the fact that hospitality organizations must compete by adopting technologies that influence the learning environment of employees, the burden of self-directed learning has grown, as well as technology. These attributes which celebrate self directed learning and those which represent an obstacle towards self directed learning should be explored to see the willingness of an employee to embrace self directed learning. Companies and employees encouraging the presence of self-directioned learning will assist in equipping the employees to join self-directed working teams and facilitate a learning company approach. The issues of this research are to examine the connections that the chosen variables to determine on the self-directed learning readiness. The establishment of a wider knowledge foundation of self-directed learning preparedness with the use of targeted variables may actually not only advantage a people-based business that is highly focused in hiring and serving people, but other scientists and practitioners may take advantage of the exclusive social learning setting the hospitality business may present (Ansari et al., 2024; Ansari & Singh, 2023; Berry & Singh, 2024). Organizations and employers that encourage the self-directed learning preparedness will assist in equipping their employees to engage in the self-directed work groups and aid a learning organisational strategy. Encouragement of the employees to bring change should be prepared and practiced. When training an employee to be self-directed, the assumptions include taking into account the self-directed learning techniques. Awareness of the theoretical study of the variables that determine the self-directed learning readiness will enable a hospitality organization to develop effective and efficient programmes and practices to maximize on the talents of employees (Singh & Kumar, 2025a; Singh & Kumar, 2025b). A benefits of self-directed learning do not only benefit a people-based business that is highly focused on employing and serving people, but it also takes advantage of the social learning environment that is unique to hospitality industry (Berry & Singh, 2024; Bhalla et al., 2023).There are some similar and distinct views that are shared in definitions of self directed learning. In spite of some common aspects in the definition of self-directed learning, there is as well an alarming ambiguity in the exact definition in the research. The self-directed learning theory might have been given more focus compared to the practical aspect which has not been developed with an equivalent focus. Self-directedness is an attribute of adult learning that goes hand in hand with self directed learning, and entails some degree of choice making and personal control during the learning process. Self-efficacy theory is grounded on two forms of expectations, discussed earlier as efficacy expectations and outcome expectations, as well as the characteristics, behavior, and the results of the behavior of the individual. Self-efficacy could be a better predictor of performance because the outcome expectations are reliant on self-efficacy. Employees, in their turn, can be more encouraged to act in a way they see that will deliver desirable results. Additional insight into the manner people operate in terms of the decisions they take (selection processes), effort (time and persistence), motivation (initiation), thought patterns (cognitive processes) and emotional responses (affective processes) to different situations (Bhalla et al., 2023; Chourasia et al., 2025; Deri et al., 2024; Francis et al., 2024). The popularity of self-service technology (SST) is on the rise because of the COVID-19 epidemic that resulted in radically different user habits under its influence and the resulting alteration of the user behavior. As a matter of fact, COVID-19 pandemic has created enormous demand of SSTs. To provide an example, the present forms of segregation have resulted in the creation of contactless consumption where people do not desire to have physical contact with another but rather adopt self-service and healthier services provided by technology, therefore, the do-it-myself style of interaction has become more desirable. This chapter tries to critically analyze the advantages of SST, the issues of adopting SST and some implementation ideas of SST techniques. The customers embraced Kiosks as easy to use, fast to execute, fun to operate, yet without human interaction and counter language concerns were both discussed at the same time. The kiosks have been coming up as self-service technologies in hotels and are significant in helping to minimize bottlenecks in hotel operations (Al-Romeedy et al., 2025a; Ansari et al., 2024; Ansari & Singh, 2023). The results indicated that customers would prefer to interact with an employee as compared to relying on a technology-based self-service in most of their service encounters; the preference of customers to use technology-based self-service as compared to human encounters varies across different points throughout their occupancy cycle; customers choose to be served in terms of speed and easy service is the least important point in determining their preference to use technology-based self-service as compared to human contact encounters; the preference of customers to receive direct person contact is the most significant factor in determining preference to be served in terms of human-based encounters.

Guest Autonomy and Technology Acceptance Model

Self-service technologies (SSTs) can transform hospitality, which can be better explained using the body of existing theoretical explanations of the previously mentioned factors of guest behavior, motivation, and technological innovation acceptance (Singh, 2024; Singh & Singh, 2024; Singh & Bathla, 2023). Two of the most applicable ones, namely, the Guest Autonomy Theory and the Technology Acceptance Model (TAM), provide useful insights into the way technology provides guests with power and, at the same time, reinvigorates the relations of service provision (Chourasia et al., 2025; Deri et al., 2024).These frameworks may be used to explain why guests embrace technology as well as why the balance between convenience, control, and emotional experience determines their satisfaction and loyalty.

Guest Autonomy Theory

Hospitality has been based on service dependency where customers are dependent on employees to provide comfort, convenience, and care. The advent of SSTs, however, mobile check-ins, digital concierge applications, automated payment systems, etc., are putting guests in charge of their experiences. This change can be associated with the Guest Autonomy Theory that accentuates the human aspiration to independence, competence, and self-determination in consumption experience. Guest autonomy is a psychological process of empowerment, and the technology is used as a means of self-control and personalization (Deri et al., 2024; Francis et al., 2024; Harit et al., 2025).To illustrate, being in charge of room options, check-in without lines, and manage in-room comfort with the help of mobile-powered tools makes one feel more in control. Consumer psychology studies indicate that people are more satisfied and involved when they feel that they have control over the environment. In this way, SSTs affect autonomy through reducing friction and making them customize their stay in the way they precisely need. The correlation between autonomy and satisfaction is however not straight (Singh & Kumar, 2025a; Singh &

Kumar, 2025b). Over-automation may lead to so-called autonomy fatigue, the state when the guest feels overburdened by decision-making or depersonalized when he/she does not see a human face (Harit et al., 2025; Maheshwari et al., 2024). This means that as much as autonomy promotes empowerment, it has to be accompanied by optional human interaction to ensure emotional security as well as inclusivity. Therefore, efficient digital hospitality is not complete automation but moderate autonomy, in which the guests are free in their selection between self-service and personal service.

Technology Acceptance Model (TAM)

The Technology Acceptance Model (TAM) was developed by Davis (1989) to offer a framework through which we can understand the state of acceptance and adoption of new technologies by users. TAM suggests that individuals who adopt or reject technological systems depend on two important variables, which are Perceived Usefulness (PU) and Perceived Ease of Use (PEOU). These constructs are crucial in the hospitality industry in terms of influence over the reception of self-service technologies by guests (Singh, 2024; Singh & Singh, 2024). When the digital kiosk or mobile application is perceived to be beneficial (i.e. saves time, makes a guest more comfortable, or is convenient), the guests are more likely to embrace it. Likewise, when the technology is user-friendly, accessible, and easy to use and the demographics, the perceived ease of use increase its acceptance. Hotels that do not match these expectations may lose less technological-savvy guests, which will either become digitally excluded or frustrated (Maheshwari et al., 2024; Islam & Singh, 2025).Social influence and enabling conditions are also taken into account in recent extensions of TAM, including the Unified Theory of Acceptance and Use of Technology (UTAUT). In the hospitality industry, this is expressed in the contribution of brand reputation, word of mouth and infrastructural provisions (e.g. Wi-Fi connectivity, clear signs, staff service) to the perceived validity and trustworthiness of SSTs. Notably, TAM emphasizes the fact that technology adoption is not strictly rational, it is also experiential. Although a system can be valuable, it can decrease adoption due to negative emotional reactions (e.g., anxiety or distrust in the safety of data). Thus, hotel managers should be able to create SSTs that work efficiently but are also safe, intuitive, and humanized.

Generalizing Theory and Practice

The integration of the Guest Autonomy Theory and TAM raises one crucial point: empowerment via technology should be in line with the psychological satisfaction and functionality. Guests are not afraid of technology that improves control and convenience, but they are retaining when the same is backed with empathy, privacy and inclusivity.

This theoretical underpinning acts as a guide to hospitality practitioners:

Apply the Guest Autonomy Theory in order to create experiences that support freedom and personalization.
Use TAM concept to make technologies helpful, available, and emotionally compelling.

Finally, the theory is a reminder that technological empowerment in the hospitality industry is not all about innovation but also about human-centered design where all clicks, swipes and automated actions remain to be hospitable.

LITERATURE REVIEW

Another current development in the hospitality sector is that some organizations are shifting out of traditional services provided by human workers to the adoption of self-service technology (SST) in which the services are co-created by customers. Cashless touch-screen kiosks or self-order and pay tablets, which are used in restaurants, are used to illustrate this trend (Bhalla et al., 2023; Chourasia et al., 2025). This type of self-service is growing more popular following the introduction of the two most famous casual dining chains, Chili's and Applebee's, that offered the option to place orders and pay services on the tablet-based touch-screen gadgets in 2014. The resultant savings on costs in the form of reducing the human cost factor can be quite a relief to organizations in the hospitality industry since the hospitality industry is a human intensive industry as compared to other sectors.

Service Excellence

Service excellence can be described as an act or provision of service level that yields delight. Delight is explained as a manifestation of very high satisfaction, extreme expression of positive affect due to unexpectedly good performance. Satisfaction is a judgment of meeting expectations, and delight is a customer affects due to such a judgment, an emotion. In the qualitative study of service excellence as mentioned by Johnston, the customers had a notion of service excellence which they termed as being easy to conduct business. Johnston suggests that there are four major components of service excellence, namely delivering the promise, delivering a personal touch, going the extra mile, and responding to problems and queries appropriately (Maheshwari et al., 2024; Islam & Singh, 2025; Nadda et al., 2024).

The central component is the fulfilment of the promise or the performance of what the organization claims that it would perform or do are to meet the expectations (not outdo them). Personal touch can be termed as service and shows how customers are being attended to and this may create a measure of delight to the customers. Going the extra mile is also known as anticipation of needs of customers, and simply as giving a little extra which is actually always welcome. The slightest extra effort that companies can apply to their clients is mutually advantageous to the customers and businesses.

SST in Luxury Hotel

With the advent of technology making such serious infiltrations into business and marketing strategy, self-service technologies (SSTs) become an option or even a substitute to service transactions made by human beings. The advantages of using SST are associated with labor costs savings in the long term, establishment of competitive niches and differentiation, and enhanced consumer service and operational efficiency (Singh & Kumar, 2025b; Singh, 2024; Singh & Singh, 2024). The operators and managers are still reluctant to move towards adoption of SSTs because it has a heavy burden on the resources in case the consumers reject SSTs. Service organizations are afraid that the loss of customer-employee relationship would deter service recovery processes, destroy social networks between consumers and organization, diminish up-selling chances, and make service staff develop resentment of the technology (Chourasia et al., 2025; Francis et al., 2024). Given that there are two possible alternatives, SSTs vs. service staff to complete the same task objective, a consumer would probably choose between them based on the perceived power of extrinsic or intrinsic motivation either alternative offers in fulfilling the task objective. A hotel check-in service is an example of such a situation, and a fast check-in service will be considered as a quality one. The options available to the travelers are to check in at the front desk, or a recently installed SST kiosk. The front desk is also an option that would provide me with a chance to communicate with a service staff. Hotel selecting the kiosk can be driven more by the extrinsic desires of speedy check-in, whereas those guests that choose the service desk fulfil intrinsic desires of communication with the staff. SST intention is operationalized as the probability of using SSTs in making a service transaction as opposed to the service staff. Extrinsic and intrinsic motivations have a direct relation to attempting SSTs. TAM and similar research give grounds to believe that perceived usefulness is a direct antecedent of intention to use SSTs. The usefulness of SST enhances the probability of usage, particularly when it has an alternative service transaction system like staffed counters (Berry & Singh, 2024; Bhalla et al., 2023; Chourasia et al., 2025). Data security and privacy of personal information is also a significant concern in a few service transactions. According to the surveys, 59-68 percent of the consumers decided to use SST kiosks to secure their privacy during hotel registration. The aspect of consumer security and privacy is also of great concern in the study of consumer research and the policy and law in the consumer sector, particularly in a transaction involving technology. show that the level of privacy concerns demanded of online transactions has a negative relationship with the purchase behavior of consumers.

AI and Hospitality

Artificial intelligence can be explained to mean systems that blend advanced hardware and software with extravagant databases and knowledge-based processing schemes to exhibit attributes of successful human decision making. Besides hotels, restaurants, events, and travel agencies are also rapidly adopting and implementing technologies, including service automation and robots, the first one being kiosks.

Chatbot

A chatbot is an application that mimics human-to-human communication by use of text or voice messages. The original chatbot is ELIZA created by a professor at MIT, Joseph Weizenbaum in 1966. A chatbot is one of the significant social problems because companies are eager to make bot-enabled services available to users. This study claimed that a chatbot can save labor costs and respond to the demand of a client in a short time. A chatbot, in turn, can create some extra expenses, including installation or maintenance, and can lead to the complaint of employees or customers regarding the nonperfect services.

Service Failure

It entails that service provision falls outside the perceived patience zone of a customer, and the customer is not satisfied. the offline studies of service failure has been replicated on-line. Moreover, under the banner of service encounters, it has been broadened to approach service failures associated with technology-based problems and self-service, and employees.

Empowerment and Experience

The advent of self-service technologies (SSTs) in hospitality, including mobile check-ins, digital kiosks, voice assistants and AI-based concierge apps, has turned the interaction with service environments into a completely new form (Nadda et al., 2024; Prakash et al., 2024; Sharma & Singh, 2024a).The innovations are expected to be convenient, personal and empowering and allow guests to have more control over their experiences (Singh & Kumar, 2025a; Singh & Kumar, 2025b; Singh, 2024). But with these advantages come a set of three interconnected issues: data security, digital exclusion of not-so-technologically-advanced guests, and loss of human touch, all with substantial consequences of guest satisfaction, ethical responsibility, and brand loyalty. As much as technology has taken a new frontier of the modern hospitality industry, the issue of making sure that the empowerment is not at the expense of inclusiveness, trust, or emotional bonding has continued to be a key focus in the contemporary hospitality management.

Information security and privacy issues

Hotels, resorts and restaurants are part of a hyper-connected hospitality ecosystem that gathers large amounts of personal data on the guest. Passport scans, payment information, and behavioral data, as well as location information, all of which are captured by digital systems, have turned personal information into an asset and an exposure. Self-service technologies, in particular, mobile apps and automated check-in kiosks usually need to be authenticated by credit cards, identification uploads, or biometric data, and all of these are the areas that can potentially be compromised (Singh, 2024; Singh & Singh, 2024; Singh & Bathla, 2023; Srivastava et al., 2025). The hospitality industry has been repeatedly jeopardized when it comes to cyber-security and has been exposed to data breaches. To take an example, the event of high-profile breaches like the breach of Marriott International impacted millions of clients exposing sensitive personal information about them, including credit card numbers and passport details. These incidents have pointed to how hospitality organizations are vulnerable to cybercriminals because of the obsolete infrastructure, integrations with third parties, and ad hoc adherence to data protection laws such as the GDPR or the Digital Personal Data Protection Act of 2023 in India (DPDP 2023). The same technologies that are intended to give more power to the guests such as mobile key systems, contactless payments, and personalized digital profiles broaden the scope of attack by hackers. Data theft, ransomware, malware, and phishing can destroy confidence in the guest and distrust. In addition, misuse of data is not just an external threat. Privacy can also be compromised by internal mishandling such as illegal access by employees or lack of proper encryption (Islam & Singh, 2025; Nadda et al., 2024; Prakash et al., 2024; Sharma & Singh, 2024b). Hospitality organizations need to take data protection as part of service quality to protect the empowerment of guests. It is necessary to have transparent privacy policies, informed consent systems, frequent cybersecurity audits, and staff education in data ethics. Such technologies as blockchain can be used to provide immutable records of transactions as well, and AI-based anomaly detection systems may be used to detect suspicious actions in real-time. Finally, data protection should not only become a technical compliance matter but also a strategic digital hospitality pillar, which should be directly correlated with the level of guest trust and brand image.

Digital Imbalance

Although SSTs are intended to be autonomous and efficient, it is possible that they will unintentionally exclude a large proportion of guests, especially older people, people with disabilities or belonging to low-digital-literacy groups. Digital divide that has been experienced in the hospitality industry can be seen in situations where hotels make the assumption that all guests are technologically competent enough to work with mobile applications or automated teller machines. As an illustration, visitors who do not know QR code menus or mobile check-in applications will be alienated or embarrassed when they are forced to request the staff to assist them in their simple processes. On the same note, international travelers in areas with low internet penetration or other technology platforms might not be able to interact with some of these systems. This virtual marginalization negatively affects the very concept that hospitality is all about inclusivity to which the hospitality is dedicated the philosophy of ensuring that all the guests feel desired and looked after. Access is not only a problem but also design bias. Numerous online interfaces are focused on the aesthetic of the surface, being sleek and minimalistic, neglecting users with visual impairments, non-native languages, or cognitive disabilities. Moreover, over use of technology may unwillingly discriminate against those guests who want to interact with others or are just in digital detox holidays (Al-Romeedy et al., 2025a; Prakash et al., 2024; Sharma & Singh, 2024c; Sharma et al., 2024).To deal with this the hospitality managers have to adopt the concept of universal design which includes designing interfaces and processes that are easy to use, multilingual and demographically accessible. Alternative to full automation Assisted digital service options like the hybrid check-in counters, where staff assist guests with the digital systems, can help fill the gap between digital and traditional service. Furthermore, the digital literacy courses and in-premises technology ambassadors will enable to train all guests and the employees so that there should be no one who feels the excluded party of innovation. Hospitality must also change the meaning of empowerment as forcing digital adoption to mean provide meaningful choice. The real guest empowerment is to be flexible, to give everyone the choice on whether they want to engage digitally, personally or by a combination of the two.

Human Brokenness and Emotional Stability

Welcoming, understanding, and human relationship has been the longstanding definition of hospitality. Yet, with the shift toward automation, the problem of the emotional authenticity loss has become even more acute. Although digital kiosks are efficient in processing a check-in, a smile, a sincere greeting, or an understanding gesture, which gives a feeling of belonging, will not be duplicated. Dehumanization is a risk that threatens to happen in the event that efficiency is put ahead of emotion. Visitors can find tech-enhanced service at the place to be chilly or impersonal, especially in the culture that places much emphasis on interpersonal warmth. Human touch is not only about interaction it is emotional labor, intangible competence that leaves a memorable experience and emotional commitment. It is what hotels cannot afford to do because it will make hospitality become a transactional process instead of a relationship process. However, it does not mean that the division between technology and humanity has to be complete. Well considered SSTs can add to human service and not take the place of it. As an example, online platforms will liberate employees by moving them out of the usual administrative tasks to higher-order interactions with the guests, such as anticipating their needs, giving local suggestions, or offering personalized care (Deri et al., 2024; Francis et al., 2024; Harit et al., 2025; Maheshwari et al., 2024).Simple questions and queries can be addressed by chatbots and AI concierges, whereas human employees will be engaged in the area of emotionally demanding cases when empathy and intuition cannot be substituted. The trick is to create the balance between automation and authenticity. The issue that the hospitality managers have to redefine roles of the employees not to be redundant but to be redefined. The employees must evolve into more of an experience curator rather than a service provider and use technology as an extension to, instead of replacing, emotional involvement. In addition, digital empathy, or learning to feel the emotions of guests despite mediated communication, may be used to assist the staff in feeling connected to other people even in the technology-saturated setting.

Ethics and Experience in Web-based Hospitality

The ethical basis of self-service transformation in hospitality is the convergence of data security, digital inclusion, and emotional authenticity. Guest empowerment goes beyond providing guests with technological control, it is also about making sure that their dignity, safety, and emotional gratification are the focal points of all innovations. Ethical hospitality requires openness in gathering and utilization of guest information, diversity in system design and compassion in service provision. A digitally enhanced guest experience is not just an efficient experience but rather a fair and emotionally intelligent one (Al-Romeedy et al., 2025b; Ansari et al., 2024; Ansari & Singh, 2023; Berry & Singh, 2024; Sharma et al., 2024b). There are no signs of digital convenience becoming the sole measure of excellence as hospitality brands that manage to bring these values together will stand out in an increasingly crowded market. The aftermath of the pandemic has supported the concept of automation in the name of safety and efficiency. Nevertheless, it also has brought a new thought back: how can technology enhance the fundamental values of hospitality and not blur them? Phygital hospitality is the future, a harmonious combination of both physical and digital service when technology serves as a convenience factor, and human beings can increase the level of emotion. In managerial perspective, it requires a strategic foresight and ongoing review of the comments made by guests, data protection procedures, and inclusivity. The interaction between technology and man has to be handled with sensitivity where the idea of empowerment should never become one-sided and impersonal. Self-service technologies have certainly transformed the hospitality industry, providing unprecedented comfort and flexibility of the guest. However, empowerment without empathy will lead to alienation; innovation without inclusion will lead to inequity; and personalization based on data without privacy will lead to suspicion (Islam & Singh, 2025; Nadda et al., 2024; Prakash et al., 2024).The reason is that, as the hospitality industry moves on with the digitalization process, adopting more technology is not necessarily the solution, but adopting it with a purpose, by ensuring that any innovation implemented is rooted in morality, accessibility, and human warmth. The emergence of self-service hospitality is a turning point: it may transform the concept of service excellence by harmonizing empowerment or destroy it by becoming disconnected with technology. The future is those who are capable of keeping guest data as safe as they keep guest feelings, design inclusively as much as they design efficiently and maintain the spirit of hospitality in the loops of automation.

DISCUSSION AND INTERPRETATION

Customer activities that are self service must be undertaken by their customers. SST (Self-Service Technology) or TBSS (Technology Based Self-Service) is referred to as an activity or benefit which relies on hard technologies which are provided by service providers in such a way that the customers may themselves handle the service or even some sections of the service. SSTs (Self-Service Technologies) are the technology interfaces, which allow customers to benefit in a service without the service employee participation (Bhalla et al., 2023; Chourasia et al., 2025). Some of the examples of Self-service technologies are conducting bank transactions and using Automated Teller Machines (ATM), self-service gas stations, self-health diagnosis and shopping with the use of the Internet, making reservations and buying tickets through kiosks, checking out of hotel rooms using interactive television, self-scanning systems at retail stores, vending machines, internet banking, phone banking, kiosks, self scanning check out, and self-service health diagnosis (Al-Romeedy et al., 2025a; Al-Romeedy et al., 2025b). The totality of self-service technologies contributes to the convenience of the customers and allows them time as well. Self service technology can result in better customer services, empowered customers and employees and greater efficiency. The high pace of technology allows service firms to integrate technology and their activities more and more. The new technologies are transforming the environment of services but not everyone is of the opinion that the use of technology in the service delivery is beneficial. Self-service technologies can be the cause of the success as well as the cause of failure. The customers are usually able to remember the failures as compared to the successes of the self-service technologies. Implicates that customers in almost all situations (96 percent) attribute these disappointing experiences to the company and its technology rather than the customer. The customers never lack any need of improved purchasing opportunities. However, in most cases companies are unable to see the vision of the customer. The difference between what the customer expects and what he perceives to receive has a tremendous influence in determining whether he is satisfied or not and their likelihood to go back to service or to find substitute products of other competitors. Companies seek SSTs because of a lot of reasons. Quality as a concept is highly significant to marketers since quality influences the formulation of any marketing strategy (Harit et al., 2025; Sharma et al., 2024).It implies that all service providers who take advantage of self-service technologies ought to put a lot of attention to operational excellence, quality, and the quality of customer service. This particularly with technology-based self-service channels has infiltrated the 21 st century service economy. It was concluded in the paper that companies should not redirect customers to self-service channels totally, particularly at the start of a relationship. The research has noted the relevance of the current context in knowing when and how self-service technologies form valuable customer experiences and emphasize. Further study would need to incorporate other factors that may influence the self-direction although self-efficacy was the variable that was most closely related with the self-directed learning preparedness of the hospitality workers. The hotels have integrated kiosks on self-services, the other hotels are not that much at ease with self-service.The chapter will discuss opportunities, challenges and issues that hoteliers experience when handling self-service kiosks, it also examines the viewpoint of the customer and the advantages of self-service kiosk to the end-user (Francis et al., 2024; Harit et al., 2025). Dwelling upon an SST kiosk system installed in one of the most luxurious hotel/casinos of Macau, this study demonstrates that no matter how technologically ready a customer may feel, in general, customers consider most of the SST options to be insignificant. The findings indicate that the guests of the hotel consider the SST option that assists them in gaining access to discounts on entertainment and eating facilities as the most important of all the other SST options (Nadda et al., 2024; Prakash et al., 2024; Sharma & Singh, 2024a). Therefore, fun can be identified as a precursor to SST. SST is an incredible transformation to service providers since the usage can satisfy the quest of customers to live an efficient life, research lacks enough information on the impacts of the usage of SST to customers experience of the services offered in the tourism and hospitality environments. The results showed that technology readiness and technology characteristics were significant in correlation with customer acceptance of SST in hotels. Moreover, the adoption of SST by customers and their desire to use it were strongly related. The study provides an extensive understanding of SST in five-star hotels in Egypt. Moreover, numerous applications can assist in helping hotel managers and decision-makers to know what helps customers embrace SST.self-service technology (SST) has become more of welcome in the business and society, especially with the onset of the COVID-19 pandemic. SST is an essential instrument to most businesses in their endeavor of achieving efficiency in operations and maximization of profits. Self-service technology does impact the hotel guest experience in a significant manner. We should have a glimpse of some of its effects.

Expedite the Service Request:

Although individuals are more relaxed and tolerant when they are on vacation, it should be mentioned that nobody wants to be put on hold to make a request or seek assistance in solving a problem. Using self-service technology guests will be served faster through requesting a service through the virtual concierge who in turn informs the front desk or respective department to have that request implemented. This enhances customer care and also allows the front desk to deal with the clients face to face. Quick service also enhances satisfaction of the guests, as they are aware that they can have what they desire, when they want it.

Make a good initial impression

Check-in is among the initial contacts between a guest and your hotel, hence it must be an effective and smooth procedure (Srivastava et al., 2025). A strong first impression can be made by using technology such as a kiosk to check-in since it is more efficient and the customer will not be required to stand in line. Such a hassle-free check-in process will make your guests appreciate your hotel and form a good opinion about it in the beginning.

Put guests at ease

Contactless technology use in hotels has increased after the pandemic to minimize physical contact and assist guests in feeling comfortable. Making contact-free will make guests feel safer, and make it so that all they need to do is be in close proximity with the hotel employees should they wish to. This implies that they have no fear of travelling regardless.

Provide convenience

You are able to give convenience to your guests with self-service technology. This implies that it does not matter when they check in or check out since they are able to check in using mobile check-in and check in ahead of their arrival (Prakash et al., 2024; Sharma & Singh, 2024c; Sharma et al., 2024). They can go into their room after a long flight only to unlock their rooms using digital room keys which can unlock their rooms using their smartphones. This is highly convenient to your guests as well as your hotel since you do not always need someone on duty round the clock. To the visitors who would like to order the room service, the same mobile app that you have used to check in can be utilized as a means of ordering the room service expounding that you can order the room service directly via the bed without even picking the phone to make the call. Easy and convenient service can hardly be matched to improve guest satisfaction.

Effortless Payment Experience

By utilizing technology, you can establish seamless payment process through a kiosk or internet-based payments so that the guest can take care of their own payment using a payment option of their choice, either PayPal or Apple Pay or reward credit. Through digital payments, guests can make payments in the manner and at the time they desire. The lighter and easier it is to make a transaction, the better the guest experience.

Autonomy

One of the most effective methods that a hotel can employ to increase guest satisfaction is self-service. Provided with the right tools, guests are now able to design and have their own experience, checking in at any time of the day, calling a room service on their smartphones, and requesting services without addressing any person (Bhalla et al., 2023; Prakash et al., 2024; Sharma et al., 2024b).They are able to select the degree of relation they desire with employees and address their own issues using the available means. Customers should be left to their own devices, which is a significant element of establishing a good brand experience and ensuring that the clientele feels that they have had a pleasant stay, in large part because they are essentially left to their own devices. Naturally, this is coupled with a great customer service that would be available in case they require it.

24 Hours Support

With the help of the tools which enable the visitors to arrange service orders themselves, your hotel will be able to provide 24/7 care without having to have employees available all the time. Chatbots, virtual assistants, self check-in kiosks and other contactless technologies collaborate to provide a smooth process of communication with your brand, and guests can be served at any time of the day and night (Singh & Bathla, 2023; Srivastava et al., 2025). The fact that they can attend to their requests 24/7 is what you need to make you feel that the brand is concerned about them.

CONCLUSION

Self-Service Technology (SST) is a disruptive technology that has transformed the way customers relate to each other, made them more efficient and allowed making decisions that are data-driven. Its influence keeps shifting with the development of technology and with the shift in customer demands and expectations, and therefore, it is one of the major factors that businesses ought to consider in a changing environment. Self-efficacy role concerning self-directed learning readiness will be a subject of keen attention. The improvement in the degree of self-directed learning readiness of the employees will have to be accompanied with the measures that enhance self-efficacy. the prevalence of focused facilitation to supplement self-directed processes can be used to offer positive and successful experiences with self-service technologies and with enhanced self-efficacy. Chatbot studies of service failure have not been fully studied to date, our research is significant since it examines the relationship between type of service and service condition. It is remarkable that this study proposes which business strategies must be emphasized, in line with the characteristics of customers with technology. Furthermore, our study presents research impact due to being the first article to empirically analyze the service failures of chatbot and how users respond to them. In this way, after the additional analysis, we were able to discuss not only the reaction of customers to service failures associated with AI, but also the guidelines that might be used to institute the recovery strategies. establish what challenges hotels are facing and the opportunities that they observe in the installation of kiosks; thus, the findings can be of great use The trend in society and business that we are currently experiencing, the international SST market is set to expand by many folds. One report projected that the SST market would blow up to a surpassing 72 billion dollars by the year 2030 and that the rate of growth will be more than 11 times the compound. Due to the rapid influence of technology on society and hospitality industry, emergence of new tools that will enhance the framework of already existing self-directed learning readiness instruments will aid in offering better insight on emerging processes of self-directed learning. With technology constantly rapidly defining society and the hospitality industry, new instruments that build upon the framework of already existing self-directed learning readiness instruments, will assist in offering more insight into new self-directed learning processes. self-service processes will require assistance to formulate strategies capable of delivering the most desirable experiences to their employees and customers. Self-service technology (SST) is changing the hospitality industry at a very rapid rate, be it hotels or restaurants. Giving guests the chance to conduct their business in the digital sphere, through the implementation of digital technology, including kiosks, mobile applications, and various online platforms, SST poses both a great opportunity and a great challenge to businesses in the industry.

OPPORTUNITIES

Improved Productivity and Lessened Expenditures:

SST automates the repetitive activities such as check in, check out and taking of orders. This will help to ease the staff work load, particularly at the time of peak and can help to save a lot of labor expenses. It is possible to redirect staff to more high-value products, including delivering individualized guest services or specific requests to enhance the overall quality of the service.

Improved Guest Experience:

Convenience, speed, and privacy are important to a number of travelers in the modern world. SST serves such needs by eradicating long queues at the front desk or counter. Visitors can arrive or leave at their convenience, and anytime, in the day or night, which comes in handy especially when their arrivals or departures are late. These systems can also provide personalized recommendations that are based on past preferences which result in a more personalized and fulfilling journey.

Enhanced Data Gathering and Insights:

The SST platforms will be able to gather quality data about the guests and their preferences and behaviors. This information helps hotels to develop specific marketing campaigns, make personalized offers and make data-driven decisions to enhance services. In the case of restaurants, this information can aid in the refinement of menus and streamlining of activities.

Higher Revenue:

Self-service systems will frequently result in the growth of ancillary revenue. It is simple to offer guests upsell and cross-sell items in kiosks and apps like room upgrade, late check-out, or more. These choices might be more attractive to the customers to choose these options at their pace without the strain of a face-to-face encounter. It has been established that whether in self-service kiosks, customers tend to spend more.

24/7 Availability:

With SST, there is 24/7 availability as opposed to a front desk manned by human beings or service counter. This gives the guest the option of using the services whenever they require it, in checking into the rooms when they are late or ordering room services in the late hours.

CHALLENGES

First Investment and Maintenance Costs: The adoption of SST will involve a large initial investment in equipment (kiosks, tablets, etc.) and software. Moreover, these systems need constant maintenance, technical support and frequent software upgrade so that they are operating properly and safely.

Change and User Adoption:

The resistance to change and user adoption is not universal among the guests. There might be those who like the traditional, human to human interaction as they see this as a necessary experience with the hospitality. Companies must be able to guarantee the seamless transition and provide the combination of self-service and conventional services to satisfy all clients. The staff should also be trained to provide support to the guest who might not be conversant with the technology.

Security and Privacy Issues:

Self-service systems process sensitive guest data, such as personal details and credit card details. That exposes them to cyberattack and data breaches. Companies need to have strong security systems to keep the details of their guests safe and retain their confidence.

Technological Integration and Interoperability:

The integration of new SST platforms with the current property management systems (PMS) and point-of-sale (POS) systems among other hotel software may be complicated. The compatibility is important to ensure that there are no errors and the efficiency remains operational.

Risk of Lost Human Interaction:

As much as SST has the potential of liberating staff to concentrate on more valuable interactions, there is a possibility of losing the general human aspect of the hospitality industry. The welcome on the front desk or the polite conversation with a waiter is also included in the special guest experience. Companies need to create a balance that does not compromise on the personal touch but rather utilizes technology.

To sum up, self-service technology in hospitality is a potent instrument that can potentially automatize the process, minimize expenses, and improve the experience of guests. Nonetheless, its effective adoption hinges on the strategic approach that would counter the problems of user adoption, the security of the data, and the necessity to support the high-quality human factor in service.

FUTURE RESEARCH OUTLOOK

Future study of Self-Service Technology (SST) in hospitality industry ought to explore further the utilization of finding a balance between technological effectiveness and human touch. Although a lot has been done on the adoption and advantages of SSTs, it is important to see the differences between the preference of guests in various demographics and trip marketing (Singh & Hassan, 2024; Singh & Kumar, 2025a). As an example, how are any business traveler needs of speed and autonomy compared to leisure traveler needs of personalised interactions, and how can SSTs be fitted in response to these different expectations? The long-term psychological effect of a lesser human interaction on the satisfaction and loyalty of the guests should also be studied further. Will there be a feeling of detachment or a loss of perception of quality of service even though practical needs will be satisfied with time-lived dependency on SST? This may include longitudinal research of the behavior of guests and their attitude during more than one visit. Also, the best application of Artificial Intelligence (AI) and Machine Learning (ML) in SSTs deserves consideration. What will it take to enable AI-powered chatbots and virtual assistants to go beyond simple query resolution to provide genuinely proactive, understanding and personalized suggestions, to the extent that the border between human and machine service is broken? The second important area is the contribution of SST in ensuring sustainability. One of the studies that researchers may undertake is how digital check-in/out, smart room controls, and AI-based energy management systems can prove their negative impact on the environment and how these sustainable features affect guests and hotel brands (Singh & Kumar, 2025a; Singh & Kumar, 2025b). Lastly, since SST involves a large-scale initial investment and a continuous maintenance fee, the future research should center on the establishment of robust systems and methods of appraising the ROI of different SST applications with emphasis on not only the cost savings but also on the intangible functions such as increased brand recognition and better understanding of the guests. This would offer good advice to hospitality businesses in the changing technological world.

REFERENCES

Al-Romeedy, B. S., Hussein, H., Emam, M. E., & Singh, A. (2025b). From Potential to Peril: Evaluating AI and the Metaverse in Modern Research. In M. Al Aqad, A. SorayyaeiAzar, A. Albattat, & A. Singh (Eds.), Navigating AI and the Metaverse in Scientific Research (pp. 563-584). IGI Global Scientific Publishing. DOI: 10.4018/979-8-3373-0340-6.ch027

Al-Romeedy, B. S., Hussein, H., & Singh, A. (2025a). From Data to Delight: Using Wearable Technology to Tailor Tourist Experiences. In Rana, V., Bathla, G., Raina, A., & Chhabra, D. (Eds.), *Transforming the Service Sector With New Technology* (pp. 381–398). IGI Global Scientific Publishing., DOI: 10.4018/979-8-3693-7447-4.ch022

Ansari, A. I., & Singh, A. (2023), "Application of Augmented Reality (AR) and Virtual Reality (VR) in Promoting Guest Room Sales: A Critical Review", Tučkovά, Z., Dey, S.K., Thai, H.H. and Hoang, S.D. (Ed.) Impact of Industry 4.0 on Sustainable Tourism, Emerald Publishing Limited, Leeds, pp. 95-104. DOI: 10.1108/978-1-80455-157-820231006

Ansari, A. I., Singh, A., Pattanaik, S., Patnaik, R., & Suryawanshi, U. (2024). Exploring the Role of AI-Powered Virtual Assistants in Optimizing Self-Service Experiences for Customers Within the Service Sector. In Nadda, V., Tyagi, P., Singh, A., & Singh, V. (Eds.), *Integrating AI-Driven Technologies Into Service Marketing* (pp. 225–238). IGI Global., DOI: 10.4018/979-8-3693-7122-0.ch013

Berry, K., & Singh, A. (2024). AI-Driven Service Marketing: Transforming Customer Experience and Operational Efficiency. In Nadda, V., Tyagi, P., Singh, A., & Singh, V. (Eds.), *AI Innovations in Service and Tourism Marketing* (pp. 35–56). IGI Global., DOI: 10.4018/979-8-3693-7909-7.ch003

Bhalla, A., Singh, P., & Singh, A. (2023). Technological Advancement and Mechanization of the Hotel Industry. In Tailor, R. (Ed.), *Application and Adoption of Robotic Process Automation for Smart Cities* (pp. 57–76). IGI Global., DOI: 10.4018/978-1-6684-7193-7.ch004

Chourasia, S., Kaur, S., & Singh, A. (2025). The Future of Indian Restaurants: The Role of Technology in the Success of Cloud Kitchens. In Singh, K., Kumar, S., & Talukder, M. (Eds.), *Impact of AI and the Evolution of Future Ghost Kitchens* (pp. 15–34). IGI Global Scientific Publishing., DOI: 10.4018/979-8-3373-0432-8.ch002

Deri, M. N., Zaazie, P., & Singh, A. (2024). Digital Future of the Hospitality Industry and Hospitality Education Globally. In Sharma, A. (Ed.), *International Handbook of Skill, Education, Learning, and Research Development in Tourism and Hospitality. Springer International Handbooks of Education*. Springer., DOI: 10.1007/978-981-97-4318-6_14

Francis, R. S., Anantharajah, S., Sengupta, S., & Singh, A. (2024). Leveraging ChatGPT and Digital Marketing for Enhanced Customer Engagement in the Hotel Industry. In Bansal, R., Ngah, A., Chakir, A., & Pruthi, N. (Eds.), *Leveraging ChatGPT and Artificial Intelligence for Effective Customer Engagement* (pp. 55–68). IGI Global., DOI: 10.4018/979-8-3693-0815-8.ch004

Harit, P., Singh, A., Agrawal, K., Verma, D., Bathla, G., & Raina, A. (2025). Exploring the Growing Threat of Cyber Risks and Vulnerabilities in the Hospitality Industry. In Srivastava, R., Rana, V., & Faiz, M. (Eds.), *Exploring the World With-Blockchain Through Cryptotravel* (pp. 299–316). IGI Global Scientific Publishing., DOI: 10.4018/979-8-3693-6562-5.ch017

Islam, N., & Singh, A. (2025). Role of Digitalization into Hospitality Businesses: Innovations and Trends. In Rana, V., Bathla, G., Raina, A., & Chhabra, D. (Eds.), *Transforming the Service Sector With New Technology* (pp. 447–458). IGI Global Scientific Publishing., DOI: 10.4018/979-8-3693-7447-4.ch027

Maheshwari, N., Singh, V., Singh, A., & Ansari, A. I. (2024). Understanding Artificial Intelligence Adoption in Tourism Services and Marketing: Boon or Bane? In Nadda, V., Tyagi, P., Singh, A., & Singh, V. (Eds.), *AI Innovations in Service and Tourism Marketing* (pp. 359–374). IGI Global., DOI: 10.4018/979-8-3693-7909-7.ch018

Nadda, V., Tyagi, P. K., Singh, A., & Singh, V. (Eds.). (2024). *Integrating AI-Driven Technologies Into Service Marketing* (Vols. 1–2). IGI Global., DOI: 10.4018/979-8-3693-7122-0

Prakash, J., Singh, V., Singh, A., & Kumari, U. (2024). Investigating the Synergistic Impact of AI and Marketing Within Virtual Tourism Platforms for Millennials: A Review. In Nadda, V., Tyagi, P., Singh, A., & Singh, V. (Eds.), *Integrating AI-Driven Technologies Into Service Marketing* (pp. 419–438). IGI Global., DOI: 10.4018/979-8-3693-7122-0.ch020

Sharma, M., & Singh, A. (2024a). Enhancing Competitive Advantages Through Virtual Reality Technology in the Hotels of India. In Kumar, S., Talukder, M., & Pego, A. (Eds.), *Utilizing Smart Technology and AI in Hybrid Tourism and Hospitality* (pp. 243–256). IGI Global., DOI: 10.4018/979-8-3693-1978-9.ch011

Sharma, M., & Singh, A. (2024b). Embracing Technological Innovation: A Review of Hi-Tech Services in Hospitality Industry. *Evergreen, 11*(4), 2818–2830. DOI: 10.5109/7326926

Sharma, M., Singh, A., & Rana, S. (2024). Empowering Hospitality Brands AI Strategies for Enhanced Customer Experiences in Emerging Markets. In Nadda, V., Tyagi, P., Singh, A., & Singh, V. (Eds.), *Integrating AI-Driven Technologies Into Service Marketing* (pp. 173–190). IGI Global., DOI: 10.4018/979-8-3693-7122-0.ch010

Sharma, R., Sachdeva, T., Singh, A., & Nadda, V. (2024b). A New Era of Engagement and Satisfaction: Transforming Customer Experience With AI-Driven Technologies. In Nadda, V., Tyagi, P., Singh, A., & Singh, V. (Eds.), *AI Innovations in Service and Tourism Marketing* (pp. 1–16). IGI Global., DOI: 10.4018/979-8-3693-7909-7.ch001

Sharma, R., & Singh, A. (2024c). Blockchain Technologies and Call for an Open Financial System: Decentralised Finance. In Vardari, L., & Qabrati, I. (Eds.), *Decentralized Finance and Tokenization in FinTech* (pp. 21–32). IGI Global., DOI: 10.4018/979-8-3693-3346-4.ch002

Singh, A. (2024). Virtual Research Collaboration and Technology Application: Drivers, Motivations, and Constraints. In Chakraborty, S. (Ed.), *Challenges of Globalization and Inclusivity in Academic Research* (pp. 250–258). IGI Global., DOI: 10.4018/979-8-3693-1371-8.ch016

Singh, A., & Bathla, G. (2023). Fostering Creativity and Innovation: Tourism and Hospitality Perspective. In P. Tyagi, V. Nadda, V. Bharti, & E. Kemer (Eds.), Embracing Business Sustainability through Innovation and Creativity in the Service Sector (pp. 70-83). IGI Global.

Singh, A., & Hassan, S. C. (2024). Service Innovation Through Blockchain Technology in the Tourism and Hospitality Industry: Applications, Trends, and Benefits. In Singh, S. (Ed.), *Service Innovations in Tourism: Metaverse, Immersive Technologies, and Digital Twin* (pp. 205–214). IGI Global., DOI: 10.4018/979-8-3693-1103-5.ch010

Singh, A., & Kumar, V. (2025a). Organizational Innovation and Adaptation in Hospitality and Tourism: Strategies for Resilience and Leadership. In C. Ramos, T. Costa, R. Madeira, & J. Ferreira do Rosário (Eds.), Open Innovation and Technology in Tourism and Hospitality (pp. 355-372). IGI Global Scientific Publishing. DOI: 10.4018/979-8-3693-8633-0.ch018

Singh, A., & Kumar, V. (2025b). User-Centered Hospitality: UI/UX as the New Frontier in Hotel and Restaurant Innovation. In C. Ramos, T. Costa, R. Madeira, & J. Ferreira do Rosário (Eds.), Open Innovation and Technology in Tourism and Hospitality (pp. 395-410). IGI Global Scientific Publishing. https://doi.org/DOI: 10.4018/979-8-3693-8633-0.ch020

Singh, V., & Singh, A. (2024). Revolutionizing the Hospitality Industry: How chatGPT Empowers Future Hoteliers. In Bansal, R., Ngah, A., Chakir, A., & Pruthi, N. (Eds.), *Leveraging ChatGPT and Artificial Intelligence for Effective Customer Engagement* (pp. 192–203). IGI Global., DOI: 10.4018/979-8-3693-0815-8.ch011

Srivastava, A., Singh, B. M., Singh, A., & Singh, A. (2025). Customer Immersiveness and Customer Retention: A Metaverse-Driven Study for Brands. In M. Al Aqad, A. SorayyaeiAzar, A. Albattat, & A. Singh (Eds.), Navigating AI and the Metaverse in Scientific Research (pp. 475-500). IGI Global Scientific Publishing. DOI: 10.4018/979-8-3373-0340-6.ch023

Chapter 11
AI and Metaverse-Powered Immersive Hotel Experiences for Smart Tourism Transformation

Rupam Konar
https://orcid.org/0000-0002-3235-3842
Taylor's University, Malaysia

Chen Weimei
https://orcid.org/0009-0007-8563-0587
Taylor's University, Malaysia

ABSTRACT

This chapter discusses how artificial intelligence (AI) and Metaverse technologies are transforming smart tourism, especially virtual hotel tours and personalized room previews. It examines how these innovations change tourist experiences, planning, and engagement with hospitality services before travel. The chapter begins by explaining AI-driven personalization and virtual environments, then covers intelligent service design, data-driven hotel operations, and interactive marketing through virtual platforms. Using examples and analysis, it shows how digital tech enhances bookings, customer satisfaction, and destination appeal. It concludes with managerial insights and research directions on user experience, ethics, and cross-sector collaboration. This content benefits tourism scholars, industry professionals, and digital strategists interested in innovative travel services.

DOI: 10.4018/979-8-3373-6541-1.ch011

1. INTRODUCTION

With the rapid evolution of technology, artificial intelligence (AI) and Metaverse technologies are reshaping the way industries operate at an unprecedented rate, and the tourism industry is no exception (Basheer et al., 2024). The application of AI in the tourism industry has been widely validated and covers a wide range of dimensions from personalized recommendations to operational optimization (Doğan and Niyet, 2024; Dwivedi et al., 2024; Wang and Uysal, 2024). Specifically, AI can provide customized hotel recommendations and service solutions based on tourists' historical behaviors, preferences, and search paths (Tuo et al., 2024). Combined with technologies such as virtual assistants (VAs), augmented reality (AR), and virtual reality (VR), AI can also enhance the interactive experience and linguistic suitability, thus improving the overall quality of travel decisions and customer satisfaction (Ma, 2024). According to related research, the commercial application of AI has brought significant results for tourism enterprises, such as the booking conversion rate increased by about 30 percent (Samala et al., 2020).

Meanwhile, the Metaverse, as a comprehensive platform integrating VR, real-time interaction, and immersive environment, is also increasingly showing great potential in the application of tourism scenarios (Zhong et al., 2024). With VR technology, tourists can realize panoramic virtual hotel tours, including spatial structure, decorative styles, and window views, and even simulate specific behaviors such as rest and entertainment (Nair et al., 2023), which significantly reduces information asymmetry and decision-making uncertainty in the actual booking process.

Although AI and Metaverse have demonstrated many advantages in enhancing customer service and optimizing experience design, there are still key gaps in existing research. First, while academics have separately explored the efficacy of AI and the Metaverse in tourism (Basheer et al., 2022), there is a lack of in-depth exploration of how the two can synergize to provide systematic and sustainable immersive solutions (Jia et al., 2023; Laachach et al., 2024; Youssef et al., 2024). For example, how AI tour guides, behavior-driven recommender systems, and immersive technologies such as AR and VR can work together in the tourism decision-making process remains to be further analyzed (Ma, 2024). Second, systematic empirical research on the role of these technologies in improving overall tourist experience, satisfaction, and brand interaction is still lacking (García-Madurga & Grilló-Méndez, 2023; Sousa et al., 2024).

Furthermore, while the application of AI in customer experience management such as chatbots, personalized suggestions, and itinerary planning is expanding (Aljizawi, 2024; Avula and Sithole, 2024; Babu Reddy et al., 2024; Ivanova, 2019; Ray et al., 2024), the further integration of big data, ML, and immersive interactions to achieve a truly personalized service experience is still a core challenge for current research and practice (Kazak et al., 2020; Mishra et al., 2024).

Therefore, to cope with the evolving needs of smart tourism, this paper explores the integrated application of AI and Metaverse technology in virtual hotel tours and personalized room previews, explores its comprehensive impact on improving user experience, enhancing the level of smart services, and enhancing the competitiveness of destinations, and proposes a forward-looking theoretical basis for promoting innovation and digital transformation in the tourism industry.

2. THEORETICAL FOUNDATION OF AI AND METAVERSE IN THE CONTEXT OF SMART TOURISM

2.1 Application of AI in Smart Tourism

2.1.1 Personalized Tourism Experience Empowerment

Yang et al. (2024) and Hong et al. (2023) highlighted the growing importance of personalized travel recommender systems in tourism marketing. Studies have shown that consumer perceptions of such systems are influenced by factors such as personalization, appearance, and message quality (Yang et al., 2024). Intelligent service robots utilizing deep learning can further enhance recommendation accuracy (Huang, 2022). Several studies have adopted personalized algorithms based on user behavior to recommend attractions and routes, integrating AR into virtual tours to promote user interaction and community engagement (Chen, 2024; Kesorn et al., 2017; Rane et al., 2023; Sharma, 2024; Torres-Ruiz et al., 2020; Yang et al., 2024).

Moreover, rich social media content can be leveraged through multimodal data mining to improve recommendation quality (Liu, 2023). Personalized travel recommendation systems typically prioritize user interests while considering constraints such as time and cost (Kumar et al., 2025). For instance, preference-based systems employ latent dirichlet allocation (LDA) and Jaccard coefficients to refine user profiles and combine A* algorithms with heuristic shortest path methods to optimize travel routes (Kanimozhi et al., 2021). Research has also focused on developing more intuitive user interfaces using natural language processing (NLP) techniques (Jang and Park, 2022; Mishra et al., 2024; Shelke et al., 2024). Huang (2022) proposed a personalized travel route recommendation model that employs deep learning and a self-attention mechanism to capture user preferences and generate customized travel route suggestions.

2.1.2 Predictive Analytics and Dynamic Pricing Strategies

Hopper is a company that focuses on AI technology, making travel price predictions by automatically recommending the best time and place to travel through advanced ML algorithms, with airfare prediction accuracy reaching up to 95 percent (Saurabh, 2024; Thea, 2018). This high level of accuracy is largely attributed to the ability of ML models to analyze and learn from large volumes of historical data to identify complex patterns in price fluctuations (Choudhary et al., 2023). Hopper's system not only predicts airfares but also assists travelers in determining the optimal time to purchase tickets by continuously monitoring and notifying users about price changes (Saurabh, 2024; Thea, 2018). Relevant data indicate that each customer can save an average of approximately 50 dollars per ticket (Choudhary et al., 2023).

This dynamic pricing strategy not only helps travelers reduce their expenses but also enables airlines and hotels to enhance their operational revenues. For instance, hotels can raise room rates during peak seasons to maximize income and lower them during off-seasons to attract more guests, based on predictive analytics (Choudhary et al., 2023). On average, tourism enterprises adopting AI-driven predictive analytics have seen a 15 percent increase in operating revenue (Thea, 2018). These findings suggest that the use of predictive analytics and dynamic pricing can effectively respond to market demand while improving the operational efficiency of tourism-related businesses (Choudhary et al., 2023). Empirical research has demonstrated that fare trends predicted by ML models can significantly reduce travel costs for consumers by identifying optimal purchasing times. The implementation of such technological advancements not only improves the consumer experience but also reinforces the competitive advantage of travel companies (Kalampokas et al., 2023).

2.2 Core features of the Metaverse and its integration with tourism scenarios

The Metaverse is a virtual digital environment that integrates advanced technologies including VR, AR, AI, and blockchain, offering an open, immersive, and interactive three-dimensional space that transcends the temporal and spatial constraints of the physical world (Xiaoxin et al., 2026; Wang et al., 2023). Originating in the 1980s as a conceptual framework (Evans et al., 2022), the Metaverse has evolved into practical applications due to the rapid advancement and convergence of core technologies such as VR, AR, mixed reality, blockchain, and AI (Fan et al., 2023).

This cross-technology integrated virtual ecosystem offers highly realistic environments for social, entertainment, and commercial interactions and presents transformative opportunities for service innovation and experiential redesign in the tourism sector. Within the framework of smart tourism, the Metaverse enables immersive virtual travel experiences, allowing users to explore museums, heritage sites, and natural attractions remotely through VR and AR technologies (Aktaş et al., 2023). Moreover, personalized recommendation systems can be developed on Metaverse platforms using user preferences and historical behavior data to deliver tailored travel services (Park and Kim, 2022). Tourism enterprises can also leverage the Metaverse to create virtual tourism products and scenarios, expand service boundaries and market reach, and enhance tourists' brand perception and engagement (Zhou and Yan, 2023).

2.2.1 Virtual World Properties in the Metaverse

The Metaverse is a digital realm that integrates VR and AR technologies, which have profoundly transformed the tourism industry (Basheer et al., 2025; Buhalis et al., 2023). It enables users to engage in immersive virtual tourism experiences anytime and anywhere, eliminating constraints related to physical location and fixed schedules (Aktaş et al., 2023; Zhou and Yan, 2023). This flexibility facilitates on-demand travel participation and expands access to a wide range of experiential scenarios. Beyond conventional sightseeing, Metaverse tourism includes exploratory journeys through ancient ruins, participation in themed activities, and simulations of interstellar travel, offering highly diversified and novel experiences that cater to evolving consumer demands for uniqueness and innovation (Damjanov and Crouch, 2019).

VR applications that simulate space environments without physical displacement exemplify a paradigm shift in tourism production and practice, influencing perceptions of future space exploration (Fan et al., 2023). In addition, the integration of Metaverse technologies into cultural tourism presents substantial opportunities. Users can engage with heritage content through immersive and interactive digital experiences, which not only enrich the tourism experience but also contribute to the digitization and modernization of the cultural tourism sector (Tiwari, 2021). This transformation supports the diversification of cultural offerings and enhances sectoral development through technological innovation.

2.2.2 Metaverse in Hotel Presentation and Marketing

With the rapid advancement of VR technologies, the concept of Metaverse hotels is fundamentally reshaping how the hospitality industry presents, markets, and delivers its services. Through the construction of three-dimensional (3D) immersive environments, hotels can offer potential guests interactive virtual experiences that surpass traditional two-dimensional visuals. Utilizing VR headsets, users can virtually explore hotel spaces prior to booking, gaining a detailed understanding of room layouts, interior design, amenities, and overall ambiance. Simulated real-life scenarios—such as resting, watching media, or using facilities—further support sensory-based evaluations, thereby reducing perceived risk and enhancing booking confidence (McLean & Barhorst, 2022). These immersive pre-stay experiences have been shown to increase customer trust and conversion rates.

Moreover, the Metaverse facilitates hyper-personalized marketing strategies, enabling hotels to co-create tailored virtual environments aligned with user preferences. Within these platforms, users can customize room themes, background music, and lighting to reflect their aesthetic tastes and emotional states (Chen, 2023). The Metaverse also supports a broad range of interactive digital events—such as virtual concerts, digital art exhibitions, and metaverse-based retail activities—which extend the service scope and enrich the brand experience.

Practically, numerous global hotel groups have begun prototyping virtual properties in the Metaverse, reporting encouraging market responses. Notably, the proposed "space hotel" concept is considered a radical manifestation of Metaverse hospitality, simulating extraterrestrial lodging environments to create novel, immersive experiences. Similarly, the idea of the "Quantum Hilbert Hotel," inspired by quantum theory and proposing a theoretically infinite virtual accommodation system, offers speculative but insightful guidance for the spatial and algorithmic design of future Metaverse hotels.

Empirical studies affirm that immersive virtual displays significantly enhance consumers' understanding of hotel offerings, increase booking intentions, and foster loyalty (Demir, 2022). Consequently, the Metaverse is not only reframing hotel marketing as a technologically mediated platform but also reconfiguring the customer journey by foregrounding experiential engagement, strengthening brand awareness, and improving decision-making efficiency. This underscores the transformative potential of immersive displays and interactions in the context of smart tourism.

3. INTELLIGENT SERVICE APPLICATION IN VIRTUAL TOUR AND ROOM PREVIEW

3.1 Personalized Recommendation Mechanism Based on AI

3.1.1 User Behavior-Driven Content Recommendation

The widespread application of AI in personalized recommendation systems enables travel platforms to generate more targeted travel recommendations based on user behavior data. By analyzing users' web browsing history, search keywords, click frequency, and booking history, AI systems can accurately identify users' interests and potential needs (Bhuiyan, 2024; Karthiyayini & Anandhi, 2024; Londhe et al., 2024; Yang et al., 2024). For example, when a user frequently searches for "beach vacation" related content, the system can automatically recognize his/her preference and push virtual hotels located in the coastal area with immersive room previews (Konar et al., 2024).

Such behavior driven recommendation mechanisms not only improve the matching efficiency between users and content but also enhance users' interactive experience and satisfaction during the virtual tour. Empirical studies have shown that personalized recommendations based on user behavioral analysis can increase booking conversions by approximately 20% (Konar et al., 2024), demonstrating their effectiveness in driving marketing conversions and enhancing user experience.

3.1.2 Intelligent Customer Service and Differentiated Service Design

AI chatbots, as a key interaction interface in smart tourism services, rely on NLP and ML technologies to understand and respond to users' natural language queries, thereby offering instant and personalized interactive services (Rane et al., 2024). These systems are commonly deployed on travel platforms and hotel booking websites to enable two-way interactions with users via voice or text channels, addressing frequently asked questions and proactively delivering recommendations based on user needs (Athikkal and Jenq, 2022).

During virtual tours and room previews, users often encounter complex informational choices and diverse service scenarios. AI chatbots can provide real-time assistance throughout the journey, offering explanatory support and operational guidance for tasks such as selecting room types, learning about amenities, comparing prices, or customizing room designs (Mei et al., 2024; Dobbala and Lingolu, 2024). Particularly when users face excessive options or lack clear preferences, AI systems can propose feasible decision paths through inference mechanisms and behavioral prediction algorithms (Nica et al., 2018), thereby reducing decision fatigue and minimizing user churn risk.

In addition, based on the recommendation capabilities of ML models, AI chatbots can identify user interests and behavioral patterns to suggest relevant local attractions, dining options, and leisure activities aligned with user preferences (Yan et al., 2025b; Bhuiyan, 2024). For instance, the system can generate customized city itineraries based on search behavior and historical data or guide users to participate in virtual interactive activities within the hotel environment. Through such integrated services, AI systems not only enhance the variety and depth of user experience but also increase user engagement and loyalty to the platform.

Multiple studies have confirmed the effectiveness of AI chatbots in enhancing service efficiency, optimizing interaction flows, and boosting conversion rates (Avula and Sithole, 2024; Dwivedi et al., 2024; Kannan, 2024; Ray et al., 2024).

For example, Athikkal and Jenq (2022) developed a voice interaction system that facilitates efficient communication between hotels and guests, while Car et al. (2024) introduced AI-driven personalized recommendations and intelligent marketing strategies on online travel agency (OTA) platforms, significantly improving booking conversions.

3.2 Application of Intelligent Technology in Hotel Operation and Service Management

3.2.1 Data-driven operation optimization

Under the development framework of smart tourism, AI is becoming the core driving force in promoting the digital and intelligent transformation of hotel management systems. Its application in operations management spans from data processing and process automation to intelligent decision support, significantly enhancing operational efficiency and service responsiveness in the hotel sector (Xiaoxin et al., 2025; Shirzad and Rahmani, 2024).

AI can integrate diverse data sources such as invoices, orders, sensory devices, and customer behaviors, and employ algorithms to automatically perform data cleansing and categorization, thereby improving the real-time accuracy of management information systems. For example, Ctrip's hotel AI system can monitor room status in real time and track customer behavior across the entire service cycle, from check-in to check-out, enabling managers to dynamically optimize room allocation and service strategies (Yan et al., 2024; Yan et al., 2025a; Gajić et al., 2024). On this basis, AI systems have been widely adopted to automate check-in and check-out procedures. Self-service devices equipped with facial recognition and biometric technologies help reduce waiting times and enhance check-in convenience (Kiptanui et al., 2024; Lee et al., 2024; Thenahandi, 2023).

Moreover, by dynamically processing customer preferences and booking data, AI can complete room assignments and service preparations in advance, thus elevating personalized service delivery and customer satisfaction (Acharya and Mahapatra, 2024; Al-Hyari et al., 2023). At the revenue management level, AI systems incorporate internal and external variables including room configuration, market demand, traffic conditions, weather patterns, and local events into predictive models to generate precise pricing strategies and inventory control plans (Ivanov, 2023; Rekiek et al., 2024). Empirical studies suggest that the implementation of AI-based revenue optimization systems can increase average hotel revenue by up to 15 percent (Imad, 2019).

Furthermore, the deep integration of AI and IoT is driving hotel resource management toward intelligence and sustainability. Real-time regulation of lighting, air conditioning, and energy consumption through sensing technologies not only lowers operational costs but also supports the goals of sustainable tourism, such as energy conservation and meeting green certification standards (Gajić et al., 2024; Milton, 2024). AI-enabled operational systems not only enhance hotel competitiveness in the digital landscape but also provide a robust infrastructure for immersive travel experience development (Kannan, 2024; M'hamed and Idrissi, 2024).

3.2.2 Response Mechanisms in Interruption Scenarios

In the smart tourism system, the continuity and stability of tourism services are increasingly emphasized, and AI technology provides unprecedented support for hotels' emergency response and risk control in unexpected events (Mohanty et al., 2025). Especially in the case of natural disasters, equipment failures, loss of luggage, or service interruptions, AI technology greatly improves the operational resilience and service recovery of hotels through real-time early warning, resource deployment, and process adaptive capabilities.

The intelligent early warning function of AI can predict disasters and deploy response strategies in advance based on big data analysis. For example, hotels can receive upcoming severe weather forecasts, such as storms and heavy rains, through the AI system, and accordingly trigger emergency operation processes, including adjusting check-in plans, evacuation route planning, and customer information notification (Ma, 2024; Mohanty et al., 2025). This mechanism ensures that hotels are adequately prepared to respond to disasters before they occur, thereby reducing the negative impact on the visitor experience.

The integration of AI with the IoT is equally powerful in disruption scenarios. In the case of lost luggage, for example, a smart luggage tracking system that integrates Global Positioning System (GPS) modules with devices such as Arduino can monitor the location of luggage in real time and accurately label it on a map (Jain et al., 2023). When unexpected events lead to delayed or misdirected baggage, the AI system can automatically identify the problematic links and send status updates to the back office of the tourists and hotels through the visualization interface, which dramatically shortens the problem resolution cycle.

In the case of service disruption or abnormal customer paths, the AI-based intelligent navigation system can automatically adjust the service process. For example, a robot or a guided tour system that recognizes a service failure or abnormal customer behavior can immediately adjust the recommended path, initiate an alternative service, or direct to human assistance, ensuring minimal disruption to the customer (Asmoro et al., 2023; Booranakittipinyo et al., 2024; Soori et al., 2023).

The value of AI in risk assessment and decision support is also increasingly evident in higher-level management perspectives. Through ML algorithms, AI can integrate information from multiple channels, such as customer feedback, complaint data, and platform behavioral records, to dynamically analyze the stability of service systems and user satisfaction (Qi et al., 2024). For example, when detecting frequent customer dissatisfaction or operational interruptions triggered by a specific service module, the AI system can proactively suggest improvements or automatically optimize process parameters.

In addition, AI systems can assist managers in identifying potential operational risks and formulating intervention strategies to improve resource scheduling efficiency and revenue management capabilities (Imad, 2019). By integrating intelligent warning, automatic response, and risk identification mechanisms, AI technology not only improves the hotel's resilience in unexpected situations but also provides key technical support for realizing resilient management of service systems (Ramachandran et al., 2023; Zhang, 2022).

4. IMMERSIVE TOURISM MARKETING EXPERIENCE EMPOWERED BY THE METAVERSE

4.1 Immersive Construction and Interaction Design of Virtual Hotel

4.1.1 Experience Mechanism of Virtual Hotel

The in-depth application of Metaverse technology in the tourism industry has facilitated the development of virtual hotel immersive experiences, emerging as a key approach to enhancing customers' perceived value and strengthening marketing effectiveness (Sun et al., 2025). By constructing highly realistic virtual environments, this technology allows tourists to perceive the spatial atmosphere of various destinations and hotels without physical presence, thereby forming clearer psychological expectations and perceptual judgments during the booking process (Koohang et al., 2023). The combination of immersion and interactivity enables users to explore different hotel options prior to booking, which increases their engagement and decision-making confidence (Koohang et al., 2023).

In practice, VR and AR technologies are widely applied to create interactive virtual tourism experiences (Syarul, 2023). Techniques such as 360-degree video tours, 3D spatial modeling, and real-time navigation enable tourists to explore hotel layouts, interior design, and amenities from a first-person perspective. Compared with traditional text and image displays, these technologies provide a more engaging and informative visual experience (Syarul, 2023). Research indicates that such immersive virtual tours can significantly enhance users' interest and positive attitudes toward hotel brands, thereby promoting booking intentions (Almeida et al., 2024).

The commercial application of Metaverse-based hospitality has begun to show promising results, exemplified by M Social Decentral, the world's first Metaverse hotel (Emily, 2022). This project offers a roamable and interactive hotel experience within a virtual environment, strengthening the perceived connection between users and hotel brands. Through virtual identity registration, avatar-based exploration, and participation in digital activities, users can experience a comprehensive virtual representation of hotel services (Almeida et al., 2024).

Notably, the immersive characteristics of Metaverse hotels are gradually being extended to the design and operation of physical spaces. For instance, at the M Social Hotel in Singapore, Philippe Starck's open kitchen integrates an artistic installation composed of 28 LCD screens to produce dynamic lighting effects, blending technology and art (Tian and Song, 2023). Similarly, the M Social Hotel in Auckland incorporates New Zealand's nautical culture, featuring a boat-shaped chandelier and a projection wall showcasing historical and cultural narratives, creating an immersive and story-rich environment (Dewi Andriana and Damayanti, 2023; Millennium, 2024).

The Horigin platform, developed by Haitian Hotel Management Co., Ltd., represents an emerging approach to immersive hotel experience design in the Metaverse. This platform provides customized virtual environments specifically designed for small and medium-sized boutique hotels (Ocean, 2024a). For example, Perhui Hotel, as a pilot project, grants guests access to Horigin upon check-in. Guests can explore five distinct virtual spaces, including the lobby, fitness center, meeting room, restaurant, and rooftop garden. These virtual spaces replicate the physical layout of the hotel and enable interactive exploration, such as browsing hotel information, engaging with other users, and participating in virtual wellness activities. Gamified features also allow guests to earn HoriCoin, a virtual currency redeemable for tangible benefits such as room upgrades or extended check-out services (Ocean, 2024b).

This type of immersive environment reflects a broader trend in hotel design that integrates virtual and physical elements to enhance guest experience. The use of mixed reality technologies enables hotels to extend experiential offerings beyond physical limitations, fostering emotional engagement, personalization, and brand differentiation. These mechanisms not only enhance the perceived service value but also underscore the growing importance of virtual engagement in the future of hospitality.

4.1.2 Personalized Interaction and Customer Engagement

Metaverse technology is transforming the way customers interact with hotels by emphasizing the delivery of highly personalized and immersive service experiences that promote deeper customer engagement and brand loyalty. As an early innovator in this field, international hotel group Marriott was among the first hospitality brands to integrate non-fungible tokens (NFTs) with the Metaverse through its Marriott Bonvoy campaign. This loyalty initiative combined digital art with immersive storytelling, enabling guests to engage with the brand in virtual environments that reflect the convergence of hospitality, technology, and next-generation customer experiences. By offering users exclusive digital artworks and virtual identities, the program expanded the expression of brand value and introduced a new approach for customer relationship building in the hotel industry (James, 2021; Ledger, 2021).

Within Metaverse environments, hotel brands can design personalized service scenarios based on user behavior and preferences. Guests can customize room decor, background music, lighting color and brightness, and even the virtual window view to create spaces that align with their aesthetics and needs (Filippova, 2023). This level of scenario customization enhances emotional connection and satisfaction while contributing to brand differentiation and increased customer loyalty.

Additionally, Metaverse platforms offer VR tours that provide users with a comprehensive understanding of room structures, amenities, and spatial layouts before booking (Xiaoxin et al., 2026; Solakis et al., 2024). For example, guests can simulate the experience of visiting fitness centers, spa facilities, or restaurants, gaining concrete spatial awareness and reducing decision-making uncertainty caused by information asymmetry.

Hotels also integrate a variety of interactive entertainment offerings into virtual platforms, including virtual concerts, immersive art exhibitions, e-commerce events, and social spaces, thereby expanding the channels of user engagement within Metaverse contexts (Cynthia and Kiruthiga, 2024). These services not only enhance the entertainment dimension of the guest experience but also contribute to a cross-temporal and cross-cultural interactive ecosystem, enriching the diversity and innovation of hotel services.

4.2 The Role of Metaverse in Tourism Marketing and Planning

4.2.1 The Impact of Virtual Scene Visualization on Decision Making

Metaverse introduces an innovative path centered on virtual scene visualization, which provides travelers with an immersive and interactive environmental experience before making decisions. With the help of VR and AR technologies, tourists can "walk into" the hotel scene before booking to gain a deeper understanding of its spatial layout, décor, amenities, and service details, which can significantly reduce information asymmetry and decision-making uncertainty (Koohang et al., 2023).

Specifically, users can freely navigate the room layout, furniture configuration, and landscape view in the virtual environment, and even simulate daily activities in the room, such as resting, watching movies, or using the bathroom facilities. This highly realistic interactive experience breaks through the limitations of traditional promotional methods and significantly improves the perceived quality and trust of tourists in hotel products (Koohang et al., 2023).

Studies have shown that Metaverse-based virtual venue presentations have a significant advantage in increasing customer booking conversion rates by an average of about 30% compared to traditional static images or text descriptions (Almeida et al., 2024; Koohang et al., 2023). This visual comparison mechanism enables customers to compare different hotel options more clearly and thus make choices that are more in line with individual needs (Koohang et al., 2023).

From a marketing perspective, the Metaverse is not only a cutting-edge medium for showcasing hotel products but also becomes a new platform for brand communication and distribution strategies. On the one hand, hotel companies can use virtual visualization scenarios to enhance brand impressions and users' emotional connection, and improve online marketing effects; on the other hand, the Metaverse itself can be used as an incremental sales channel for tourism products, creating new revenue sources by providing immersive virtual experiences (e.g., virtual tours, immersive activities, etc.) and indirectly promoting the growth of physical hotel bookings (Koohang et al., 2023; Pratisto et al., 2022; Zaman et al., 2025).

4.2.2 Contextualized Design of Pre-Trip Experiences and Brand Interaction

The introduction of Metaverse technologies has enabled travelers to engage in more immersive and interactive contextualized experiences prior to their trips, thereby reshaping the relationship between destination brands and tourists. Within the Metaverse, tourists can explore natural landscapes, cultural heritage, and historical narratives of destinations through virtual experiences before visiting in person. These pre-experiences facilitate deeper cognitive understanding and emotional attachment, enhancing psychological engagement with the destination. In addition to improving satisfaction during the physical visit, such virtual engagements contribute to the formation of a more memorable and emotionally resonant destination image, thereby reinforcing brand attachment and recall (Aktaş et al., 2023; Almeida et al., 2024).

Moreover, the integration of Metaverse technologies supports personalized travel planning, allowing tourists to customize itineraries and simulate activities based on individual interests and preferences. Supported by AI and user profiling systems, tourism enterprises and hotel operators can deliver tailored content and services that align with specific user expectations. This level of personalization increases the relevance and value of pre-travel engagement, enhances customer satisfaction, and strengthens brand loyalty in competitive tourism markets (Damaševičius, 2023; Kouroupi and Metaxas, 2023).

In addition, the Metaverse expands the social dimension of tourism by offering interactive platforms where users can communicate, share experiences, and build virtual communities. This form of virtual socialization not only enhances the enjoyment and participation in tourism activities but also improves brand visibility and social recognition (Damaševičius, 2023).

Importantly, the Metaverse also presents a novel pathway for promoting sustainable tourism. By reducing the frequency of physical travel and the consumption of on-site resources, it alleviates environmental pressure and supports the development of green and low-carbon tourism models (Kouroupi and Metaxas, 2023). Therefore, the Metaverse redefines the contextualized pre-travel experience while offering transformative opportunities in destination branding, user co-creation, and sustainable tourism development.

5. CONCLUSION

5.1 Key Findings and Practical Implications

This study reveals the synergy between AI and Metaverse technologies in virtual hotel tours and personalized room preview scenarios, demonstrating a significant impact on service innovation and the digital transformation of smart tourism (Koohang et al., 2023; Mishra et al., 2024; Konar et al., 2024). AI enhances personalized service delivery through precise data analysis and NLP technologies, providing tourists with behavior-based recommendations that improve service responsiveness and customer satisfaction. For example, recommendation mechanisms based on user search and booking history have been shown to increase hotel booking conversion rates by an average of 20% (Konar et al., 2024). Intelligent customer service systems leveraging semantic recognition and ML technologies support instant interactions in booking, consultation, and complaints, significantly enhancing user experience (Avula and Sithole, 2024; Dobbala and Lingolu, 2024). In hotel operations, AI has also improved efficiency and financial performance through automated check-in and check-out systems, dynamic pricing, and optimized revenue management (Rekiek et al., 2024; Xiaoxin et al., 2025), with some hotels achieving up to 15 percent revenue growth from AI-based revenue optimization systems (Imad, 2019). Furthermore, AI improves risk identification and emergency responsiveness through data-driven early warning systems and IoT integration, thereby enhancing organizational resilience and guest safety (Mohanty et al., 2025; Qi et al., 2024; Ramachandran et al., 2023).

Simultaneously, Metaverse technologies offer highly immersive three-dimensional virtual experiences that transcend the physical limitations of tourism marketing and consumption (Almeida et al., 2024; Xiaoxin et al., 2026). Through virtual scene visualization and 3D interactive tours, tourists can gain detailed insights into hotel layouts, decor, and amenities prior to booking, significantly reducing decision uncertainty and improving booking conversion rates by approximately 30 percent compared to static visual presentations (Koohang et al., 2023; Tul-Krzyszczuk, 2023). Benchmark cases such as M Social Decentral and Marriott Bonvoy illustrate how Metaverse hotels combine immersive visualization with interactive content, including NFT art, customizable room configurations, and virtual concerts, to create personalized experiential paths for guests (Cynthia and Kiruthiga, 2024). Additionally, the Metaverse enables travelers to explore historical heritage and natural landscapes virtually during the pre-trip stage, enhancing emotional resonance and increasing travel intentions (Aktaş et al., 2023; Kouroupi and Metaxas, 2023). This multi-dimensional scenario design not only enhances engagement during the tourist journey but also offers opportunities for brand communication, customer loyalty building, and secondary marketing (Damaševičius, 2023; Filippova, 2023).

From a managerial perspective, this study proposes four strategic insights. First, tourism enterprises should systematically integrate AI and Metaverse technologies into service design and operational management to achieve dual goals of service personalization and operational intelligence (Sousa et al., 2024). This strategy enhances intelligent responsiveness at customer touchpoints while improving service chain automation and efficiency. Second, in customer relationship management, firms should leverage the communicative value of virtual tours, immersive experiences, and digital content such as NFTs to strengthen brand narratives and emotional engagement throughout the customer journey (Ledger, 2021). Third, the Metaverse should be positioned not only as a marketing tool but also as a long-term interface for customer interaction, supporting customer acquisition, retention, and re-engagement by extending the value of customer lifecycle management. Finally, in deploying such technologies, enterprises must address data ethics and platform governance, with particular focus on privacy protection, algorithmic transparency, and system reliability. Establishing a trust-based technological framework is essential for fostering a secure, dependable, and sustainable smart tourism ecosystem (Gajić et al., 2024; Zhou and Yan, 2023).

5.2 Future Research Directions

With the continuous progress of science and technology, the application of AI and Metaverse technology in the tourism industry is promising. Future research can further explore the direction of the integration of the two and their sustained impact on the tourism industry from the following aspects.

5.2.1 Evolutionary Path of Intelligent Technology

In the continuous evolution of smart tourism, technological innovation serves as the primary catalyst for transforming immersive experiences and services. Within the Metaverse context, the integration of AR and VR technologies is progressively overcoming the spatial-temporal constraints of traditional tourism, offering more immersive and interactive experience pathways. AR superimposes virtual layers onto real-world environments, enabling features such as 3D reconstructions of historical architecture and virtual tour guides that provide real-time interpretation, thereby enhancing contextual immersion and the educational value of cultural experiences. Concurrently, VR supports the creation of highly realistic and reproducible virtual environments, allowing users to explore historically or culturally significant scenarios, which strengthens their sense of presence, engagement, and emotional resonance. The incorporation of MR further extends the perceptual and cognitive boundaries of tourism by enriching spatial storytelling and enabling multi-sensory interactions, thus increasing the symbolic and commercial value of destinations.

The advancement of AI also plays a pivotal role in deepening personalized tourism services. Leveraging ML and deep learning algorithms, AI systems can accurately identify and anticipate user preferences, dynamically generating tailored recommendations. Reinforcement learning (RL) mechanisms further enhance these systems by continuously optimizing recommendation strategies based on behavioral feedback, fostering seamless coordination among intelligent customer service, itinerary planning, and real-time responsiveness. This data-driven service logic significantly boosts user satisfaction and supports the construction of a closed-loop smart tourism experience.

Additionally, the application of digital twin (DT) technology in the tourism sector is gaining prominence. By generating high-fidelity virtual replicas of destinations and hotel infrastructures, DT enables real-time monitoring, predictive maintenance, and improved resource allocation efficiency. For tourists, DT facilitates pre-trip familiarization with destination environments and spatial layouts, while offering simulated decision-making scenarios that improve planning efficiency and reduce uncertainty. Overall, the deep convergence of AR/VR, AI, and DT technologies is reshaping the service logic and experience architecture of tourism, steering smart tourism systems toward greater intelligence, immersion, and sustainability.

5.2.2 Optimization Trend of User Experience Design

The development of Metaverse tourism is increasingly advancing toward the synergistic integration of multi-sensory interactive experiences and deeply personalized services, signifying a new phase in tourism experience design characterized by heightened immersion and emotional engagement. Beyond traditional visual and auditory stimuli, the incorporation of multi-dimensional sensory technologies such as olfactory and tactile perception is expected to expand the scope of user sensory participation and enhance the realism and presence of virtual tourism environments. Through specialized equipment that simulates wind, water, terrain undulations, and emits spatially oriented scents such as forest or ocean aromas, the fidelity of natural environmental representation in virtual settings has been significantly improved. These sensory enhancements not only elevate the immersive quality of virtual experiences but also reduce the cognitive dissonance between real and virtual environments, thereby increasing the perceived value and emotional resonance of digital tourism.

Simultaneously, the enhancement of social interaction mechanisms has emerged as a critical trend in optimizing user experience. By establishing multi-user collaboration platforms and virtual social spaces, tourists can engage in real-time interactions with companions or other virtual travelers, participate in shared explorations, complete collaborative tasks, and join virtual social events. Such immersive and collective engagement mechanisms strengthen emotional bonds among users and facilitate cross-cultural exchanges and co-created tourism meanings (Konar et al., 2018). Furthermore, the integration of Metaverse tourism platforms with mainstream social media enables users to share travel experiences, upload multimedia content, and participate in virtual community interactions, thereby amplifying user engagement, content dissemination, and the sense of community belonging.

The evolution toward in-depth personalization within Metaverse tourism supports the construction of a user-centric experiential architecture. Tourists are empowered to customize environmental elements such as accommodation styles, lighting, and music according to personal preferences and exercise autonomy in itinerary planning, food selection, and activity arrangements. For instance, systems may offer dietary suggestions based on users' health data and eating habits or recommend themed experiences such as cultural immersion or adventure activities tailored to interest profiles. This service model, supported by behavioral data and semantic analysis, facilitates the delivery of personalized and differentiated tourism experiences aligned with individual user characteristics. By responding to users' implicit preferences and situational needs, such a model significantly enhances satisfaction, engagement, and brand loyalty across the entire tourism lifecycle.

5.2.3 Synergy and Governance in Smart Tourism Ecosystem

In the context of the accelerated integration of AI and Metaverse technologies, constructing a synergistic and symbiotic smart tourism ecosystem has become a critical task for advancing innovation and sustainable industry transformation. Cross-sector collaboration between tourism enterprises and technology firms has emerged as a key model for enabling technological empowerment. Tourism enterprises, with their accumulated destination resources, operational experience, and user insights, can form complementary partnerships with technology providers possessing core capabilities in fields such as VR and AI. For example, hotel groups can develop immersive virtual tour systems with the support of VR companies, allowing users to explore room types, facilities, and environments through 3D visualization before making bookings. Similarly, destination managers can implement intelligent tour systems based on semantic recognition and real-time recommendation mechanisms in cooperation with AI solution providers, enabling personalized service delivery and improved visitor satisfaction. Such cross-disciplinary integration enhances service quality while laying the groundwork for the diversification and scalability of tourism technology applications.

Nevertheless, the widespread commercialization of emerging technologies has raised growing concerns regarding system ethics and platform governance. In virtual tourism, data-driven marketing, and digital identity scenarios, critical issues such as user data privacy, algorithmic transparency, and the delineation of rights and responsibilities within virtual environments require regulatory attention. To protect user rights and foster trust, it is essential to establish comprehensive industry standards and compliance frameworks addressing virtual experience quality, data sovereignty, information security, and user behavior regulation. These frameworks are instrumental not only in mitigating platform risks and curbing unfair competition but also in providing institutional safeguards for the responsible deployment of technologies and supporting sustainable innovation in smart tourism within a credible and trustworthy environment.

Furthermore, technological infrastructure upgrades must be accompanied by the advancement of human capital. In a smart tourism ecosystem where intelligence and virtualization are increasingly converging, the demand for interdisciplinary talent is becoming more pressing. Higher education and vocational training systems in fields such as tourism management, information technology, and service design should enhance AI and XR (Extended Reality) curricula, equipping students with competencies in big data analytics, contextualized experience design, and platform governance. In parallel, enterprises should establish cross-sector talent development mechanisms through retraining programs and industry-academia partnerships to facilitate the integration of technological proficiency, customer insight, and service innovation. Such efforts are crucial for cultivating a future-oriented knowledge and skill framework suited to the evolving demands of smart tourism.

REFERENCES

Acharya, P., & Mahapatra, S. S. (2024). Exploring the Impact of Artificial Intelligence Integration on Guest Experience in the Hotel Industry. *Geo Journal of Tourism and Geosites*, *54*(2 supplement), 802–810. DOI: 10.30892/gtg.542spl04-1255

Aktaş, E., Gürsoy, I. T., & Kurgun, A. (2023). The future of metaverse in the tourism industry: Tool or purpose? *International Journal of Contemporary Tourism Research*, *7*(2), 143–152.

Al-Hyari, H. S., Al-Smadi, H. M., & Weshah, S. R. (2023). The Impact of Artificial Intelligence (AI) on Guest Satisfaction in Hotel Management: An Empirical Study of Luxury Hotels. *Geojournal of Tourism and Geosites, 48*(2spl), 810–819.

Aljizawi, J. (2024). *Personalized Travel Recommendations and Marketing Automation for Saudi Arabia: Harnessing AI for Enhanced User Experience and Business Growth.*

Almeida, G. G. F., Almeida, P., & Cardoso, L. (2024). Scope extension of the metaverse tourism concept: Proposing a research agenda. *Journal of Infrastructure. Policy and Development*, *8*(4), 3483.

Asmoro, A. Y., Butler, G., & Szili, G. (2023). Exploring the Current Status and Future Potential of Robot, Artificial Intelligence, and Service Automation in the Indonesian Tourism Industry. *Jurnal Kepariwisataan: Destinasi. Hospitalitas Dan Perjalanan*, *7*(2), 133–153.

Athikkal, S., & Jenq, J. (2022). Voice chatbot for Hospitality. *ArXiv Preprint ArXiv:2208.10926*. DOI: 10.5121/csit.2022.121315

Avula, M., & Sithole, T. (2024). Artificial Intelligence's Potential to Improve Operational Efficiency and Customer Experience in the Hospitality and Tourism Sectors. In *Impact of AI and Tech-Driven Solutions in Hospitality and Tourism* (pp. 351–382). IGI Global. DOI: 10.4018/979-8-3693-6755-1.ch018

Babu Reddy, S., Kathpalia, R., Sil, R., & Nag, A. (2024). Tourism Companion: Enhancing Travel Experiences with AI Chatbot and Soft Computing. In C. K. K. Reddy, T. Sithole, M. Ouaissa, Ö. ÖZER, & M. M. Hanafiah (Eds.), *Soft Computing in Industry 5.0 for Sustainability* (pp. 301–317). Springer Nature Switzerland.

Basheer, S., Farooq, S., & Reshi, M. A. (2022). Tourism, the Metaverse, Artificial Intelligence, and Travel: Striking a Balance between Innovation and Concerns. *Journal of Social Responsibility. Tourism and Hospitality*, *26*, 19–30.

Basheer, S., Walia, S., & Farooq, S. (2025). Metaverse Explosion and Its Consequences for the Travel Industry, AI, and VR. In Chhabra, G., & Kaushik, K. (Eds.), *Understanding the Metaverse: Applications, Challenges, and the Future* (pp. 307–322). Springer Nature Singapore. DOI: 10.1007/978-981-97-2278-5_14

Ben Youssef, A., Dutta, P. K., Doshi, R., & Sajnani, M. (2024). *AI, Blockchain, and Metaverse in Hospitality and Tourism Industry 4.0: Case Studies and Analysis*. CRC Press.

Bhuiyan, M. S. (2024). The role of AI-Enhanced personalization in customer experiences. *Journal of Computer Science and Technology Studies*, *6*(1), 162–169. DOI: 10.32996/jcsts.2024.6.1.17

Booranakittipinyo, A., Li, R. Y. M., & Phakdeephirot, N. (2024). Travelers' perception of smart airport facilities: An X (Twitter) sentiment analysis. *Journal of Air Transport Management*, *118*, 102600. DOI: 10.1016/j.jairtraman.2024.102600

Buhalis, D., Leung, D., & Lin, M. (2023). Metaverse as a disruptive technology revolutionising tourism management and marketing. *Tourism Management*, *97*, 104724. DOI: 10.1016/j.tourman.2023.104724

Car, T., Šuman, S., & Kliman, A. (2024). Customer Booking Habits and Attitudes Towards AI-Powered Features in Online Travel Agencies. *International Journal (Toronto, Ont.)*, *13*(9).

Chen, Z. (2023). Beyond reality: Examining the opportunities and challenges of cross-border integration between metaverse and hospitality industries. *Journal of Hospitality Marketing & Management*, *32*(7), 967–980. DOI: 10.1080/19368623.2023.2222029

Chen, Z. (2024). Beyond boundaries: Exploring the Metaverse in tourism. *International Journal of Contemporary Hospitality Management*. Advance online publication. DOI: 10.1108/IJCHM-06-2023-0900

Choudhary, A. K., Jagadeesh, R. P., Girija, E., Madhuri, M., & Shravani, N. (2023). Flyhigh:Machine Learning Based Airline Fare Prediction Model. *2023 6th International Conference on Information Systems and Computer Networks (ISCON)*, 1–8.

Cynthia, J. P. B., & Kiruthiga, V. (2024). Impression of Metaverse on Gen Z Customers in the Hotel Industry. *Shanlax International Journal of Management*, *12*(1), 24–30. DOI: 10.34293/management.v12i1.7780

Damaševičius, R. (2023). From E-commerce to V-commerce: Understanding the Impact of Virtual Reality and Metaverse on Economic Activities. *Journal of Information Economics*, *1*(3), 16. DOI: 10.58567/jie01030005

Damjanov, K., & Crouch, D. (2019). Virtual reality and space tourism. In *Space tourism* (Vol. 25, pp. 117–137). Emerald Publishing Limited. DOI: 10.1108/S1571-504320190000025007

Demir, Ç. (2022). Metaverse Teknolojisinin Otel Sektörünün Geleceğine Etkileri Üzerine Bir İnceleme (An Investigation into the Impact of Metaverse Technology on The Hotel Industry's Future). *Journal of Tourism & Gastronomy Studies, 10*(1), 542–555.

Dewi Andriana, S., & Damayanti, F. (2023). Desain Metaverse. *The Future of Tourism Pariban Berbasis Augmented Reality Menggunakan Metode Markerless User Defined Target., 16*(2), 172–182.

Dobbala, M. K., & Lingolu, M. S. S. (2024). Conversational AI and Chatbots: Enhancing User Experience on Websites. *American Journal of Computer Science and Technology, 10*(3), 62–70. DOI: 10.11648/j.ajcst.20240703.11

Doğan, S., & Niyet, İ. Z. (2024). Artificial Intelligence (AI) in Tourism. In *Future Tourism Trends Volume 2: Technology Advancement, Trends and Innovations for the Future in Tourism* (pp. 3–21). Emerald Publishing Limited.

Dwivedi, Y. K., Pandey, N., Currie, W., & Micu, A. (2024). Leveraging ChatGPT and other generative artificial intelligence (AI)-based applications in the hospitality and tourism industry: Practices, challenges and research agenda. *International Journal of Contemporary Hospitality Management, 36*(1), 1–12. DOI: 10.1108/IJCHM-05-2023-0686

Emily, S.-D. (2022, May 11). *Metaverse hospitality*. VML. https://www.vml.com/insight/metaverse-hospitality

Evans, L., Frith, J., & Saker, M. (2022). The roots of the metaverse. In *From Microverse to Metaverse: Modelling the Future through Today's Virtual Worlds* (pp. 15–24). Emerald Publishing Limited. DOI: 10.1108/978-1-80455-021-220221002

Fan, X., Wang, H., & Wang, L. (2023). P-2.8: Metaverse: Origin, Current Applications and Prospects for Future Development. *SID Symposium Digest of Technical Papers, 54*, 504–506. DOI: 10.1002/sdtp.16342

Filippova, O. Yu. (2023). International Hotel Service Industry Trends. *Visnyk of Donetsk National University of Economics and Trade Named after Mykhailo Tugan-Baranovsky, 2(79) 2023*, 63–68.

Gajić, T., Petrović, M. D., Pešić, A. M., Conić, M., & Gligorijević, N. (2024). Innovative Approaches in Hotel Management: Integrating Artificial Intelligence (AI) and the Internet of Things (IoT) to Enhance Operational Efficiency and Sustainability. *Sustainability (Basel)*, *16*(17), 7279. DOI: 10.3390/su16177279

García-Madurga, M.-Á., & Grilló-Méndez, A.-J. (2023). Artificial Intelligence in the tourism industry: An overview of reviews. *Administrative Sciences*, *13*(8), 172. DOI: 10.3390/admsci13080172

Hong, M., Chung, N., Koo, C., & Koh, S.-Y. (2023). TPEDTR: Temporal preference embedding-based deep tourism recommendation with card transaction data. *International Journal of Data Science and Analytics*, *16*(2), 147–162. DOI: 10.1007/s41060-022-00380-7

Huang, X. (2022). Personalized travel route recommendation model of intelligent service robot using deep learning in big data environment. *Journal of Robotics*, *2022*(1), 7778592. DOI: 10.1155/2022/7778592

Imad, A. S. (2019). Remodeling hospitality industry through artificial intelligence. *2019 13th International Conference on Software, Knowledge, Information Management and Applications (SKIMA)*, 1–7.

Ivanov, S. (2023). *Tourism and Hotel Technologies*. INSETE. https://books.google.com.my/books?id=hBy1EAAAQBAJ

Ivanova, M. (2019). Robots, Artificial Intelligence, and Service Automation in Travel Agencies and Tourist Information Centers. In S. Ivanov & C. Webster (Eds.), *Robots, Artificial Intelligence, and Service Automation in Travel, Tourism and Hospitality* (pp. 221–237). Emerald Publishing Limited.

Jain, S., Aithal, S., Allamsetty, V., K S, S., Gobi, Dr. N., & Rathi, Dr. R. (2023). Smart Luggage Tracking using IoT and GPS Technology. *IJARCCE, 12*(4).

James, B. (2021, December 4). *Marriott Bonvoy logs into the metaverse with debut of travel-inspired NFTs*. Marriott News Center. https://news.marriott.com/news/2021/12/04/marriott-bonvoy-logs-into-the-metaverse-with-debut-of-travel-inspired-nfts

Jang, Y., & Park, E. (2022). Satisfied or not: User experience of mobile augmented reality in using natural language processing techniques on review comments. *Virtual Reality (Waltham Cross)*, *26*(3), 839–848. DOI: 10.1007/s10055-021-00599-y

Jia, S., Chi, O. H., Martinez, S. D., & Lu, L. (2023). When "Old" Meets "New": Unlocking the Future of Innovative Technology Implementation in Heritage Tourism. *Journal of Hospitality & Tourism Research (Washington, D.C.)*, •••, 10963480231205768.

Kalampokas, T., Tziridis, K., Kalampokas, N., Nikolaou, A., Vrochidou, E., & Papakostas, G. A. (2023). A holistic approach on airfare price prediction using machine learning techniques. *IEEE Access : Practical Innovations, Open Solutions*, *11*, 46627–46643. DOI: 10.1109/ACCESS.2023.3274669

Kanimozhi, U., Ganapathy, S. D. M., & Kannan, A. (2021). *A User Preference Tree based Personalized Route Recommendation System for Constraint Tourism and Travel*. DOI: 10.21203/rs.3.rs-755856/v1

Kannan, R. (2024). Revolutionizing the Tourism Industry through Artificial Intelligence: A Comprehensive Review of AI Integration, Impact on Customer Experience, Operational Efficiency, and Future Trends. *International Journal for Multidimensional Research Perspectives*, *2*(2), 1–14. https://www.chandigarhphilosophers.com/index.php/ijmrp/article/view/115

Karthiyayini, J., & Anandhi, R. J. (2024). To Analyze the Various Machine Learning Algorithms That Can Effectively Process Large Volumes of Data and Extract Relevant Information for Personalized Travel Recommendations. *SN Computer Science*, *5*(4), 336. DOI: 10.1007/s42979-024-02667-x

Kazak, A. N., Chetyrbok, P. V., & Oleinikov, N. N. (2020). Artificial intelligence in the tourism sphere. *IOP Conference Series. Earth and Environmental Science*, *421*(4), 042020. DOI: 10.1088/1755-1315/421/4/042020

Kesorn, K., Juraphanthong, W., & Salaiwarakul, A. (2017). Personalized attraction recommendation system for tourists through check-in data. *IEEE Access : Practical Innovations, Open Solutions*, *5*, 26703–26721. DOI: 10.1109/ACCESS.2017.2778293

Kiptanui, M. J., Karei, R., & Makomere, J. (2024). Effect of Automated Check in on Service Delivery in Star-Rated Hotels in Uasin Gishu County, Kenya. *Journal of Frontiers in Humanities and Social Sciences*, *2*(3), 68–79.

Konar, R., Bhutia, L. D., Fuchs, K., & Balasubramanian, K. (2024). Role of Virtual Reality Technology in Sustainable Travel Behaviour and Engagement Among Millennials. In El Archi, Y., Benbba, B., Dávid, L., & Cardoso, L. (Eds.), *Promoting Responsible Tourism With Digital Platforms* (pp. 1–19). IGI Global Scientific Publishing. DOI: 10.4018/979-8-3693-3286-3.ch001

Koohang, A., Nord, J., Ooi, K., Tan, G., Al-Emran, M., Aw, E., Baabdullah, A., Buhalis, D., Cham, T., Dennis, C., Dutot, V., Dwivedi, Y. K., Hughes, L., Mogaji, E., Pandey, N., Phau, I., Raman, R., Sharma, A., Sigala, M., & Wong, L.-W. (2023). Shaping the metaverse into reality: A holistic multidisciplinary understanding of opportunities, challenges, and avenues for future investigation. *Journal of Computer Information Systems*, *63*(3), 735–765. DOI: 10.1080/08874417.2023.2165197

Kouroupi, N., & Metaxas, T. (2023). Can the Metaverse and Its Associated Digital Tools and Technologies Provide an Opportunity for Destinations to Address the Vulnerability of Overtourism? In *Tourism and Hospitality* (Vol. 4, Issue 2, pp. 355–373). DOI: 10.3390/tourhosp4020022

Kumar, A. V. S., Mosikan, M., Dutta, A., & Musirin, I. Bin, Akula, V. S. G., Khalid, S., Yalavarthi, S. B., DV, C. S., Suganya, R. V, & Dulhare, U. N. (2025). AI-Driven Personalization in Tourism: Balancing Innovation with Ethical Challenges. In *AI Technologies for Personalized and Sustainable Tourism* (pp. 31–58). IGI Global.

Laachach, A., Mumtaz, S., & Andaloussi, B. Z. (2024). Unveiling opportunities and challenges of the metaverse in the tourism and the hospitality sector: A qualitative investigation. *Tourism and Hospitality Management*, *30*(4). Advance online publication. DOI: 10.20867/thm.30.4.1

Ledger, I. (2021, December 6). *Marriott launches NFT Collection at contemporary art fair*. Ledger Insights - Blockchain for Enterprise. https://www.ledgerinsights.com/marriott-launches-nft-collection-at-contemporary-art-fair/

Lee, S., Shulga, L. V., Busser, J. A., & Hwang, J. (2024). Self-Service Hotel Check-In: Is There Still a Role for the In-Person Experience? *International Journal of Hospitality & Tourism Administration*, •••, 1–28.

Liu, R. (2023). Development of a travel recommendation algorithm based on multi-modal and multi-vector data mining. *PeerJ. Computer Science*, *9*, e1436. DOI: 10.7717/peerj-cs.1436

Londhe, K., Dharmadhikari, N., Zaveri, P., & Sakoglu, U. (2024). Enhanced Travel Experience using Artificial Intelligence: A Data-driven Approach. *Procedia Computer Science*, *235*, 1920–1928. DOI: 10.1016/j.procs.2024.04.182

M'hamed, E. G., & El Idrissi, R. (2024). Transforming Hospitality: Harnessing Artificial Intelligence for Enhanced Guest Experience and Operational Efficiency. *The International Workshop on Big Data and Business Intelligence*, 173–185. DOI: 10.1007/978-3-031-65014-7_17

Ma, S. (2024). Enhancing Tourists' Satisfaction: Leveraging Artificial Intelligence in the Tourism Sector. *Pacific International Journal*, *7*(3), 89–98. DOI: 10.55014/pij.v7i3.624

McLean, G., & Barhorst, J. B. (2022). Living the experience before you go... but did it meet expectations? The role of virtual reality during hotel bookings. *Journal of Travel Research*, *61*(6), 1233–1251. DOI: 10.1177/00472875211028313

Mei, C. W., Konar, R., & Kumar, J. (2024). The Role of AI Chatbots in Transforming Guest Engagement and Marketing in Hospitality. In Nadda, V., Tyagi, P., Singh, A., & Singh, V. (Eds.), *Integrating AI-Driven Technologies Into Service Marketing* (pp. 595–620). IGI Global Scientific Publishing., DOI: 10.4018/979-8-3693-7122-0.ch029

Millennium. (2024). *M Social Hotel Auckland*. Millennium Hotels and Resorts. https://www.millenniumhotels.com/en/auckland/m-social-auckland/

Milton, T.Dr.T. Milton. (2024). Artificial Intelligence Transforming Hotel Gastronomy: An In-depth Review of AI-driven Innovations in Menu Design, Food Preparation, and Customer Interaction, with a Focus on Sustainability and Future Trends in the Hospitality Industry. *International Journal for Multidimensional Research Perspectives*, *2*(3), 47–61. DOI: 10.61877/ijmrp.v2i3.126

Mishra, D., Das, S., & Patnaik, R. (2024). Application of AI Technology for the Development of Destination Tourism towards an Intelligent Information System. *Economic Affairs*, *69*(02), 1083–1095.

Mohanty, P., Singh, K., & Lakshmi, R. B. (2025). Can Artificial Intelligence Revolutionize Crisis Management in Tourism?: Exploring the Implications and Challenges. In *Tracking Tourism Patterns and Improving Travel Experiences With Innovative Technologies* (pp. 83–104). IGI Global.

Nair, A. J., Manohar, S., Mittal, A., & Khanna, V. (2023). Revolutionizing Tourism and Hospitality Services: Integrating AI in the Metaverse. *2023 6th International Conference on Contemporary Computing and Informatics (IC3I)*, *6*, 1206–1212.

Nica, I., Tazl, O. A., & Wotawa, F. (2018). Chatbot-based Tourist Recommendations Using Model-based Reasoning. *ConfWS*, 25–30.

Ocean. (2024a). *"horigin" metaverse hotel platform*. Ocean Sky Hotel Management Limited. https://www.oceansky.com.hk/en/horigin-metaverse-hotel-platform-en/

Ocean. (2024b). *Popway hotel hong kong*. Ocean Sky Hotel Management Limited. https://www.oceansky.com.hk/en/popway-hotel-hong-kong-en/

Park, S.-M., & Kim, Y.-G. (2022). A metaverse: Taxonomy, components, applications, and open challenges. *IEEE Access : Practical Innovations, Open Solutions*, *10*, 4209–4251. DOI: 10.1109/ACCESS.2021.3140175

Pratisto, E. H., Thompson, N., & Potdar, V. (2022). Immersive technologies for tourism: A systematic review. *Information Technology & Tourism*, *24*(2), 181–219. DOI: 10.1007/s40558-022-00228-7

Qi, H., Han, Z., & Feng, X. (2024). Research on Hotel Service Robot Management System Based on Artificial Intelligence. *2024 3rd International Conference on Artificial Intelligence and Autonomous Robot Systems (AIARS)*, 265–269.

Ramachandran, K. K., Vinjamuri, L. P., Ramesh, R., Al-Taee, M., & Alazzam, M. B. (2023). Using AI for risk management and improved business resilience. *2023 3rd International Conference on Advance Computing and Innovative Technologies in Engineering (ICACITE)*, 978–982.

Rane, N., Choudhary, S., & Rane, J. (2023, October 31). Sustainable tourism development using leading-edge Artificial Intelligence (AI), Blockchain, Internet of Things (IoT), Augmented Reality (AR) and Virtual Reality (VR) technologies. *Blockchain, Internet of Things (IoT), Augmented Reality (AR) and Virtual Reality (VR). Technologies.*

Rane, N., Choudhary, S., & Rane, J. (2024). Artificial Intelligence (AI), Internet of Things (IoT), and blockchain-powered chatbots for improved customer satisfaction, experience, and loyalty. *Internet of Things (IoT), and Blockchain-Powered Chatbots for Improved Customer Satisfaction, Experience, and Loyalty (May 29, 2024).*

Ray, N., Dey, P., Chakraborty, M., & Bandyopadhyay, M. A. (2024). A Study on AI Application in Travel and Tourism Industry. *Redefining Commerce and Management: New Paradigms for the Digital Age (Volume 1)*, 222.

Rekiek, S., Jebari, H., & Reklaoui, K. (2024). Prediction of Booking Trends and Customer Demand in the Tourism and Hospitality Sector Using AIBased Models: A Case Study of Major Hotel Chain. *International Journal of Advanced Computer Science and Applications, 15*(10). Advance online publication. DOI: 10.14569/IJACSA.2024.0151043

Samala, N., Katkam, B. S., Bellamkonda, R. S., & Rodriguez, R. V. (2020). Impact of AI and robotics in the tourism sector: A critical insight. *Journal of Tourism Futures, 8*(1), 73–87. DOI: 10.1108/JTF-07-2019-0065

Saurabh, D. (2024). *Hopper Travel app: Business Model, Development Cost, & features*. https://www.addwebsolution.com/blog/hopper-travel-app-business-model-app-development-cost-features

Sharma, R. (2024). Genetic Algorithm Based Personalized Travel Recommendation System. *2024 2nd International Conference on Intelligent Data Communication Technologies and Internet of Things (IDCIoT)*, 867–874.

Shelke, P. M., Dedgaonkar, S., & Bhimanpallewar, R. N. (2024). Powering User Interface Design of Tourism Recommendation System with AI and ML. In *Artificial Intelligence, Machine Learning and User Interface Design* (pp. 108–135). Bentham Science Publishers. DOI: 10.2174/9789815179606124010008

Shirzad, A., & Rahmani, A. (2024). Smart Treasury: Leveraging Artificial Intelligence and Robotic Process Automation for Financial Excellence. *Knowledge Economy Studies, 1*(1), 65–86.

Solakis, K., Katsoni, V., Mahmoud, A. B., & Grigoriou, N. (2024). Factors affecting value co-creation through artificial intelligence in tourism: A general literature review. *Journal of Tourism Futures, 10*(1), 116–130. DOI: 10.1108/JTF-06-2021-0157

Soori, M., Arezoo, B., & Dastres, R. (2023). Artificial intelligence, machine learning and deep learning in advanced robotics, a review. *Cognitive Robotics, 3*, 54–70. DOI: 10.1016/j.cogr.2023.04.001

Sousa, A. E., Cardoso, P., & Dias, F. (2024). The Use of Artificial Intelligence Systems in Tourism and Hospitality: The Tourists' Perspective. *Administrative Sciences, 14*(8), 165. DOI: 10.3390/admsci14080165

Sun, Y., Konar, R., & Wei, C. K. (2025). Unlocking the power of employee advocacy: The role of internal communication and engagement in five-star hotels. *International Journal of Hospitality Management, 126*, 104086. DOI: 10.1016/j.ijhm.2025.104086

Syarul, M. (2023). Potensi Dan Hambatan Metaverse Di Era Terkini. *Jurnal Mahasiswa Teknik Informatika, 7*(3).

Thea, S. (2018). *Hopper's AI-driven prediction models can help you save big on airfare.* https://outsideinsight.com/insights/hoppers-ai-driven-prediction-models-can-help-you-save-big-on-airfare/

Thenahandi, H. (2023). *The effects of self-service automation on customer experience during the check-in stage at Helsinki Vantaa airport.* https://urn.fi/URN:NBN:fi:amk-202302082195

Tian, S., & Song, Q. (2023). Study on the Technical Model Design of a Tourist Attraction from the Metaverse Perspective. *Business. Economics and Management BEMS, 6*, 371–380.

Tiwari, S. (2021). Space tourism: An initiative pushing limits. *Journal of Tourism Leisure and Hospitality, 3*(1), 38–46. DOI: 10.48119/toleho.862636

Torres-Ruiz, M., Mata, F., Zagal, R., Guzmán, G., Quintero, R., & Moreno-Ibarra, M. (2020). A recommender system to generate museum itineraries applying augmented reality and social-sensor mining techniques. *Virtual Reality (Waltham Cross), 24*(1), 175–189. DOI: 10.1007/s10055-018-0366-z

Tul-Krzyszczuk, A. (2023). The metaverse in hospitality management in the age of digital transformation. *Zeszyty Naukowe Szkoły Głównej Gospodarstwa Wiejskiego w Warszawie. Ekonomika i Organizacja Logistyki, 8*(4), 105–120. DOI: 10.22630/EIOL.2022.7.4.31

Tuo, Y., Wu, J., Zhao, J., & Si, X. (2024). Artificial intelligence in tourism: insights and future research agenda. *Tourism Review, ahead-of-print*(ahead-of-print). DOI: 10.1108/TR-03-2024-0180

Wang, H., Ning, H., Lin, Y., Wang, W., Dhelim, S., Farha, F., Ding, J., & Daneshmand, M. (2023). A survey on the metaverse: The state-of-the-art, technologies, applications, and challenges. *IEEE Internet of Things Journal, 10*(16), 14671–14688. DOI: 10.1109/JIOT.2023.3278329

Wang, Y.-C., & Uysal, M. (2024). Artificial intelligence-assisted mindfulness in tourism, hospitality, and events. *International Journal of Contemporary Hospitality Management, 36*(4), 1262–1278. DOI: 10.1108/IJCHM-11-2022-1444

Xiaoxin, L., Konar, R., & Kim, M. J. (2026). The impact of human-robot collaboration on hotel employees: A meta-analytic exploration of behavioral intentions and influencing factors from the UTAUT perspective. *International Journal of Hospitality Management, 133*, 104492. DOI: 10.1016/j.ijhm.2025.104492

Xiaoxin, L., Liqun, C., Konar, R., Sthapit, E., Garrod, B., & Weimei, C. (2025). Towards a better understanding of enhancing task performance in resort hotels. *Tourism and Hospitality Research*, •••, 14673584251314282. DOI: 10.1177/14673584251314282

Yan, Z., Balasubramanian, K., Konar, R., Chen, L., & Wei, Y. (2024). Smart hotel in the eyes of the silver: Developing and testing the silver tourists' behavioural intention scale. *Current Issues in Tourism*, •••, 1–20. DOI: 10.1080/13683500.2024.2381249

Yan, Z., Konar, R., & Balasubramanian, K. (2025b). The Future of Hotels Robotics, AI, and Service Automation in Practice. In Ozen, E., Singh, A., Taneja, S., Rajaram, R., & Davim, J. (Eds.), *Sustainability, Innovation, and Consumer Preference* (pp. 249–278). IGI Global Scientific Publishing., DOI: 10.4018/979-8-3693-9699-5.ch011

Yan, Z., Konar, R., Sthapit, E., Balasubramanian, K., Chen, L. and Prentice, C. (2025a), Opting for smart hotels: do digital immigrants perceive differently than digital natives do?, *Journal of Hospitality and Tourism Insights*, Vol. ahead-of-print No. ahead-of-print. https://doi.org/DOI: 10.1108/JHTI-08-2024-0785

Yang, X., Zhang, L., & Feng, Z. (2024). Personalized tourism recommendations and the E-tourism user experience. *Journal of Travel Research, 63*(5), 1183–1200. DOI: 10.1177/00472875231187332

Zaman, M., Hasan, P. R., Vo-Thanh, T., Shams, R., Rahman, M., & Jasim, K. M. (2025). Adopting the metaverse in the luxury hotel business: A cost–benefit perspective. *International Journal of Contemporary Hospitality Management, 37*(4), 1309–1331. DOI: 10.1108/IJCHM-08-2023-1265

Zhang, W. (2022). Research on the Application of Computer Artificial Intelligence Technology in the Response Mechanism of Emergency Logistics System. *2022 IEEE International Conference on Advances in Electrical Engineering and Computer Applications (AEECA)*, 676–680. DOI: 10.1109/AEECA55500.2022.9918994

Zhong, L., Xu, Z., Morrison, A. M., Li, Y., & Zhu, M. (2024). Metaverse customer journeys in tourism: Building viable virtual worlds. *Tourism Review, 79*(8), 1409–1426. DOI: 10.1108/TR-07-2023-0492

Zhou, M., & Yan, Y. (2023). The Research on the Application of Metaverse in the Cultural and Tourism Industry. *Advances in Economics. Management and Political Sciences, 47*(1), 109–113.

Chapter 12
Examining the Development of Cloud Kitchens:
Consequences for the Hospitality and Tourism Industries' Future

Anamika Ahirwar
https://orcid.org/0000-0003-3003-3105
Compucom Institute of Technology and Management, Jaipur, India

ABSTRACT

Cloud kitchens are changing the nature of hospitality and tourism because they are a digitally enabled form of restaurant that requires shared infrastructure to operate. They are cost-effective, efficient, and scalable by eliminating the dine-in and relying on AI, IoT, and analytics. They open up virtual brands and small businesses as well. However, they are threatening the business of traditional restaurants because they are concerned with the quality of food, monotony, regulations, and the loss of cultural dining. In this chapter, the cloud kitchen and traditional kitchen models are compared based on sustainability, customer experience, and policy. Stakeholders, operations, and technology share a framework with each other. The findings highlight the point that cloud kitchens encourage innovation; however, the balance between convenience and experience and policy integration requires moderation to make the developments sustainable in the long run.

DOI: 10.4018/979-8-3373-6541-1.ch012

1. INTRODUCTION

Cloud kitchens, also called ghost, dark, or virtual kitchens, is a disruptive business approach to food service, involving no physical restaurants, and only delivery. Under this type of format, food is cooked at centrally located kitchens to be only served in the form of takeaway or delivery, with the use of digital platforms and applications to take orders (Švancár et al., 2024; Singh, Chourasia, and Kaur, 2025). The on-demand food service based on this model of delivery has spread quickly across the global market due to growing consumer demand in terms of convenience and choice (Singh, Chourasia, and Kaur, 2025). Cloud kitchens have lower overhead costs (there is no front-of-house/dine-in space) and can be occupied by a variety of virtual brands in the same facility unlike traditional restaurants, which leads to lower start-up costs and better scalability (Švancár et al., 2024; Singh, Chourasia, and Kaur, 2025).

Previously, cloud kitchens used to be the integration of food delivery apps and redesigned hospitality business models. Before 2020, the model was still in its early days, and the COVID-19 pandemic turned out to be the stimulating factor toward its use (Švancár et al., 2024; Singh, Chourasia, and Kaur, 2025). Most of the old operations converted old kitchens into delivery-only establishments to be able to survive lockdowns and varying demand. Researchers state that cloud kitchens have a low-cost and demand-oriented business model that is entirely driven by online ordering, and that served as an essential element in the foodservice recovery program, in the event of pandemic restrictions. In fact, the study findings it that the COVID-19 spurred the rise of cloud kitchens directly because of the rates of home delivery because consumers and businesses dine out remotely (Nagadeepa, C. et al. 2023). Consequently, a niche virtual restaurant idea has been developed and evolved to become a new industry paradigm.

Historically, cloud kitchens evolved from the convergence of food delivery apps and changing business models in hospitality. The model existed in nascent form before 2020, but the COVID-19 pandemic served as a catalyst for its widespread adoption (Švancár et al., 2024; Singh, Chourasia, & Kaur, 2025). Many traditional operators converted existing kitchens to delivery-only operations to survive lockdowns and shifting demand. Scholars note that cloud kitchens operate on a low-cost, demand-driven model supported entirely by online ordering, and they became an essential element of foodservice resilience during pandemic restrictions. Indeed, studies report that the surge in home delivery during COVID-19 directly led to the rise of cloud kitchens, as consumers and businesses embraced remote dining solutions (Nagadeepa, C. et al. 2023). Consequently, what began as a niche "virtual restaurant" concept has matured into a new paradigm for the industry.

From a technological perspective, cloud kitchens epitomize the digital transformation of food service. These kitchens integrate advanced information and communication technologies at every stage of operations. For example, operators use data analytics and predictive modeling to tailor menus to local demand patterns and optimize inventory. One case study highlights that cloud kitchens "analyze enormous amounts of data" to anticipate which menu items (e.g. specific cuisines or dishes) will be in greatest demand at particular times and locations, enabling near-real-time menu adjustments (Chi, O. H. et al., 2022). In practice, cloud kitchens rely heavily on online ordering platforms and mobile apps (including global players like Uber Eats or regional ones like Swiggy and Zomato) to reach customers. Technology also improves backend processes: digital kitchen-management systems, automated order tracking, and even robotics or AI-driven cooking tools can increase throughput and consistency. In the hospitality context, this digital-first approach aligns with broader industry trends toward smart restaurants and IoT-enabled operations, allowing cloud kitchens to achieve high operational efficiency and flexibility.

From the hospitality and tourism perspective, cloud kitchens are reshaping service delivery and business models. By decoupling food production from physical restaurant space, they enable restaurants and hotel food outlets to extend their reach without costly real estate or additional staff. For example, hospitality businesses can use cloud kitchens to offer new virtual restaurant brands or cuisines to guests and local markets simultaneously. In this way, cloud kitchens serve as a form of operational innovation that addresses evolving guest expectations for convenience. As one study notes, the cloud kitchen model "capitaliz[es] on the increasing demand for food delivery services," creating a growing business model within the foodservice industry. This model also dovetails with trends in tourism, where travelers increasingly rely on digital platforms for food recommendations and ordering. Even for hotels, partnering with off-site cloud kitchens allows them to offer 24/7 food service or diverse menu options with lower overhead. In sum, cloud kitchens exemplify how digital food service models are blurring the lines between hospitality, technology, and urban delivery ecosystems.

Consumer behavior trends are tightly interwoven with the rise of cloud kitchens. Studies show that consumer awareness of cloud kitchens is expanding rapidly through social media and online marketing, and key drivers of purchase decisions include the visibility of a brand online, perceived food quality, and price or affordability. Ray et al. (2019) emphasized that the growing use of food delivery applications, driven by convenience and digital trust, has been a key enabler of the cloud kitchen business model. Importantly, cloud kitchens promise benefits valued by modern consumers: faster delivery, lower costs, and sometimes a perceived sustainability advantage by consolidating resources. These preferences were magnified in the post-pandemic era; consumers of all ages report greater willingness to use delivery-only food services and expect hygienic, contactless experiences. As one analysis concludes, shifting consumer patterns—especially among younger cohorts—have made cloud kitchens a "game-changing innovation" in foodservice, and cloud models can fulfill demand for variety and convenience without in-person contact (Nagadeepa, C. et al.2023).

In summary, cloud kitchens have emerged as a theoretically and practically significant phenomenon at the intersection of hospitality, tourism, and technology. Rooted in digital-platform economics and service-dominant logic, they represent a response to changing market conditions and consumer behaviors. Early literature emphasizes their disruptive potential: reducing costs and entry barriers, exploiting shared infrastructure, and leveraging data-driven operations (Singh, Chourasia, & Kaur, 2025). At the same time, scholars note challenges such as dependency on third-party platforms and the need to maintain food quality and brand trust in an intangible environment. This chapter, therefore examines cloud kitchens' development with full consideration of these trends, drawing on recent empirical studies to ground claims about their economic impact, operational innovation, and implications for the future of hospitality and tourism.

1.1. Research Methodology

This study adopts a qualitative, descriptive research design based on secondary data analysis. Peer-reviewed journal articles, industry reports, and government regulations were examined to assess the evolution and impact of cloud kitchens on hospitality and tourism. A systematic literature review approach was followed to identify thematic areas including technological enablers, business models, sustainability challenges, and policy frameworks. Comparative analysis between traditional and cloud-based restaurant models was used to highlight operational and cultural implications. The methodological framework ensures theoretical depth and practical relevance.

1.2. Review of Related Literature

Previous research has explored the emergence of cloud kitchens from operational, technological, and consumer behavior perspectives and emphasized how digital transformation enabled low-cost food service expansion. Gursoy and Chi (2022) discussed consumer acceptance and post-pandemic behavior shifts.

1.3. Key Definitions

- **Cloud Kitchen:** A delivery-only restaurant model operating without dine-in facilities.

- **Ghost Kitchen / Virtual Kitchen:** Synonyms for cloud kitchens focused on digital delivery through online platforms.
- **Hybrid Model:** Combines dine-in services with delivery or virtual brands.
- **Hospitality Industry:** Businesses providing lodging, food, and leisure services to guests.
- **Tourism Industry:** The collection of services supporting travel, including accommodations, food, and entertainment.

2. THE EMERGENCE OF CLOUD KITCHENS

Cloud kitchens are also known as ghost/ virtual kitchens, and they are adding to becoming a disruptive factor in the food service industry, offering a total alteration in the way food is produced and delivered. Cloud kitchens are only utilized to deliver foods as compared to traditional restaurants. They seize the opportunity of online delivery and third-party services like DoorDash, Grubhub, and Uber Eats and send the growing demand of fast and convenient food. As these kitchens have no actual dine-in seating places, companies will be able to save a considerable amount of money that would otherwise be used on front-of-house personnel and equipment. This type of lean mode of operation has made cloud kitchens develop rapidly, particularly in metropolitan cities where delivery is in demand.

The development of meal delivery applications, changing consumer demands, and technologies are some of the primary factors that enable the development of cloud kitchens. The development of meal delivery applications, changing consumer demands, and technologies are some of the primary factors that enable the development of cloud kitchens (Ahuja, Chandra, Lord, & Peens, 2021).

The meals-on-demand keeps increasing among customers as they can have them in the consecration of their homes or anywhere, they visit when they visit them through the convenience of ordering meals using apps and the growing usage of smartphones. The COVID-19 pandemic has hastened the trend and limited the dine in options, and increased the demand using food delivery services (Kulshreshtha & Sharma, 2022). It is at this stage that the cloud kitchens displayed the fact that they could turn around the traditional restaurant model by operating in a leaner and more efficient manner. A recent systematic literature review supports the observation that cloud kitchen is a technology-based sustainability that has developed in accordance with the trends of consumer preference and food entrepreneurship in the face of post-pandemic (Harini, P. et al., 2025) observes that the model has enabled small and mid-scale entrepreneurs to advance with minimal initial investment.

Cloud kitchens offer a fresh pool of opportunities to the old and new business-oriented individuals. They provide a low-cost entry point into the world of the entrepreneur aspiring to launch a food brand without incurring huge amounts of money to launch a full-service restaurant. Even the famous restaurants do not view the cloud kitchen idea as a big deal when aiming to experiment and increase the scope of delivery. The decision to run several brands under a single location is not unique, so as to maximally utilize the space of the kitchen and satisfy the needs of a more varied clientele.

However, the expansion of cloud kitchens poses a threat to the conventional sectors of hospitality and tourism. Food and beverages represent a significant component of the visitor experience in the hospitality industry, and above all in the hotel, resorts, and other event providers. Cloud kitchens change all this and may end up lowering the demand of on-site food by providing them with the option of external delivery. Since it is parallel to this, cloud kitchens can make consumers choose the comfort over the uniqueness of going out and eating in the tourism sector, where individuals visit to sample some local delicacies.

Traditional hospitality and tourism sectors have no choice but to fight back since there is mushrooming of cloud kitchens. The hospitality industry is the one that is central in the delivery of food and beverage services to the visitors especially in the case of hotels, resorts and in the events facilities. Cloud kitchens provide alternative methods of delivering outside meals which are known to reverse this, and may even decrease the on-site meals. Like with such cloud kitchens, it can divert the consumers to the concept of convenience over a unique experience of the food experience in the tourism industry since local food is one of the tourist sites.

However, the conventional hospitality and tourism sectors are worn out as cloud kitchens become decentralized. The roles of the restaurant and beverage services are very important to the hospitality industry; more so, in the hotels and resorts and the event facilities. Cloud kitchens reverse this situation and can potentially decrease the existing demand of on-site meals as there is an option of delivery off-site. Similar to this scenario, cloud kitchens may force individuals out of the one-of-a-kind dining out experience in the tourism sector where the local food forms a key draw in the sites among tourists.

2.1. The Influence of Technology

The development of technology is directly connected with the emergence of cloud kitchens, in particular due to the popularity of food delivery services, including Uber Eats, DoorDash, and Deliveroo. These online outlets enable cloud kitchens to serve a substantial number of customers without necessarily having a location with prime real estate, which results in a more flexible and cost-efficient operation (Ahuja, Chandra, Lord, & Peens, 2021).

Cloud kitchens are also able to streamline their operations by anticipating customer demand, controlling inventory, and creating more accurate menus because of data analytics and artificial intelligence (AI). Cloud kitchens can employ technology to ensure that their operations and reactions to market dynamics are high in efficiency.

2.2. Demand and Behavior of the Customers.

The consumers have been using cloud kitchens more due to the necessity to have their orders delivered promptly and in a convenient manner. Convenience food has substituted the traditional plans of eating and this is particularly in the urban circles where foods can be delivered easily. Also, there is more food delivery due to remote occupation and lifestyle change.

2.3. Role of Internet of Things (IoT) in Cloud Kitchen Operations

Internet of Things (IoT) is becoming a major enabler in digital transformation of cloud kitchen across the world. IoT, through the integration of sensors, devices, and real-time monitoring systems, as part of kitchen infrastructure, allows a higher level of automation, data collection, and distance-managing capabilities. IoT can be used in cloud kitchens to observe the temperature of cooking equipment, the state of the refrigeration, the indicators of hygiene, and energy savings. As an illustration, smart sensors will allow staff members to be notified of an unusual temperature change in storage facilities to keep food at the highest possible safety standards, whereas motion sensors and usage analytics will help to streamline work processes and minimize energy costs in the kitchen.

Also, IoT devices that will be synced with the predictive maintenance will be able to predict equipment failure, thus decreasing downtimes and preventing expensive shutdowns during times of high demand. Kitchen appliances and inventory systems can be connected in a way so that the amount of ingredients is tracked easily minimizing food waste and automatically replenishing their stock. As well, real-time GPS tracking of food orders with the help of IoT-based delivery tracking systems results in added transparency and customer satisfaction.

In case of a cloud kitchen that is strongly positioned around a remote management and delivery system, IoT can guarantee the appropriate quality, efficiency, and compliance without a need to define and control by human interaction constantly. The development of the industry will make it mandatory to integrate IoT to maintain scalable, safe, and customer-focused operations of cloud kitchens.

3. CONSEQUENCES FOR THE TOURISM AND HOSPITALITY SECTOR

The rapid transformation of cloud kitchens will be a major factor in the hospitality and tourist industries as both rely primarily on dining experiences to enhance the happiness of visitors and introduce local culture. The cloud kitchens are making a fundamental change in food distribution by its delivery-only format that disrupts the existing restaurant-based operations and altering the eating patterns in both businesses.

Cloud kitchens also provide hotels and resorts with the opportunity to experiment with new culinary concepts in order to capture the varied tastes of their customers. By exploiting the effectiveness and scalability of cloud kitchens, hospitality ventures have the capacity to accommodate the changing dining trends, seasonal demands and customer reviews, therefore enhancing total visitor experience. This strategy will also enable them to compete better with external meal delivery providers as they will keep the customers interested in their in-house dining plans.

Nevertheless, the conventional restaurant format of the hospitality sector encounters issues with the development of cloud kitchens. Hotel restaurants could also experience reduced traffic as people prefer to get their food delivered rather than eat at restaurants. The resulting shift in the role of the food and beverage services in the hospitality industry can reduce the importance of on-site restaurants as a hotel brand differentiation. Also, the reliance on external cloud kitchen can restrict the direct management of the quality of the gourmet and the service, two highly important variables of the visitor experience.

The impacts of cloud kitchens are also complex in the traveling and tourism industry. Eating and food are some of the key aspects of a cultural tourism and the travellers always seek out local restaurants to taste authentic food. Nevertheless, such an emphasis on efficiency and convenience in the realm of cloud kitchens can deny the traditional eating experience that the majority of traveling people would want to have. In such cities where food delivery is prevalent, cloud kitchens could thus pose an even greater challenge to corner restaurants.

The idea of cloud kitchen is very convenient, yet in case it gets more common, it will disrupt the local food industry as well and diminish the authenticity of food tourism. Nevertheless, restaurants can provide access to new target groups of tourists who can be business people or people seeking unfastened food, attracting both cloud kitchens without having to rely on real spaces in restaurants. The cloud kitchen will offer more choices and can attract a larger number of tourists as destinations attract more tourists who have more choices, a factor that can boost tourism.

Finally, turning the hospitality and tourism sphere upside down, cloud kitchens offer opportunities to break the traditional pattern of eating and become innovative.

The long-term effects of this shift with respect to the degree of integration of cloud kitchens into the processes combined with the retention of cultural and experiential elements which define respective industries will define the outcomes of this development. The phenomenon of cloud kitchen is a threat and a promise of thing to come in relation to travel and hospitality sector.

3.1. The Effect on Conventional Dining Establishments

The invention of cloud kitchen has brought a disrupting change in the conventional restaurant business model. Removing the food production process and dine in factor, these delivery focused kitchens have been able to reduce the requirement to provide high investment stores, waiters, and ambience design. This model of value appeals to both startups and large chains because it is low-cost to reach a wide consumer base sooner when they need to accumulate less capital investment.

With food delivery platforms such as Zomato, Swiggy, Uber Eats and Grubhub gaining momentum on consuming food in the cities, pressure is mounting on the traditional restaurants. Consumer behavioral change (i.e., the preference to convenience, a wide range of options, and ordering via apps) has caused a steep decrease in both foot traffic and in-house dining sales that can be measured. Customers no longer prioritize dining out but instead choose to have multi-cuisine delivery services that are catered efficiently by cloud kitchens.

This change in behavior has wide-ranging consequences on any heritage food places. Most restaurants have been forced into either becoming hybrid (where they offer dine in services as well as delivery practices) or to shift the entire practice (perhaps to a virtual brand using shared kitchen facilities). Besides, aggregators impose commissions and staff the data about customers, building an unequal ratio in which traditional operators do not keep up with profitability. The pandemic also fast-tracked this shift, and the necessity to stay healthy and operationally efficient in lockdowns has made delivery the accepted state of interaction between restaurants and customers.

Altogether, as much as cloud kitchens are its benefit, including operational flexibility and operation versatility, they also threaten the very identity and economics of traditional dining. Highly competitive dynamics have forced conventional restaurants to rethink their value propositions with more focus on experiential eating, brand, and a direct address to the consumers as the industry continues to shift and become delivery-focused.

3.2. Operational Efficiency and Scalability

Cloud kitchens are different to the traditional restaurant concepts in that they focus on operational efficiency and growth. Cloud kitchens have the capacity to accommodate the growing demand by streamlining operations and use of technology thereby enabling efficiency and cost-cutting. This business model has altogether revolutionized the way the food services industry operated as it provides its blueprint on how restaurants can simplify their functionality so as to become delivery-oriented.

The primary determinant of the effectiveness of the operation of cloud kitchens is the delivery-only model. Cloud kitchens may operate with cheaper and smaller rooms without having physical dine-in facilities; which appear to be common in urban or industrial locations where real estate costs will be cheaper. The overhead is also reduced, making the workflow very streamlined and leaving more funds that are allocated to marketing, technology and culinary operations. To enhance productivity, cloud kitchens often use shared facilities, and the latter enable many brands to be able to operate under the same kitchen.

To enhance the effectiveness of the operations in cloud kitchens, technology is important. Most cloud kitchen operators make use of advanced software tools to organize the delivery, track inventory, and receive orders. These solutions will ensure that the cooking of meals is done accurately and quickly through the optimization of the ordering process. Moreover, automation is becoming popular. Other cloud kitchens have robotic machines to cook and pack the food, a factor that reduces manpower expenditure and reduces mistakes (Kim & Lee, 2023). Data analytics can also be employed in cloud kitchens to keep a track of customer tastc and partner with any customer service and make on-the-fly alterations to the menu, thereby enabling the company to respond rapidly to the fluctuation in demand and reduce food wastage.

As demonstrated by Švancár, Chrpa, Dvořak, and Balyo (2024), composite AI models based on planning can be implemented to optimize the routes to deliveries, inventory, and staff scheduling in real time, which greatly enhances the efficiency of the work of cloud kitchens.

The scalability of cloud kitchens is the other factor that contributes to its success. The coverage and growth opportunities of the traditional restaurants are usually restricted by the location. Unlike the case of a restaurant, cloud kitchens do not need to be based in a specific area, hence can expand quite easily. A single cloud cooking facility can host several virtual brands in different market niches. Under this plan, cloud kitchens will have an opportunity to experiment with new concepts and dishes without incurring the cost and risk of locating new physical outlets. Moreover, cloud kitchens have the opportunity to grow their business quickly, establishing more restaurant kitchens in new areas and replicating their successful processes as well as leaping on the growth in the need of delivery services.

The connectivity that they have with third-party delivery services, such as Uber Eats, Grubhub, and DoorDash, also makes cloud kitchens more scalable. The platforms provide infrastructure necessary in accessing a huge client base without the restaurants having to develop their own delivery networks. By connecting with these systems, cloud kitchens can multiply their presence in multiple locations, often without making much more infrastructure investment. This allows them to rival conventional restaurants in a larger scale and enter into new markets with much speed. Nonetheless, achieving scale and bringing efficiency to the table entails some challenges to cloud kitchens. At times the quality gets compromised in the quest to ensure that production and delivery are made as fast as possible especially where there are multiple brands involved in the kitchens at one time. It can be also difficult to handle a lot of orders and maintain consistency across locations, especially when things are getting hectic. Moreover, growth in cloud kitchens may lead to staffing shortages, issues with organization of delivery, and control over the supply chain that might hurt their ability to maintain their level of performance.

Nonetheless, some challenges do come with the quest of cloud kitchens and scale and efficiency. On other occasions, the pressure of timely production and delivery may cause a compromise of this quality in the name of making-up in time on the factory or kitchen level because they may be dealing with multiple brands at any given time. It is sometimes difficult to handle a large number of orders and maintain consistency across different locations, especially when the orders are concentrated. In addition, as cloud kitchens expand, they may face staffing and delivery planning issues, as well as problems with supply chains, which may result in loss of ability to conduct their activities with the same degree of efficiency.

These difficulties notwithstanding, scalability and operational efficiency of cloud kitchens have seen them become an important player in the food service industry. They are in a great position to compete fairly well with the traditional restaurants because of the ability to update their operations, use technology and grow at high speed. There is a high likelihood that the cloud kitchen model to move on as the demand in the field of food delivery expands and more effective and innovative standards are established in the functioning of the business.

3.3. Eco-friendliness and Ecological Impact.

Since cloud kitchens become more trendy these days, it is actively concerned with their environmental effects and sustainability (Sultana, Nikhat, & Ahmed, 2025). Although having an extensive mix of advantages when it comes to scalability and cost effectiveness, these delivery-only services have cast serious doubts on their capacity to generate waste material, expend resources and the real impact of these businesses on the environmental footprint of the whole food service industry.

Since the viability of business operations has been highly being appreciated in any business, there is a desperate need to know how cloud kitchens affect the business sustainability.

The waste during packing has soared up, and this is one of the critical environmental solutions that comes with cloud kitchens (Chan, Cheung, Choi, & Sheu, 2023). In cloud kitchens, food is delivered exclusively, which is why every dish must be packed in single-use materials, containers, cutlery, and often dozens of protective layers of wrapping to ensure food is shipped in a secure condition. Both of these relationships have led to the growing garbage predicament at a global scale, given that complete dependence on these disposable resources is based on plastic. One of the vital environmental concerns the industry is experiencing is vast non-sustainability of packaging, although multiple cloud kitchens and delivery agents have begun to implement recyclable or biodegradable packaging alternatives.

Another environmental issue that cloud kitchens possess, besides the waste generated by packaging, is food waste. Just like the typical restaurant, cloud kitchens must minimize overstock by controlling inventory and food preparation. The danger of excess food production and food going bad is happening more and that is due to the numerous brands occupying the same kitchen and the concerns over a wide range of menu choices as well by customers. Besides its adverse influence on nature, including the increase in the level of greenhouse gases produced due to the decomposition of organic matter, excess food waste is an indicator of resource waste. Some cloud restaurants are also creating artificial intelligence (AI) and data analytics to manage inventory more efficiently and even more accurately forecast demand, which makes it less likely that any surplus food will waste.

Another factor that would impact how cloud kitchens may impact the environment is the carbon footprint associated with delivery logistics. The success of cloud kitchens is also directly connected with the expansion of food delivery services that need cars as their primary means of delivering food to their clients. The extensive unavoidable use of motorized transport, especially in the usage of transport in metropolitan areas, which are primarily subject to cloud kitchens, contributes to both air pollution and poor traffic rates. Even though some companies started investigating more environmentally friendly means of delivery, such as e-bikes or e-scooters, the total environmental harm of delivery fleets is a critical issue.

The obstacles notwithstanding, the cloud kitchen could enhance sustainability in a number of ways. In this instance, the centralized design of cloud kitchen operation means more resources, such as power and water, can be utilized more efficiently than in most normal restaurants that must have designated areas to eat and various sets of kitchen facilities. Streamlining will allow cloud kitchens to reduce the number of resources they use. Also, the need to pay for high-energy-intensive real estate changes as cloud kitchens are often located in urban or industrial areas. This has the potential to lead to a reduced energy footprint of every cooking facility as compared to big, full-service restaurants.

Such challenges notwithstanding, cloud kitchens may enhance sustainability in the following respects. As an illustration, cloud kitchen operations have a centralized format compared to regular restaurants, which require different dining rooms and sets of kitchens, which means that resources, such as energy and water, are used more effectively. Cloud kitchens are capable of reducing their overall resources through streamlining. Also, this curbs the need to spend heavy energy-using real estate due to the high availability of cloud kitchens in urban or industrial areas. This may mean fewer energy emissions per cooking facility relative to large full-service restaurants.

Some cloud kitchens are also beginning to incorporate sustainable practices in their daily routines as well. Some of these activities include using renewable power, and/or solar electricity, and less food waste by partnering with local food banks or composting programs. Cloud kitchens can reduce the impact on the environment when long-distance transportation is replaced through more sustainable supply chain options and due to sourcing ingredients from the same season and by using regional sources.

Both operations and environmental concerns are, therefore, grave even though cloud kitchens offer apparent operational advantages. Both the high level of reliance of the industry on single-use packaging, food delivery logistics (including food waste potential) may cause increased environmental impact. In their turn, cloud kitchens can reduce their environmental impact through the implementation of eco-friendly packing solutions, more sustainable ways of waste management, and delivery methods. As the industry continues to evolve, cloud kitchens should have the capacity to operate in a manner that will favour global efforts to reduce wastage and carbon emissions which will require the introduction of sustainable practices.

3.4. Implications for Tourism

Cloud kitchens have transformed the world of tourism and hospitality and have revolutionized the journey of the food experiences of travelers. Culturally, culinary tourism has always been greatly informative of cultural immersion, thus the tourists explore the cuisine in a physical space, whether it is in a fine-dining restaurant, street food vendor, or traditional restaurant. The spaces do not only provide meals but also ambience, one-on-one chef/vendor contact, and exposure to local traditions or narratives.

Nevertheless, today, tourists have a chance to enjoy the diversity of local and international food, as cloud kitchens and food delivery apps have become more and more widespread, and people do not even need to leave their accommodation to order food. The COVID-19 pandemic, in its turn, has caused this shift especially rapidly because it made travellers even more sensitive to issues of hygiene, safety, and social distancing. Cloud kitchens offer a hands-free, convenient, and personalized dining experience that would appeal to new-age travellers who value time and ease.

This is even improved with mobile applications supported by AI and location-based services which provide recommendations, transparent reviews, and ease of ordering. Gastronomic tourism has meant that tourists no longer need to make much effort to get gastronomic experiences as prepared packages, but their exposure to local food is extended, without necessarily having to interact directly with people or have to go to unknown territory. This is especially convenient in a metropolitan city or in a foreign location where the inability to communicate effectively with locals and negative cultural differences would be in the way of exploration.

Nevertheless, the increased presence of cloud kitchens in tourism may provoke an issue regarding losing the cultural richness that is attached to traditional food tourism. The transaction with a cloud kitchen does not have the sensory and social aspects that make up the aromas, decorating, live cooking, and community areas. Moreover, because cuisine has been commoditized with the use of algorithmic menus, there is a threat to the ability to respect regional cuisine and various culinary traditions.

Essentially, although the cloud kitchens correspond to the new demands of the digital-age tourists, they also challenge the very notion of an authentic culinary experience in destination marketing and tourism planning.

4. CHALLENGES AND OPPORTUNITIES

As a consequence of the explosive growth of cloud kitchens, the food service business has opportunities as well as challenges, particularly with reference to the established restaurant forms, the customer experience, and the market forces. It is important as operators aspire to be competitive and long-lasting once the industry transforms to a new way of doing business, which is delivery-driven and digital-first.

Challenges

1. Increased Competition for Traditional Restaurants

This trend is especially relevant in the food delivery sector, where cloud kitchens pose a different level of threat to traditional brick-and-mortar restaurants. The competing Cloud kitchens have the potential to offer similar prices and faster delivery due to their low overhead costs and high rate of scalability which may reduce the number of foot traffic and sales in the traditional dine-in restaurants. This will be more challenging for smaller, autonomous restaurants that may lack the infrastructure needed to compete with the incredibly efficient cloud kitchens.

2. Brand Dilution and Lack of Customer Loyalty

In spite of the fact that cloud kitchens allow operators to launch large amounts of virtual brands quite promptly, they might face loyalty and brand awareness issues. Creating long-lasting ties with the diners can prove to be a challenge when there are no physical locations to visit. Cloud kitchens can be easily lost among other participants of a crowded market, as customers can base their decision on price and convenience factor only, due to their invisible nature and non-existent contact with customers.

3. Quality Control and Consistency

The quality of food may be hard to keep high once scaled up to many locations or with many virtual brands. Unlike usual restaurants that often focus on certain menu items and face-to-face service, cloud kitchens need to manage a vast number of delivery orders and this aspect may have an impact on the quality of food they provide. Lack of enough resources coupled with staff may also make it difficult to maintain brand standards in a number of cloud cooking facilities.

4. Environmental Concerns

Cloud kitchens produce extensive waste when it comes to packaging, especially single-use plastics, since criticism in its use is so high. The environmental impacts of delivery vehicles that are often motorized vehicles, increase the carbon footprint attributed to cloud kitchens. Cloud kitchens are going to be pressured even more into using sustainable practices as knowledge of their eco-responsibility by consumers rises, potentially raising the level of operating costs and complexity.

5. Supply Chain and Logistics Issues

Cloud kitchens will have to manage complex delivery chains to support a variety of virtual brands. This often requires proper prediction in order to avoid food spoilage and also ensure that the ingredients become available so that various foods can be prepared. Any disruptions in the supply chain, caused by either the lack of supply, price fluctuation, and logistic delays, may considerably impede the smooth operation effectiveness in cloud kitchens, leading to inefficiency and the potential of customer dissatisfaction.

Opportunities

1. Lower Entry Barriers for New Entrepreneurs

Cloud kitchens reduce the initial capital cost of starting a food service business to a very significant extent. Entrepreneurs can also experiment with new business ideas at a faster rate and with fewer risks, as they do not require a vast staff and physical restaurant spaces to serve customers. This has enabled a larger population to get into the food business, experiment with food, and be able to operate on a large-scale operation without the aggravation of the restaurant environment, democratizing the restaurant business.

2. Expansion of Virtual Brands and Market Reach

The use of cloud kitchens enables operators to host multiple virtual brands in a single cooking location, thus gaining the benefit of reaching many more consumers and being able to address a diverse range of customer needs. It is not necessary to have multiple physical locations since one kitchen can produce different cuisines and, hence, find more customers. It is with the aid of this multi-brand approach that the company can rapidly expand into established brands, engage in innovation of the latest trends, and respond to the market demand with speed.

3. Data-Driven Operations

Cloud kitchens provide an advantage to the operators because they can contribute multiple virtual brands to the same cooking facility, which helps them serve more people and cater to many clients' needs. One does not have to come up with multiple physical locations since an individual kitchen can produce different types of food that may be of interest to many people. It is this multi-brand strategy that enables the ability to scale up popular brands, but also experiment with the latest trends and respond immediately to consumer demand.

4. Flexibility in Location and Market Penetration

Unlike traditional restaurants, which are limited to serving at a particular location, cloud kitchens have the ability to operate in locations where it would otherwise be uneconomical to operate an entire restaurant. Leveraging delivery platforms to deliver to higher-scale geographical areas is an additional benefit of cloud kitchens that have the opportunity to service markets that they previously could not reach because they could not invest in the overhead cost. Due to their flexibility, cloud kitchens can also easily enter underserved destinations or open new stores to comply with changing consumer preferences or the market economy.

5. Partnership Opportunities with Hotels and Other Sectors

Cloud kitchens would be able to collaborate with restaurants such as hotels, offices, and other organizations that lack full-fledged dining establishments to offer a wider food menu through delivery. To give an example, hotels and cloud kitchens may cooperate and provide room service, allowing visitors to have more options for excellent meals. Although cloud kitchens will benefit from an already established customer base, this will allow hotels to enhance the customer experience without building restaurants and kitchens of their own.

6. Globalization and Culinary Diversity

Unlike other restaurants that have to be limited to an area, cloud kitchens can be used to expand international cuisines. Cloud kitchens have simplified the provision of numerous cuisines, as customers who cannot access a specific cuisine locally can delight in a wide and diverse cuisine palate offering. Now with an opportunity to open up the exotic flavours in newer markets, this is also an opportunity that cloud kitchens have to collaborate with global chefs and businesses. Now with an opportunity to open up the exotic flavours in newer markets, this is also an opportunity that cloud kitchens have to collaborate with global chefs and businesses (John, 2021).

Cloud kitchens surpass conventional restaurants both in terms of its low cost of operation, convenience and scalability, and efficiency, and there is the potential that it could change the food service industry forever. However, the merits are associated with shortfalls such as the need to organize logistics, ensuring quality, and addressing environmental problems. So that they can survive, cloud kitchens should prevail over these barriers, taking advantage of opportunities to evolve, expand, and deliver new experiences to customers. The manner in which the traditional restaurants and the cloud kitchens will blend together, evolve, and accommodate to evolving customer demands is likely to impact the future of the food service industry.

5. FUTURE DIRECTIONS

The fast evolution of cloud kitchens keeps opening new horizons of innovation as well. New technologies like blockchain will be more transparent in food supply chains, which will help to make them traceable and food safe. The delivery of food by drones is on the rise and can indeed lead to speedier deliveries as well as alleviate urban traffic. Moreover, the use of artificial intelligence could be paramount regarding menu planning, demand forecasting, and customer preferences personalization. It should be studied in the future how these disruptive trends can be ethically, regulatory, and technologically implemented in sustainable integration into the food service ecosystem.

6. CONCEPTUAL FRAMEWORK FOR CLOUD KITCHEN INTEGRATION

A conceptual framework could be drawn to enable integration of technological enablers (AI, IoT, Data Analytics), operational functions (inventory, delivery, quality control), and external drivers (customer behavior, regulation, tourism trends) in a better understanding of how cloud kitchens fit into the hospitality and tourism ecosystem. This model, based on interconnection, would enable the stakeholders (managers, technologists, and policymakers) to understand and visualize the changing interaction between innovation and market responsiveness. An upcoming version of this chapter can feature a graphical representation that covers such relationships to drive development and scholarly research.

The conceptual framework of the cloud kitchen integration and industry transformation is meant to show the interrelations between the technological enablers, the operation strategy, and the consumer outcomes. It shows the ways in which structural change in the hospitality ecosystem is being bolstered by the digitalization of innovative platforms.

Figure 1. Conceptual Framework for Cloud Kitchen Integration and Industry Transformation.

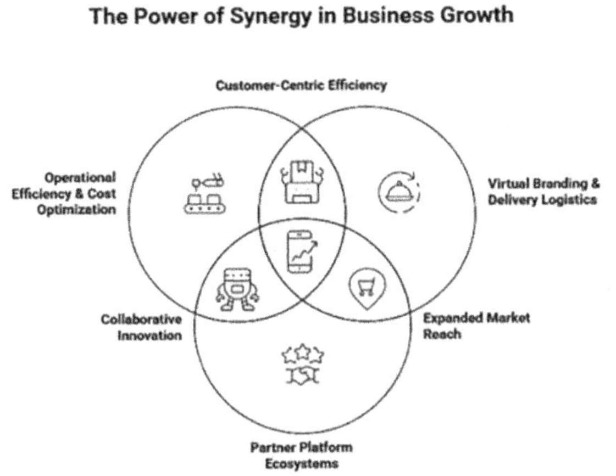

This model illustrates how technological enablers drive operational and strategic efficiencies that influence consumer behavior and overall hospitality industry transformation, with consumer feedback reinforcing further technological evolution.

As shown in Figure 1, technological enablers such as AI, cloud computing, and analytics serve as the foundation for operational efficiency and innovation in the food delivery ecosystem. These technological drivers interact with strategic management and consumer-centric approaches, leading to sustainable growth and transformation within the hospitality industry.

7. POLICY AND REGULATORY ENVIRONMENT

The control of cloud kitchens can be performed in many ways in the world. Food safety laws do exist in India, and they are in the form of licenses granted by the Food Safety and Standards Authority of India (FSSAI) and adherence to the local municipal and fire standards. To the contrary, the US is based on local health departments that are decentralized, and the guidelines are usually built on traditional restaurants, which will be unknown to virtual kitchens. Another aspect that is noted is that the UAE has embarked on a proactive or intelligent style of governance in which it has made the process of licensing and inspection straightforward by centralizing digital portals. The existence of a globally coordinated policy regime was noted to be a factor in enhancing the cross-border compliance and establishing international standards of safety, environmental friendliness and transparency in operations.

Cloud kitchen must be registered or licensed under the Food Safety and Standards Act, 2006, and must comply with the same hygiene and storage and labelling standards as their dine-in counterparts (FSSAI, 2022). The adherence to the hygienic standards and the frequency of the inspections are also highlighted as the element of the licensing system by the authority (FSSAI, 2022).

The growing interest of governments of different areas in evolving the regulatory mechanism is such that the speedy development of cloud kitchens may be corrected. These requirements are maintained through regular audits and registration in India according to the FSSAI requirements.

The food code in the United States defines the specificity of the food preparation as well as food handling off-premise preparation and handling and is implemented by the Food and Drug Administration (United States Food and Drug Administration, 2017). Ghost kitchens must follow the regulations of food storing temperatures, hygiene equipment, labeling allergens, and containers of delivery.

8. CASE STUDIES OF SUCCESSFUL CLOUD KITCHENS

There are success stories that address the successful impact of cloud kitchens in real life. Multi-brand concept that is being implemented by the Rebel Foods (India) chain is being fueled by AI-based analytics and kitchen robots that help to sustain over 30 brands across different cities. When examining Singh, Chourasia, and Kaur (2025), it is possible to highlight that, it is the technology, which has played the major role of creating scalability within Indian cloud kitchens, particularly, in the metro markets, in which cloud kitchens are in demand and becoming prohibitive (e.g., Mumbai and Delhi) (Singh et al., 2025). According to Singh, Chourasia, and Kaur (2025), technology has played a critical role in scaling the Indian cloud kitchen, as stated by the authors to include high-demand metro markets where its real estate prices are prohibitively high (Singh et al., 2025).

9. COMPARATIVE OVERVIEW OF RESTAURANT MODELS

Table 1. Comparative Analysis of Restaurant Models

Aspect	Traditional Restaurant	Cloud Kitchen	Hybrid Model
Capital Investment	High	Low	Medium
Operational Cost	High	Low	Medium
Customer Interaction	Direct	Indirect	Mixed
Scalability	Limited	High	Moderate
Sustainability	Variable	Challenging	Adaptive
Market Reach	Localized	Broad	Strategic

This contrasting discussion shows that both traditional restaurants and cloud kitchens present experiential value, but the latter ones have a higher level of scalability and cost-efficiency. As a future strategic compromise between the evolution of food service, a hybrid model can play a winning role.

10. CONCLUSION

A cloud kitchen is one of the important changes in the food services, hospitality, and tourism sectors. Supported by changing consumer needs and technological advancement, these delivery-only kitchens ensure unsurpassed flexibility and scale as well as operational efficiency. The ease of use, lower entry barriers and ability to meet the rising demand of faster food delivery has transformed the restaurant market in all aspects by introducing novel ways of serving the customers to their customers and increasing the diversity of tastes that people can experience. However, the list of critical negatives associated with the cloud kitchen model is quite long, and the escalation of the competition against traditional restaurants, the problem of food quality, supply chain logistics, and environmental sustainability could be mentioned among these issues.

Cloud kitchen entails both advantages and bits of concerns for the hospitality and tourism industries. They offer a vast variety of food to the tourists, which can enhance the accessibility and variety of gastronomic experience, but it can lead to the delusion of local cultures of food and destabilize the authenticity of the food as a significant tourism element. Another problem that should be considered as these kitchens expand globally are the issues of sustainability cast on by the packaging of the food orders and delivery arrangements.

Whether cloud kitchens will be able to achieve this goal and manage to combine environmental responsibility with operational efficiency will determine whether they can be successful and whether they can expand in the future. The two industries can coexist and thrive by ensuring that the former implements the changes by adopting the delivery model, building virtual brands, and leveraging information-driven intelligence. The emergence of cloud kitchens sees an opportunity to transform the way people eat in the future, though the implications of cloud kitchens on economic, cultural, and environmental factors of both the food and hospitality industries are an aspect that should be given close attention to.

REFERENCES

Ahuja, K., Chandra, V., Lord, V., & Peens, C. (2021, September 22). *Ordering in: The rapid evolution of food delivery*. McKinsey & Company. https://www.mckinsey.com/industries/technology-media-and-telecommunications/our-insights/ordering-in-the-rapid-evolution-of-food-delivery

Chan, H.-L., Cheung, T.-T., Choi, T.-M., & Sheu, J.-B. (2023). Sustainable successes in third-party food delivery operations in the digital platform era. *International Journal of Production Research, 61*(9), 2760–2780. DOI: 10.1080/00207543.2022.2106749 PMID: 37361094

Chi, O. H., Gursoy, D., & Chi, C. G. (2022). Tourists' attitudes toward the use of artificially intelligent (AI) devices in tourism service delivery: Moderating role of service value seeking. *Journal of Travel Research, 61*(1), 170–185. DOI: 10.1177/0047287520971054

Food Safety and Standards Authority of India. (2022). *FAQs on licensing and registration under the Food Safety and Standards (Licensing and Registration of Food Businesses) Regulations, 2011*. Food Safety Compliance System (FoSCoS). https://foscos.fssai.gov.in

Gursoy, D., & Chi, C. (2022). Impact of cloud kitchens on consumer dining preferences. *International Journal of Hospitality Management, 102*, 103157. DOI: 10.1016/j.ijhm.2022.103157

Harini, P., Rohini, A., Selvanayaki, S., Prahadeeswaran, M., Pangayar Selvi, R., & Shokila, C. (2025). A systematic literature review of cloud kitchen. *Plant Science Today, 12*(1), 1–8. DOI: 10.14719/pst.6016

John, K. T. (2021). Digital disruption: The hyperlocal delivery and cloud kitchen driven future of food services in post-COVID India. *International Hospitality Review, 37*(1), 161–187. DOI: 10.1108/IHR-06-2021-0045

Kim, J., & Lee, H. (2023). Sustainable operations of online and offline restaurants: The role of multi-brand online restaurant models and environmental performance. *Sustainability, 16*(21), 9607. DOI: 10.3390/su16219607

Kulshreshtha, K., & Sharma, G. (2022). From restaurant to cloud kitchen: Survival of the fittest during COVID-19. *Technological Forecasting and Social Change, 179*, 121629. DOI: 10.1016/j.techfore.2022.121629 PMID: 35317191

Nagadeepa, C., Jaheer Mukthar, K. P., Ramirez-Asís, E., Huaranga-Toledo, H., Espinoza-Maguiña, M., & Mory-Guarnizo, S. (2023). A reinforcer to become a foodpreneur through cloud kitchen: Sustainable technology-driven business model during pandemic. In Al Mubarak, M., & Hamdan, A. (Eds.), *Technological Sustainability and Business Competitive Advantage* (pp. 401–413). Springer., DOI: 10.1007/978-3-031-35525-7_25

Ray, A., Dhir, A., Bala, P. K., & Kaur, P. (2019). Why do people use food delivery apps (FDA)? A uses and gratification theory perspective. *Journal of Retailing and Consumer Services*, *51*, 221–230. DOI: 10.1016/j.jretconser.2019.05.025

Singh, A., Chourasia, S., & Kaur, S. (2025). The future of Indian restaurants: The role of technology in the success of cloud kitchens. In *Impact of AI and the Evolution of Future Ghost Kitchens* (pp. 15–34). IGI Global., https://www.researchgate.net/publication/390708481 DOI: 10.4018/979-8-3373-0432-8

Sultana, F., Nikhat, R., & Ahmed, B. (2025). Feeding the future: Trust, technology, and sustainability in cloud kitchens. *International Journal of Management*, *16*(3), 154–171. DOI: 10.34218/IJM_16_03_011

Švancár, S., Chrpa, L., Dvořák, F., & Balyo, T. (2024). Cloud kitchen: Using planning-based composite AI to optimize food delivery processes. *arXiv preprint* arXiv:2402.10725. https://doi.org//arXiv.2402.10725DOI: 10.48550

U.S. Food and Drug Administration. (2017). *Food Code 2017: Recommendations of the United States Public Health Service.* https://www.fda.gov/media/110822/download

Compilation of References

Acharya, P., & Mahapatra, S. S. (2024). Exploring the Impact of Artificial Intelligence Integration on Guest Experience in the Hotel Industry. *Geo Journal of Tourism and Geosites*, *54*(2 supplement), 802–810. DOI: 10.30892/gtg.542spl04-1255

Acosta, L., González, E., Rodríguez, J., Hamilton, A., Méndez, J., Hernández, S., Sigut, M., & Marichal, G. (2006). Design and implementation of a service robot for a restaurant. *International Journal of Robotics and Automation*, *21*(4). Advance online publication. DOI: 10.2316/Journal.206.2006.4.206-2909

Ahuja, K., Chandra, V., Lord, V., & Peens, C. (2021, September 22). *Ordering in: The rapid evolution of food delivery*. McKinsey & Company. https://www.mckinsey.com/industries/technology-media-and-telecommunications/our-insights/ordering-in-the-rapid-evolution-of-food-delivery

Ajzen, I. (1991). The theory of planned behavior. *Organizational Behavior and Human Decision Processes*, *50*(2), 179–211. DOI: 10.1016/0749-5978(91)90020-T

Aktaş, E., Gürsoy, I. T., & Kurgun, A. (2023). The future of metaverse in the tourism industry: Tool or purpose? *International Journal of Contemporary Tourism Research*, *7*(2), 143–152.

Alemede, V. O. (2025). Innovative process technologies: Advancing efficiency and sustainability through optimization and control. *Int J Res Publ Rev*, *6*(2), 1941–1955. DOI: 10.55248/gengpi.6.0225.0904

Al-Hyari, H. S., Al-Smadi, H. M., & Weshah, S. R. (2023). The Impact of Artificial Intelligence (AI) on Guest Satisfaction in Hotel Management: An Empirical Study of Luxury Hotels. *Geojournal of Tourism and Geosites, 48*(2spl), 810–819.

Al-Hyari, H.S., Al-Smadi, H.M., & Weshah, S.R. (2023). The impact of Artificial Intelligence (AI) on guest satisfaction in Hotel Management: an empirical study of luxury hotels. GeoJournal of Tourism and Geosites, 48(2spl), 810–819. https://doi.org/10.30892/gtg.482spl15-1081

Ali, F., Qiu, Z., & Gao, J. (2018). Luxury hotel guests' perceived authenticity and behavioral intentions: The role of traditional and modern luxury. *International Journal of Contemporary Hospitality Management, 30*(3), 1599–1616. DOI: 10.1108/IJCHM-12-2016-0683

Aljizawi, J. (2024). *Personalized Travel Recommendations and Marketing Automation for Saudi Arabia: Harnessing AI for Enhanced User Experience and Business Growth.*

Almeida, G. G. F., Almeida, P., & Cardoso, L. (2024). Scope extension of the metaverse tourism concept: Proposing a research agenda. *Journal of Infrastructure. Policy and Development, 8*(4), 3483.

Al-Oraini, B. S. (2025). Chatbot dynamics: Trust, social presence and customer satisfaction in AI-driven services. *Journal of Innovative Digital Transformation, 2*(2), 109–130. DOI: 10.1108/JIDT-08-2024-0022

Al-Romeedy, B. S., Hussein, H., Emam, M. E., & Singh, A. (2025b). From Potential to Peril: Evaluating AI and the Metaverse in Modern Research. In M. Al Aqad, A. SorayyaeiAzar, A. Albattat, & A. Singh (Eds.), Navigating AI and the Metaverse in Scientific Research (pp. 563-584). IGI Global Scientific Publishing. DOI: 10.4018/979-8-3373-0340-6.ch027

Al-Romeedy, B. S., Hussein, H., & Singh, A. (2025a). From Data to Delight: Using Wearable Technology to Tailor Tourist Experiences. In Rana, V., Bathla, G., Raina, A., & Chhabra, D. (Eds.), *Transforming the Service Sector With New Technology* (pp. 381–398). IGI Global Scientific Publishing., DOI: 10.4018/979-8-3693-7447-4.ch022

Alsheibani, S., Messom, C., & Cheung, Y. (2020). Re-thinking the competitive landscape of artificial intelligence. *Proceedings of the . . . Annual Hawaii International Conference on System Sciences/Proceedings of the Annual Hawaii International Conference on System Sciences.* DOI: 10.24251/HICSS.2020.718

Ansari, A. I., & Singh, A. (2023), "Application of Augmented Reality (AR) and Virtual Reality (VR) in Promoting Guest Room Sales: A Critical Review", Tučková, Z., Dey, S.K., Thai, H.H. and Hoang, S.D. (Ed.) Impact of Industry 4.0 on Sustainable Tourism, Emerald Publishing Limited, Leeds, pp. 95-104. DOI: 10.1108/978-1-80455-157-820231006

Ansari, A. I., Singh, A., Pattanaik, S., Patnaik, R., & Suryawanshi, U. (2024). Exploring the Role of AI-Powered Virtual Assistants in Optimizing Self-Service Experiences for Customers Within the Service Sector. In Nadda, V., Tyagi, P., Singh, A., & Singh, V. (Eds.), *Integrating AI-Driven Technologies Into Service Marketing* (pp. 225–238). IGI Global., DOI: 10.4018/979-8-3693-7122-0.ch013

Appelbaum, E., Bernhardt, A., & Murnane, R. J. (2003). *Low-wage America: An overview. Low Wage America: How Employers Are Reshaping Opportunity in the Workplace*. Russell Sage Foundation.

Arora, A., & Singh, A. (2026). Ethical considerations for emotion recognition technology in luxury hospitality: Balancing personalization and privacy. *Journal of Travel Research, 12*(1), 1–15.

Ashton, M. T. (2021). *Robotics, artificial intelligence and service automation in hotel service process design: Can customer dominant logic provide improved value?* (Doctoral dissertation, Bournemouth University).

Asmoro, A. Y., Butler, G., & Szili, G. (2023). Exploring the Current Status and Future Potential of Robot, Artificial Intelligence, and Service Automation in the Indonesian Tourism Industry. *Jurnal Kepariwisataan: Destinasi. Hospitalitas Dan Perjalanan, 7*(2), 133–153.

Athikkal, S., & Jenq, J. (2022). Voice chatbot for Hospitality. *ArXiv Preprint ArXiv:2208.10926*. DOI: 10.5121/csit.2022.121315

Avula, M., & Sithole, T. (2024). Artificial Intelligence's Potential to Improve Operational Efficiency and Customer Experience in the Hospitality and Tourism Sectors. In Impact of AI and Tech-Driven Solutions in Hospitality and Tourism (pp. 351-382). IGI Global. DOI: 10.4018/979-8-3693-6755-1.ch018

Ayyappan, V., & Ayyappan, I. (2025).Eds.Das J. & Eugene J., Predictive Analytics and IoT for Cleanliness Optimization in Indian Hotels in Sustainable Cleanliness, Teeneu Publications, ISBN 978-81-968444-3-1

Babu Reddy, S., Kathpalia, R., Sil, R., & Nag, A. (2024). Tourism Companion: Enhancing Travel Experiences with AI Chatbot and Soft Computing. In C. K. K. Reddy, T. Sithole, M. Ouaissa, Ö. ÖZER, & M. M. Hanafiah (Eds.), *Soft Computing in Industry 5.0 for Sustainability* (pp. 301–317). Springer Nature Switzerland.

Balancing Automation and Personalization in Service Industries. (n.d.). Retrieved October 7, 2025, from https://www.hospitalitynet.org/opinion/4123063.html

Baltaci, F., Başer, M. Y., & Çelik, M. (2024). Attitude towards service robots in tourism and hospitality services settings — The effect of multidimensional anthropomorphism and technology readiness. *International Journal of Tourism Research, 26*(4), e2685. Advance online publication. DOI: 10.1002/jtr.2685

Banerjee, P., & Bhat, A. (2025). Transforming Smart Hospitality: Leveraging AI and IoT for Sustainable Tourism, Enhanced Guest Experience, and Cybersecurity Resilience. *Eco-Tourism and Sustainable Development*, *1*(1), 14–26. DOI: 10.63385/etsd.v1i1.87

Banu, S., Samantaray, R. R., Ayesha, N., Fathima, A., & Tabassum, Z. (2025). Robotics and Automation in Hospitality Services. In Cutting-Edge Technologies for Business Sectors (pp. 367-400). IGI Global.

Banu, S., Samantaray, R. R., Ayesha, N., Fathima, A., & Tabassum, Z. (2025). Robotics and automation in hospitality services. In *Cutting-edge technologies for business sectors* (pp. 367–400). IGI Global.

Barnes, A. J., Zhang, Y., & Valenzuela, A. (2024). AI and culture: Culturally dependent responses to AI systems. *Current Opinion in Psychology*, *58*, 101838. DOI: 10.1016/j.copsyc.2024.101838 PMID: 39002473

Barney, J. (1991). Firm Resources and Sustained Competitive Advantage. *Journal of Management*, *17*(1), 99–120. DOI: 10.1177/014920639101700108

Basheer, S., Farooq, S., & Reshi, M. A. (2022). Tourism, the Metaverse, Artificial Intelligence, and Travel: Striking a Balance between Innovation and Concerns. *Journal of Social Responsibility. Tourism and Hospitality*, *26*, 19–30.

Basheer, S., Walia, S., & Farooq, S. (2025). Metaverse Explosion and Its Consequences for the Travel Industry, AI, and VR. In Chhabra, G., & Kaushik, K. (Eds.), *Understanding the Metaverse: Applications, Challenges, and the Future* (pp. 307–322). Springer Nature Singapore. DOI: 10.1007/978-981-97-2278-5_14

Bathla, G., Raina, A., Rana, V. S., & Bhushan, B. (2025). The Impact of Artificial Intelligence and Automation in Transforming the Hospitality Sector. In *Transforming the Service Sector with New Technology* (pp. 541–552). IGI Global Scientific Publishing. DOI: 10.4018/979-8-3693-7447-4.ch033

Baum, T., Mooney, S., Robinson, R. N. S., & Solnet, D. (2020). COVID-19's impact on the hospitality workforce—New concerns and new considerations. *International Journal of Contemporary Hospitality Management*, *32*(9), 2893–2900. DOI: 10.1108/IJCHM-04-2020-0314

Bausch, D., Kraemer, T., & Mauroner, O. (2024). Technology-induced stress and employee resistance in the context of digital transformation and identification of countermeasures. *International Journal on Management of Innovation & Technology*, *21*(04), 2450029. DOI: 10.1142/S0219877024500299

Belanche, D., Casaló, L. V., & Flavián, C. (2021). Frontline robots in tourism and hospitality: Service enhancement or cost reduction? *Electronic Markets, 31*(3), 477–492. DOI: 10.1007/s12525-020-00432-5

Belanche, D., Casaló, L. V., Flavián, C., & Schepers, J. (2019). Service robot implementation: A theoretical framework and research agenda. *Service Industries Journal, 40*(3–4), 203–225. DOI: 10.1080/02642069.2019.1672666

Ben Youssef, A., Dutta, P. K., Doshi, R., & Sajnani, M. (2024). *AI, Blockchain, and Metaverse in Hospitality and Tourism Industry 4.0: Case Studies and Analysis*. CRC Press.

Berry, K., & Singh, A. (2024). AI-Driven Service Marketing: Transforming Customer Experience and Operational Efficiency. In Nadda, V., Tyagi, P., Singh, A., & Singh, V. (Eds.), *AI Innovations in Service and Tourism Marketing* (pp. 35–56). IGI Global., DOI: 10.4018/979-8-3693-7909-7.ch003

Beyond the Front Desk: How Cloud, Mobility, and AI Are Redefining Hospitality. (n.d.). Retrieved October 7, 2025, from https://www.hospitalitynet.org/opinion/4129074.html

Bhalla, A., Singh, P., & Singh, A. (2023). Technological Advancement and Mechanization of the Hotel Industry. In Tailor, R. (Ed.), *Application and Adoption of Robotic Process Automation for Smart Cities* (pp. 57–76). IGI Global., DOI: 10.4018/978-1-6684-7193-7.ch004

Bhardwaj, P. (2018, July 5). *Robots are replacing servers and delivering fresh seafood to people's tables at Alibaba's high-tech restaurant in Shanghai*. Business Insider. https://www.businessinsider.com/alibaba-shanghai-restaurant-robothe-robots-waiters-photos-2018-7

Bhimasta, R. A., & Kuo, P. (2019). What causes the adoption failure of service robots? *UBICOMP*, 1107–1112. DOI: 10.1145/3341162.3350843

Bhinder, H. S. (n.d.). Emerging technology trends adopted in hotel guest cycle for enhancing guest stay experiences. In *Recent trends in tourism and hospitality* (p. 18).

Bhuiyan, M. S. (2024). The role of AI-Enhanced personalization in customer experiences. *Journal of Computer Science and Technology Studies, 6*(1), 162–169. DOI: 10.32996/jcsts.2024.6.1.17

Binesh, F., & Baloglu, S. (2023). Are we ready for hotel robots after the pandemic? A profile analysis. *Computers in Human Behavior, 147*, 107854. DOI: 10.1016/j.chb.2023.107854 PMID: 37389284

Blöcher, K., & Alt, R. (2020). AI and robotics in the European restaurant sector: Assessing potentials for process innovation in a high-contact service industry. *Electronic Markets*, *31*(3), 529–551. DOI: 10.1007/s12525-020-00443-2

Booking.com-sustainable-travel-report2023. (n.d.).

Booranakittipinyo, A., Li, R. Y. M., & Phakdeephirot, N. (2024). Travelers' perception of smart airport facilities: An X (Twitter) sentiment analysis. *Journal of Air Transport Management*, *118*, 102600. DOI: 10.1016/j.jairtraman.2024.102600

Bowen, J., & Morosan, C. (2018). Beware, hospitality industry: The robots are coming. *Worldwide Hospitality and Tourism Themes*, *10*(6), 726–733. DOI: 10.1108/WHATT-07-2018-0045

Breazeal, C. (2003). Toward sociable robots. *Robotics and Autonomous Systems*, *42*(3–4), 167–175. DOI: 10.1016/S0921-8890(02)00373-1

Buhalis, D., & Leung, D. (2018). Smart hospitality: Interoperability and the future of hotel operations. *International Journal of Contemporary Hospitality Management*, *30*(3), 1365–1383. DOI: 10.1108/IJCHM-08-2017-0471

Buhalis, D., Leung, D., & Lin, M. (2023). Metaverse as a disruptive technology revolutionising tourism management and marketing. *Tourism Management*, *97*, 104724. DOI: 10.1016/j.tourman.2023.104724

Buhalis, D., & Sinarta, Y. (2019). Real-time co-creation and nowness service: Lessons from tourism and hospitality. *Journal of Travel & Tourism Marketing*, *36*(5), 563–582. DOI: 10.1080/10548408.2019.1592059

Cahigas, M. M. L., Prasetyo, Y. T., Persada, S. F., & Nadlifatin, R. (2023). Examining Filipinos' intention to revisit Siargao after Super Typhoon Rai 2021 (Odette): An extension of the theory of planned behavior approach. *International Journal of Disaster Risk Reduction*, *84*, 103455. DOI: 10.1016/j.ijdrr.2022.103455

Cain, L. N., Thomas, J. H., & Alonso, M.Jr. (2019). From sci-fi to sci-fact: The state of robotics and AI in the hospitality industry. *Journal of Hospitality and Tourism Technology*, *10*(4), 624–650. DOI: 10.1108/JHTT-07-2018-0066

Cain, L., Thomas, A., & Alonso, A. (2019). Robots and the tourism industry: An analysis of current applications and future considerations. *Tourism Management Perspectives*, *31*, 218–226. DOI: 10.1016/j.tmp.2019.05.004

Canal, G., Escalera, S., & Angulo, C. (2016). A real-time Human-Robot Interaction system based on gestures for assistive scenarios. *Computer Vision and Image Understanding*, *149*, 65–77. DOI: 10.1016/j.cviu.2016.03.004

Car, T., Šuman, S., & Kliman, A. (2024). Customer Booking Habits and Attitudes Towards AI-Powered Features in Online Travel Agencies. *International Journal (Toronto, Ont.), 13*(9).

Çeltek, E. (2023). Analysis of smart technologies used in smart hotels. *Journal of Business Research – Turk, 15*(4), 2880–2892. DOI: 10.20491/isarder.2023.1754

Challoumis, C. (2024). The dawn of artificial intelligence. In *XIX International Scientific Conference*. London. Great Britain(pp. 169-205).

Chan, L., Hogaboam, L., & Cao, R. (2022). Artificial intelligence in tourism and hospitality. In *Applied innovation and technology management* (pp. 213–229). DOI: 10.1007/978-3-031-05740-3_14

Chan, E. S. W., & Hsu, C. H. C. (2016). Effects of carbon management on the financial performance of hotels. *International Journal of Hospitality Management, 57*, 172–177. DOI: 10.1016/j.ijhm.2016.06.002

Chan, H.-L., Cheung, T.-T., Choi, T.-M., & Sheu, J.-B. (2023). Sustainable successes in third-party food delivery operations in the digital platform era. *International Journal of Production Research, 61*(9), 2760–2780. DOI: 10.1080/00207543.2022.2106749 PMID: 37361094

Chatterjee, S., Chaudhari, R., Kumar, A., & Schiavone, F. (2024). Emergence of Social Robot Technology and Its Friction-Free Diffusion in Society: From Engineering Aesthetics and Compatibility Perspective. *IEEE Transactions on Engineering Management, 71*, 13179–13190. DOI: 10.1109/TEM.2023.3324149

Chen, S., Chen, H., & Chen, M. (2009). Determinants of satisfaction and continuance intention towards self-service technologies. *Industrial Management & Data Systems, 109*(9), 1248–1263. DOI: 10.1108/02635570911002306

Chen, Z. (2023). Beyond reality: Examining the opportunities and challenges of cross-border integration between metaverse and hospitality industries. *Journal of Hospitality Marketing & Management, 32*(7), 967–980. DOI: 10.1080/19368623.2023.2222029

Chen, Z. (2024). Beyond boundaries: Exploring the Metaverse in tourism. *International Journal of Contemporary Hospitality Management*. Advance online publication. DOI: 10.1108/IJCHM-06-2023-0900

Chi, O. H., Denton, G., & Gursoy, D. (2020). Artificially intelligent device use in service delivery: A systematic review, synthesis, and research agenda. *Journal of Hospitality Marketing & Management, 29*(7), 757–786. DOI: 10.1080/19368623.2020.1721394

Chi, O. H., Gursoy, D., & Chi, C. G. (2022). Tourists' attitudes toward the use of artificially intelligent (AI) devices in tourism service delivery: Moderating role of service value seeking. *Journal of Travel Research, 61*(1), 170–185. DOI: 10.1177/0047287520971054

Choi, Y., Choi, M., Oh, M., & Kim, S. S. (2020). Service robots in hotels: Understanding the service quality perceptions of human–robot interaction. *Journal of Hospitality Marketing & Management, 29*(6), 613–635. DOI: 10.1080/19368623.2020.1703871

Choudhary, A. K., Jagadeesh, R. P., Girija, E., Madhuri, M., & Shravani, N. (2023). Flyhigh:Machine Learning Based Airline Fare Prediction Model. *2023 6th International Conference on Information Systems and Computer Networks (ISCON)*, 1–8.

Chourasia, B. (2024, June). To study the technological impact and innovation on the hotel industry in relation to the employees working in the concierge section of front office. In *International Conference on Innovation and Regenerative Trends in Tourism and Hospitality Industry (IRTTHI 2024)* (pp. 158–176). Atlantis Press. DOI: 10.2991/978-94-6463-437-2_11

Chourasia, S., Kaur, S., & Singh, A. (2025). The Future of Indian Restaurants: The Role of Technology in the Success of Cloud Kitchens. In Singh, K., Kumar, S., & Talukder, M. (Eds.), *Impact of AI and the Evolution of Future Ghost Kitchens* (pp. 15–34). IGI Global Scientific Publishing., DOI: 10.4018/979-8-3373-0432-8.ch002

Cobanoglu, C., Dogan, S., Berezina, K., Collins, G., Nanu, L., Shahtakhtinskaya, K., & Parvez, O. (2021). *Hospitality and tourism information technology* (Vol. 17). University of South Florida M3 Center Publishing.

Curran, J. M., Meuter, M. L., & Surprenant, C. F. (2003). Intentions to Use Self-Service Technologies: A Confluence of Multiple Attitudes. *Journal of Service Research, 5*(3), 209–224. DOI: 10.1177/1094670502238916

Cynthia, J. P. B., & Kiruthiga, V. (2024). Impression of Metaverse on Gen Z Customers in the Hotel Industry. *Shanlax International Journal of Management, 12*(1), 24–30. DOI: 10.34293/management.v12i1.7780

Damaševičius, R. (2023). From E-commerce to V-commerce: Understanding the Impact of Virtual Reality and Metaverse on Economic Activities. *Journal of Information Economics, 1*(3), 16. DOI: 10.58567/jie01030005

Damjanov, K., & Crouch, D. (2019). Virtual reality and space tourism. In *Space tourism* (Vol. 25, pp. 117–137). Emerald Publishing Limited. DOI: 10.1108/S1571-504320190000025007

Dasgupta, D. S., & Jamader, A. R. (2024). Revolutionizing tourism and hospitality: Harnessing AI for sustainable transformation. Available at *SSRN* 4896697. DOI: 10.2139/ssrn.4896697

Davenport, T. H., & Kirby, J. (2016). *Only humans need apply: Winners and losers in the age of smart machines*. Harper Business.

Davis, F. D. (1989). Perceived usefulness, perceived ease of use, and user acceptance of information technology. *Management Information Systems Research Center, University of Minnesota, Vol. 13,*(No. 3 (Sep., 1989),), 319-340 (22 pages).

Davis, F. D. (1989). Perceived usefulness, perceived ease of use, and user acceptance of information technology. *Management Information Systems Quarterly, 13*(3), 319–340. DOI: 10.2307/249008

Delmas, M. A., & Burbano, V. C. (2011). The drivers of environmental performance: The role of "conspicuous conservation". *Journal of Environmental Economics and Management, 61*(2), 195–211. DOI: 10.1016/j.jeem.2010.10.001

Demir, Ç. (2022). Metaverse Teknolojisinin Otel Sektörünün Geleceğine Etkileri Üzerine Bir İnceleme (An Investigation into the Impact of Metaverse Technology on The Hotel Industry's Future). *Journal of Tourism & Gastronomy Studies, 10*(1), 542–555.

Deri, M. N., Zaazie, P., & Singh, A. (2024). Digital Future of the Hospitality Industry and Hospitality Education Globally. In Sharma, A. (Ed.), *International Handbook of Skill, Education, Learning, and Research Development in Tourism and Hospitality. Springer International Handbooks of Education*. Springer., DOI: 10.1007/978-981-97-4318-6_14

Dewi Andriana, S., & Damayanti, F. (2023). Desain Metaverse. *The Future of Tourism Pariban Berbasis Augmented Reality Menggunakan Metode Markerless User Defined Target., 16*(2), 172–182.

Dobbala, M. K., & Lingolu, M. S. S. (2024). Conversational AI and Chatbots: Enhancing User Experience on Websites. *American Journal of Computer Science and Technology, 10*(3), 62–70. DOI: 10.11648/j.ajcst.20240703.11

Doğan, S., & Niyet, İ. Z. (2024). Artificial Intelligence (AI) in Tourism. In *Future Tourism Trends Volume 2: Technology Advancement, Trends and Innovations for the Future in Tourism* (pp. 3–21). Emerald Publishing Limited.

Dwivedi, Y. K., Pandey, N., Currie, W., & Micu, A. (2024). Leveraging ChatGPT and other generative artificial intelligence (AI)-based applications in the hospitality and tourism industry: Practices, challenges and research agenda. *International Journal of Contemporary Hospitality Management, 36*(1), 1–12. DOI: 10.1108/IJCHM-05-2023-0686

Ejigu, A. K., & Yeshitela, K. (2024). Envisioning sustainable sanitation planning: A unified approach of diffusion of innovation and theory of planned behavior in predicting ecosan toilet adoption in Arba Minch City, Ethiopia. *Frontiers in Environmental Science, 12*, 1371659. Advance online publication. DOI: 10.3389/fenvs.2024.1371659

El-Said, O. A., & Al Tall, T. (2020). Studying the factors influencing customers' intention to use self-service kiosks in fast food restaurants. In *Information and Communication Technologies in Tourism 2020* (pp. 206–217). Springer., DOI: 10.1007/978-3-030-36737-4_17

Emily, S.-D. (2022, May 11). *Metaverse hospitality*. VML. https://www.vml.com/insight/metaverse-hospitality

European Union (EU). (2021). *Proposal for a Regulation of the European Parliament and of the Council Laying Down Harmonised Rules on Artificial Intelligence (Artificial Intelligence Act)*. European Commission.

Evans, L., Frith, J., & Saker, M. (2022). The roots of the metaverse. In *From Microverse to Metaverse: Modelling the Future through Today's Virtual Worlds* (pp. 15–24). Emerald Publishing Limited. DOI: 10.1108/978-1-80455-021-220221002

Fan, X., Wang, H., & Wang, L. (2023). P-2.8: Metaverse: Origin, Current Applications and Prospects for Future Development. *SID Symposium Digest of Technical Papers, 54*, 504–506. DOI: 10.1002/sdtp.16342

Fernandes, M. D., Wani, P. S., & Shah, G. D. (2024). The Integration of Artificial Intelligence in the Hospitality Industry. In *Hotel and Travel Management in the AI Era* (pp. 525–550). IGI Global. DOI: 10.4018/979-8-3693-7898-4.ch025

Filippova, O. Yu. (2023). International Hotel Service Industry Trends. *Visnyk of Donetsk National University of Economics and Trade Named after Mykhailo Tugan-Baranovsky, 2(79) 2023*, 63–68.

Food Safety and Standards Authority of India. (2022). *FAQs on licensing and registration under the Food Safety and Standards (Licensing and Registration of Food Businesses) Regulations, 2011*. Food Safety Compliance System (FoSCoS). https://foscos.fssai.gov.in

Fornell, C., & Larcker, D. F. (1981). Structural Equation Models with Unobservable Variables and Measurement Error: Algebra and Statistics. *JMR, Journal of Marketing Research, 18*(3), 382–388. DOI: 10.1177/002224378101800313

Francis, R. S., Anantharajah, S., Sengupta, S., & Singh, A. (2024). Leveraging ChatGPT and Digital Marketing for Enhanced Customer Engagement in the Hotel Industry. In Bansal, R., Ngah, A., Chakir, A., & Pruthi, N. (Eds.), *Leveraging ChatGPT and Artificial Intelligence for Effective Customer Engagement* (pp. 55–68). IGI Global., DOI: 10.4018/979-8-3693-0815-8.ch004

Fu, S., Zheng, X., & Wong, I. A. (2022). The perils of hotel technology: The robot usage resistance model. *International Journal of Hospitality Management, 102*, 103174. DOI: 10.1016/j.ijhm.2022.103174 PMID: 35095168

Gajić, T., Petrović, M. D., Pešić, A. M., Conić, M., & Gligorijević, N. (2024). Innovative Approaches in Hotel Management: Integrating Artificial Intelligence (AI) and the Internet of Things (IoT) to Enhance Operational Efficiency and Sustainability. *Sustainability (Basel), 16*(17), 7279. DOI: 10.3390/su16177279

García-Madurga, M.-Á., & Grilló-Méndez, A.-J. (2023). Artificial Intelligence in the tourism industry: An overview of reviews. *Administrative Sciences, 13*(8), 172. DOI: 10.3390/admsci13080172

Garrell, A., Villamizar, M., Moreno-Noguer, F., & Sanfeliu, A. (2017). Teaching robots' proactive behaviour using human assistance. *International Journal of Social Robotics, 9*(2), 231–249. DOI: 10.1007/s12369-016-0389-0

Gaur, L., Afaq, A., Singh, G., & Dwivedi, Y. K. (2021). Role of artificial intelligence and robotics to foster the touchless travel during a pandemic: A review and research agenda. *International Journal of Contemporary Hospitality Management, 33*(11), 4079–4098. DOI: 10.1108/IJCHM-11-2020-1246

Gebauer, H., Johnson, M., & Enquist, B. (2010). Value co-creation as a determinant of success in public transport services: A study of the Swiss Federal Railway operator (SBB). *Managing Service Quality, 20*(6), 511–530. DOI: 10.1108/09604521011092866

Ghosh, S., & Pherwani, J. (2015). Designing a natural voice assistant for mobile through a user-centred design approach. In *Lecture Notes in Computer Science* (pp. 320–331). DOI: 10.1007/978-3-319-20901-2_29

Gill, A., & Mathur, A. (2024). Emotional Intelligence in the Age of AI: Enhancing Workforce Development for Human-Machine Collaboration. In Sustainable Innovation for Industry 6.0 (pp. 263-285). IGI Global.

Gong, L., & Nass, C. (2007). When a Talking-Face Computer Agent is Half-Human and Half-Humanoid: Human Identity and Consistency Preference. *Human Communication Research*, *33*(2), 163–193. DOI: 10.1111/j.1468-2958.2007.00295.x

Gossling, S., & Higham, J. (2021). The low-carbon transition in tourism: A challenge for business, a necessity for humanity. *Journal of Sustainable Tourism*, *29*(6), 843–856. DOI: 10.1080/09669582.2021.1901035

Gupta, P., & Pareek, R. (2024). AI-Enabled Guest Experience: Disrupting Hospitality through Next-Generation AI Technologies. In Technology and Luxury Hospitality (pp. 165-181). Routledge.

Gursoy, D. (2025). Artificial intelligence (AI) technology, its applications and the use of AI powered devices in hospitality service experience creation and delivery. *International Journal of Hospitality Management*, *129*, 104212. DOI: 10.1016/j.ijhm.2025.104212

Gursoy, D., & Chi, C. (2022). Impact of cloud kitchens on consumer dining preferences. *International Journal of Hospitality Management*, *102*, 103157. DOI: 10.1016/j.ijhm.2022.103157

Hair, J. F., Black, W. C., Babin, B. J., & Anderson, R. E. (2019). *Multivariate Data Analysis* (8th ed.). Cengage Learning.

Han, H., Kim, S. I., Lee, J.-S., & Jung, I. (2025). Understanding the drivers of consumers' acceptance and use of service robots in the hotel industry. *International Journal of Contemporary Hospitality Management*, *37*(2), 541–559. DOI: 10.1108/IJCHM-02-2024-0163

Han, H., Yu, J., & Kim, M. (2021). The role of robot presence on customer satisfaction and behavioral intention in a hotel context. *International Journal of Hospitality Management*, *97*, 103004. DOI: 10.1016/j.ijhm.2021.103004

Harini, P., Rohini, A., Selvanayaki, S., Prahadeeswaran, M., Pangayar Selvi, R., & Shokila, C. (2025). A systematic literature review of cloud kitchen. *Plant Science Today*, *12*(1), 1–8. DOI: 10.14719/pst.6016

Harit, P., Singh, A., Agrawal, K., Verma, D., Bathla, G., & Raina, A. (2025). Exploring the Growing Threat of Cyber Risks and Vulnerabilities in the Hospitality Industry. In Srivastava, R., Rana, V., & Faiz, M. (Eds.), *Exploring the World With-Blockchain Through Cryptotravel* (pp. 299–316). IGI Global Scientific Publishing., DOI: 10.4018/979-8-3693-6562-5.ch017

Henseler, J., Ringle, C. M., & Sarstedt, M. (2015). A new criterion for assessing discriminant validity in variance-based structural equation modeling. *Journal of the Academy of Marketing Science*, *43*(1), 115–135. DOI: 10.1007/s11747-014-0403-8

Hong, M., Chung, N., Koo, C., & Koh, S.-Y. (2023). TPEDTR: Temporal preference embedding-based deep tourism recommendation with card transaction data. *International Journal of Data Science and Analytics*, *16*(2), 147–162. DOI: 10.1007/s41060-022-00380-7

Hospitality Industry in 2023: Responding to new trends. (n.d.). Retrieved October 7, 2025, from https://www.hospitalitynet.org/opinion/4114325.html

Hou, Y., Zhang, K., & Li, G. (2021). Service robots or human staff: How social crowding shapes tourist preferences. *Tourism Management*, *83*, 104242. DOI: 10.1016/j.tourman.2020.104242

Huang, D., Chen, Q., Huang, S., & Liu, X. (2024). Consumer intention to use service robots: A cognitive–affective–conative framework. *International Journal of Contemporary Hospitality Management*, *36*(6), 1893–1913. DOI: 10.1108/IJCHM-12-2022-1528

Huang, M., & Rust, R. T. (2018). Artificial intelligence in service. *Journal of Service Research*, *21*(2), 155–172. DOI: 10.1177/1094670517752459

Huang, X. (2022). Personalized travel route recommendation model of intelligent service robot using deep learning in big data environment. *Journal of Robotics*, *2022*(1), 7778592. DOI: 10.1155/2022/7778592

Hu, Y. (2021). An improvement or a gimmick? The importance of user perceived values, previous experience, and industry context in human–robot service interaction. *Journal of Destination Marketing & Management*, *21*, 100645. DOI: 10.1016/j.jdmm.2021.100645

Hwang, J., Kim, H., & Kim, W. (2019). Investigating motivated consumer innovativeness in the context of drone food delivery services. *Journal of Hospitality and Tourism Management*, *38*, 102–110. DOI: 10.1016/j.jhtm.2019.01.004

Hwang, J., Park, S., & Kim, I. (2020). Understanding motivated consumer innovativeness in the context of a robotic restaurant: The moderating role of product knowledge. *Journal of Hospitality and Tourism Management*, *44*, 272–282. DOI: 10.1016/j.jhtm.2020.06.003

Imad, A. S. (2019). Remodeling hospitality industry through artificial intelligence. *2019 13th International Conference on Software, Knowledge, Information Management and Applications (SKIMA)*, 1–7.

Irimia-Diéguez, A., Velicia-Martín, F., & Aguayo-Camacho, M. (2023). Predicting Fintech Innovation Adoption: The Mediator Role of Social Norms and Attitudes. *Financial Innovation*, *9*(1), 36. DOI: 10.1186/s40854-022-00434-6 PMID: 36687792

Ivanov, S. (2023). *Tourism and Hotel Technologies*. INSETE. https://books.google.com.my/books?id=hBy1EAAAQBAJ

Ivanov, S. H., Webster, C., & Seyyedi, P. (n.d.). Consumers' attitudes towards the introduction of robots in accommodation establishments. *ResearchGate, 66/ No. 3/ 2018/ 302-317*, 302–317.

Ivanov, S., & Webster, C. (2019). Economic fundamentals of using robots, artificial intelligence, and service automation in travel, tourism, and hospitality. In *Emerald Publishing Limited eBooks* (pp. 39–55). DOI: 10.1108/978-1-78756-687-320191002

Ivanov, S., Webster, C., & Berezina, K. (2017). *Adoption of robots and service automation by tourism and hospitality companies*. https://papers.ssrn.com/sol3/papers.cfm?abstract_id=2964308

Ivanova, M. (2019). Robots, Artificial Intelligence, and Service Automation in Travel Agencies and Tourist Information Centers. In S. Ivanov & C. Webster (Eds.), *Robots, Artificial Intelligence, and Service Automation in Travel, Tourism and Hospitality* (pp. 221–237). Emerald Publishing Limited.

Ivanov, S. H., & Webster, C. (2019). The next wave of automation in tourism, travel, and hospitality—A new field of research. *Journal of Tourism Futures*, *5*(2), 114–120. DOI: 10.1108/JTF-05-2019-048

Ivanov, S. H., & Webster, C. (2020). *The robotisation of travel, tourism and hospitality: Research, theory and practice*. CABI.

Ivanov, S. H., Webster, C., & Berezina, K. (2017). Adoption of robots and service automation by travel, tourism and hospitality companies. *International Journal of Contemporary Hospitality Management*, *29*(10), 2901–2917. DOI: 10.1108/IJCHM-08-2016-0492

Ivanov, S., Gretzel, U., Berezina, K., Sigala, M., & Webster, C. (2019). Progress on robotics in hospitality and tourism: A review of the literature. *Journal of Hospitality and Tourism Technology*, JHTT-08-2018-0087. Advance online publication. DOI: 10.1108/JHTT-08-2018-0087

Ivanov, S., Seyitoğlu, F., & Markova, M. (2020). Hotel managers' perceptions towards the use of robots: A mixed-methods approach. *Information Technology & Tourism*, *22*(4), 505–535. DOI: 10.1007/s40558-020-00187-x

Ivanov, S., Seyitoğlu, F., & Webster, C. (2022). The economics of service robots in hospitality companies: Task automatability, operational impact and external drivers. *Service Industries Journal.* Advance online publication. DOI: 10.1177/13548166251362622

Ivanov, S., & Webster, C. (Eds.). (2019). *Robots, Artificial Intelligence, and Service Automation in Travel, Tourism and Hospitality*. Emerald Publishing Limited., DOI: 10.1108/9781787566873

Ivanov, S., Webster, C., & Garanina, A. (2022). *Robots, Artificial Intelligence, and Service Automation in Travel, Tourism and Hospitality 2.0.* Emerald Publishing.

Jain, S., Aithal, S., Allamsetty, V., K S, S., Gobi, Dr. N., & Rathi, Dr. R. (2023). Smart Luggage Tracking using IoT and GPS Technology. *IJARCCE, 12*(4).

Jaipuriar, S. (2021). Role of technology in accommodation operation department of hotel: An employee's perspective. In *Hospitality and tourism: Emerging practices in human resource management* (p. 70).

James, B. (2021, December 4). *Marriott Bonvoy logs into the metaverse with debut of travel-inspired NFTs*. Marriott News Center. https://news.marriott.com/news/2021/12/04/marriott-bonvoy-logs-into-the-metaverse-with-debut-of-travel-inspired-nfts

Jang, H. W., & Lee, S. B. (2020). Serving robots: Management and applications for restaurant business sustainability. *Sustainability (Basel), 12*(10), 3998. DOI: 10.3390/su12103998

Jang, Y., & Park, E. (2022). Satisfied or not: User experience of mobile augmented reality in using natural language processing techniques on review comments. *Virtual Reality (Waltham Cross), 26*(3), 839–848. DOI: 10.1007/s10055-021-00599-y

Jiang, J., Liu, R., Zhou, H., Shi, L., & Xu, L. (2025). The influence of the perception of human–robot interaction on customer value co-creation intention: A case study of hotel service robots. *Journal of Hospitality and Tourism Technology, 16*(5), 1004–1023. DOI: 10.1108/JHTT-05-2024-0293

Jia, S., Chi, O. H., Martinez, S. D., & Lu, L. (2023). When "Old" Meets "New": Unlocking the Future of Innovative Technology Implementation in Heritage Tourism. *Journal of Hospitality & Tourism Research (Washington, D.C.)*, •••, 10963480231205768.

Jin, N., Line, N. D., & Goh, B. (2016). Experiential value, relationship quality, and customer loyalty in the context of upscale hotel brands. *Journal of Travel Research, 55*(3), 356–370. DOI: 10.1177/0047287514555848

Jogarao, M. (2024). AI in the hospitality industry: Revolutionizing guest experience. In *Technology and luxury hospitality* (pp. 199–210). Routledge. DOI: 10.4324/9781003488248-14

John, K. T. (2021). Digital disruption: The hyperlocal delivery and cloud kitchen driven future of food services in post-COVID India. *International Hospitality Review, 37*(1), 161–187. DOI: 10.1108/IHR-06-2021-0045

Juquelier, A., Poncin, I., & Hazée, S. (2025). Empathic chatbots: A double-edged sword in customer experiences. *Journal of Business Research, 188*, 115074. DOI: 10.1016/j.jbusres.2024.115074

Kabadayi, S., Ali, F., Choi, H., Joosten, H., & Lu, C. (2019). Smart service experience in hospitality and tourism services. *Journal of Service Management, 30*(3), 326–348. DOI: 10.1108/JOSM-11-2018-0377

Kadagidze, L., & Ugrelidze, E. (2023). AI in Hospitality industry: a comprehensive study on its impact on operations, customer experience, and revenue management. Conference: VII International European Conference on Interdisciplinary Scientific Research

Kalampokas, T., Tziridis, K., Kalampokas, N., Nikolaou, A., Vrochidou, E., & Papakostas, G. A. (2023). A holistic approach on airfare price prediction using machine learning techniques. *IEEE Access : Practical Innovations, Open Solutions, 11*, 46627–46643. DOI: 10.1109/ACCESS.2023.3274669

Kalıpçı, B., & Nebioğlu, O. (2020). Planlı Davranış Teorisi ve Turizm Üzerine Yapılan Uluslararası Yayınların Bibliyometrik Analizi. *Turk Turizm Arastirmalari Dergisi, 4*(1), 1–14. DOI: 10.26677/TR1010.2020.296

Kandampully, J., Zhang, T., & Bilgihan, A. (2015). Customer loyalty: A review and future directions with a special focus on the hospitality industry. *International Journal of Contemporary Hospitality Management, 27*(3), 379–414. DOI: 10.1108/IJCHM-03-2014-0151

Kanimozhi, U., Ganapathy, S. D. M., & Kannan, A. (2021). *A User Preference Tree based Personalized Route Recommendation System for Constraint Tourism and Travel.* DOI: 10.21203/rs.3.rs-755856/v1

Kannan, R. (2024). Revolutionizing the Tourism Industry through Artificial Intelligence: A Comprehensive Review of AI Integration, Impact on Customer Experience, Operational Efficiency, and Future Trends. *International Journal for Multidimensional Research Perspectives, 2*(2), 1–14. https://www.chandigarhphilosophers.com/index.php/ijmrp/article/view/115

Kapogianni, I. (2025). Enhancing the guest experience in smart hotels through technology: A glimpse into the future.

Kapoor, A., Balaji, M. S., & Kannan, P. K. (2023). Algorithmic bias in service robots: A path to ethical service delivery. *Journal of Service Research, 26*(4), 509–529. DOI: 10.1177/10946705231154561

Kapur, P., & Williams, J. D. (2025). Balancing efficiency and human touch: The role of AI and robotics in hospitality. *Artificial Intelligence, Machine Learning, &. Robotics in Business*, *1*(1), 138286. DOI: 10.32473/aimlrb.1.1.138286

Karthiyayini, J., & Anandhi, R. J. (2024). To Analyze the Various Machine Learning Algorithms That Can Effectively Process Large Volumes of Data and Extract Relevant Information for Personalized Travel Recommendations. *SN Computer Science*, *5*(4), 336. DOI: 10.1007/s42979-024-02667-x

Karwatzki, S., Dytynko, O., Trenz, M., & Veit, D. (2017). Beyond the personalization–privacy paradox: Privacy valuation, transparency features, and service personalization. *Journal of Management Information Systems*, *34*(2), 369–400. DOI: 10.1080/07421222.2017.1334467

Kattara, H. S., & El-Said, O. A. (2013). Customers' preferences for new technology-based self-services versus human interaction services in hotels. *Tourism and Hospitality Research, 13*(2), 67–82. DOI: 10.1177/1467358413519261

Kaul, D. (2023). Ai-driven real-time inventory management in hotel reservation systems: Predictive analytics, dynamic pricing, and integration for operational efficiency. *Emerging Trends in Machine Intelligence and Big Data*, *15*(10), 66–80.

Kaushik, A. K., Agrawal, A. K., & Rahman, Z. (2015). Tourist behaviour towards self-service hotel technology adoption: Trust and subjective norm as key antecedents. *Tourism Management Perspectives*, *16*, 278–289. DOI: 10.1016/j.tmp.2015.09.002

Kaye, S.-A., Lewis, I., Buckley, L., Gauld, C., & Rakotonirainy, A. (2020). To share or not to share: A theoretically guided investigation of factors predicting intentions to use fully automated shared passenger shuttles. *Transportation Research Part F: Traffic Psychology and Behaviour*, *75*, 203–213. DOI: 10.1016/j.trf.2020.10.010

Kazak, A. N., Chetyrbok, P. V., & Oleinikov, N. N. (2020). Artificial intelligence in the tourism sphere. *IOP Conference Series. Earth and Environmental Science*, *421*(4), 042020. DOI: 10.1088/1755-1315/421/4/042020

Kesorn, K., Juraphanthong, W., & Salaiwarakul, A. (2017). Personalized attraction recommendation system for tourists through check-in data. *IEEE Access : Practical Innovations, Open Solutions*, *5*, 26703–26721. DOI: 10.1109/ACCESS.2017.2778293

Kim, J., & Lee, H. (2023). Sustainable operations of online and offline restaurants: The role of multi-brand online restaurant models and environmental performance. *Sustainability, 16*(21), 9607. DOI: 10.3390/su16219607

Kim, S., Kim, J., Badu-Baiden, F., Giroux, M., & Choi, Y. (2020). Do you prefer robot service or human service in hotels? Impacts of the COVID-19 pandemic. *International Journal of Hospitality Management, 93*, 102795. DOI: 10.1016/j.ijhm.2020.102795 PMID: 36919174

King, C. A. (1995). What is hospitality? *International Journal of Hospitality Management, 14*(3–4), 219–234. DOI: 10.1016/0278-4319(95)00045-3

Kiptanui, M. J., Karei, R., & Makomere, J. (2024). Effect of Automated Check in on Service Delivery in Star-Rated Hotels in Uasin Gishu County, Kenya. *Journal of Frontiers in Humanities and Social Sciences, 2*(3), 68–79.

Konar, R., Bhutia, L. D., Fuchs, K., & Balasubramanian, K. (2024). Role of Virtual Reality Technology in Sustainable Travel Behaviour and Engagement Among Millennials. In El Archi, Y., Benbba, B., Dávid, L., & Cardoso, L. (Eds.), *Promoting Responsible Tourism With Digital Platforms* (pp. 1–19). IGI Global Scientific Publishing. DOI: 10.4018/979-8-3693-3286-3.ch001

Koohang, A., Nord, J., Ooi, K., Tan, G., Al-Emran, M., Aw, E., Baabdullah, A., Buhalis, D., Cham, T., Dennis, C., Dutot, V., Dwivedi, Y. K., Hughes, L., Mogaji, E., Pandey, N., Phau, I., Raman, R., Sharma, A., Sigala, M., & Wong, L.-W. (2023). Shaping the metaverse into reality: A holistic multidisciplinary understanding of opportunities, challenges, and avenues for future investigation. *Journal of Computer Information Systems, 63*(3), 735–765. DOI: 10.1080/08874417.2023.2165197

Kouroupi, N., & Metaxas, T. (2023). Can the Metaverse and Its Associated Digital Tools and Technologies Provide an Opportunity for Destinations to Address the Vulnerability of Overtourism? In *Tourism and Hospitality* (Vol. 4, Issue 2, pp. 355–373). DOI: 10.3390/tourhosp4020022

Kraus, J., Miller, L., Klumpp, M., Babel, F., Scholz, D., Merger, J., & Baumann, M. (2024). On the Role of Beliefs and Trust for the Intention to Use Service Robots: An Integrated Trustworthiness Beliefs Model for Robot Acceptance. *International Journal of Social Robotics, 16*(6), 1223–1246. DOI: 10.1007/s12369-022-00952-4

Kulshreshtha, K., & Sharma, G. (2022). From restaurant to cloud kitchen: Survival of the fittest during COVID-19. *Technological Forecasting and Social Change, 179*, 121629. DOI: 10.1016/j.techfore.2022.121629 PMID: 35317191

Kumar, A. V. S., Mosikan, M., Dutta, A., & Musirin, I. Bin, Akula, V. S. G., Khalid, S., Yalavarthi, S. B., DV, C. S., Suganya, R. V, & Dulhare, U. N. (2025). AI-Driven Personalization in Tourism: Balancing Innovation with Ethical Challenges. In *AI Technologies for Personalized and Sustainable Tourism* (pp. 31–58). IGI Global.

Kumari, M., Kumar, S., & Devi, N. (2025). The impact of service automation and robotics on hospitality business models. In *Metaverse and sustainable business models in SMEs* (pp. 317–338). IGI Global Scientific Publishing. DOI: 10.4018/979-8-3693-9005-4.ch013

Kumar, S., Poudyal, A., Choudhury, S., & Jha, A. (2025). Analysing customer preferences for AI-Enabled service robots' physical appearance in the tourism and hospitality industry: Insights and Exploration. *Tourism Planning & Development*, 1–18. DOI: 10.1080/21568316.2025.2461008

Kuo, M. H., Chen, C. M., & Tseng, Y. C. (2017). The effect of perceived robot usefulness, ease of use, and trust on the intention to use hotel service robots. *International Journal of Hospitality Management*, *67*, 28–39. DOI: 10.1016/j.ijhm.2017.07.001

Kuo, M. H., Chen, C. M., & Tseng, Y. C. (2022). Examining the continuous use intention of service robots in the hospitality industry: A perspective of service-dominant logic. *International Journal of Contemporary Hospitality Management*, *34*(3), 1269–1290. DOI: 10.1108/IJCHM-05-2021-0676

Laachach, A., Mumtaz, S., & Andaloussi, B. Z. (2024). Unveiling opportunities and challenges of the metaverse in the tourism and the hospitality sector: A qualitative investigation. *Tourism and Hospitality Management*, *30*(4). Advance online publication. DOI: 10.20867/thm.30.4.1

Lakshmikanth, R., Vijai, C., Krishna, S. H., Alla, P. C. R., Lourens, M., & Soni, M. (2024, March). Utilizing AI to tailor leadership styles and interventions for optimal employee engagement and performance. In *2024 International Conference on Trends in Quantum Computing and Emerging Business Technologies* (pp. 1-7). IEEE.

Lashley, C. (2008). Studying Hospitality: Insights from Social Sciences1. *Scandinavian Journal of Hospitality and Tourism*, *8*(1), 69–84. DOI: 10.1080/15022250701880745

Ledger, I. (2021, December 6). *Marriott launches NFT Collection at contemporary art fair*. Ledger Insights - Blockchain for Enterprise. https://www.ledgerinsights.com/marriott-launches-nft-collection-at-contemporary-art-fair/

Lee, C. J. (2011). Understanding bank service quality in customers' terms: An exploratory analysis of top-of-mind definition. *International Journal of Business and Social Science*, *2*(21), 1–7.

Lee, K.-J. (2022). Effects of service robots in the hospitality and tourism sectors: A critical review and future research directions. *International Journal of Tourism & Hospitality Research*, *36*(4), 21–33. DOI: 10.21298/IJTHR.2022.4.36.4.21

Lee, K.-J. (2024). Understanding service robots: A systematic literature review of hospitality employees' perspectives. *Tourism Management*. Advance online publication. DOI: 10.1016/j.tourman.2023.104456

Lee, S., Shulga, L. V., Busser, J. A., & Hwang, J. (2024). Self-Service Hotel Check-In: Is There Still a Role for the In-Person Experience? *International Journal of Hospitality & Tourism Administration*, •••, 1–28.

Lee, W. H., Lin, C. W., & Shih, K. H. (2018). A technology acceptance model for the perception of restaurant service robots for trust, interactivity, and output quality. *International Journal of Mobile Communications*, *16*(4), 361–376. DOI: 10.1504/IJMC.2018.092666

Lee, Y., Lee, S., & Kim, D. Y. (2021). Exploring hotel guests' perceptions of using robot assistants. *Tourism Management Perspectives*, *37*, 100781. DOI: 10.1016/j.tmp.2020.100781

Lei, S. S. I., Wu, D., Liu, G. G., & Law, R. (2025). Customizing customer journey in hotels: A focus on human touch. *International Journal of Hospitality Management*, *124*, 103956. DOI: 10.1016/j.ijhm.2024.103956

Le, K. B. Q., Sajtos, L., & Fernandez, K. V. (2023). Employee-(ro) bot collaboration in service: An interdependence perspective. *Journal of Service Management*, *34*(2), 176–207. DOI: 10.1108/JOSM-06-2021-0232

Li, B., Lai, E. Y., & Wang, X. (2025, June 21). (Shane). (2025). From Tools to Agents: Meta-Analytic Insights into Human Acceptance of AI. *Journal of Marketing*, 00222429251355266. Advance online publication. DOI: 10.1177/00222429251355266

Liu, R. (2023). Development of a travel recommendation algorithm based on multi-modal and multi-vector data mining. *PeerJ. Computer Science*, *9*, e1436. DOI: 10.7717/peerj-cs.1436

Liu, X., & Li, Y. (2025). Examining the double-edged sword effect of AI usage on work engagement: The moderating role of core task characteristics substitution. *Behavioral Sciences (Basel, Switzerland)*, *15*(2), 206. DOI: 10.3390/bs15020206 PMID: 40001838

Liu, X., Zhang, L., Chen, H., & Li, Y. (2024). The evolution of smart hotels: Current trends and future directions. *Tourism Management Perspectives*, *51*, 101081. DOI: 10.1016/j.tmp.2024.101081

Londhe, K., Dharmadhikari, N., Zaveri, P., & Sakoglu, U. (2024). Enhanced Travel Experience using Artificial Intelligence: A Data-driven Approach. *Procedia Computer Science*, *235*, 1920–1928. DOI: 10.1016/j.procs.2024.04.182

Lukanova, G., & Ilieva, G. (2019). Robots, artificial intelligence, and service automation in hotels. In *Robots, artificial intelligence, and service automation in travel, tourism and hospitality* (pp. 157–183). Emerald Publishing Limited. DOI: 10.1108/978-1-78756-687-320191009

Lu, L., Cai, R., & Gursoy, D. (2019). Developing and validating a service robot integration willingness scale. *International Journal of Hospitality Management*, *80*, 36–51. DOI: 10.1016/j.ijhm.2019.01.005

Luong, T.-B. (2025). Green consumption intention of tourist in tourist destinations: A moderation of destination social responsibility in the Theory of Planned Behavior model. *Tourism Recreation Research*, *50*(2), 339–355. DOI: 10.1080/02508281.2023.2281058

Lusch, R. F., & Nambisan, S. (2015). Service Innovation: A Service-Dominant Logic Perspective1. *Management Information Systems Quarterly*, *39*(1), 155–175. DOI: 10.25300/MISQ/2015/39.1.07

M'hamed, E. G., & El Idrissi, R. (2024). Transforming Hospitality: Harnessing Artificial Intelligence for Enhanced Guest Experience and Operational Efficiency. *The International Workshop on Big Data and Business Intelligence*, 173–185. DOI: 10.1007/978-3-031-65014-7_17

Maheshwari, P., Singh, S., Singh, A., & Ansari, H. M. (2024). Framing AI adoption in tourism: The role of sustainability and responsible innovation. *Journal of Vacation Marketing*, *30*(1), 1–19. DOI: 10.1177/13565381231189474

Makivić, R., Vukolić, D., Veljović, S., Bolesnikov, M., Dávid, L. D., Ivanišević, A., Silić, M., & Gajić, T. (2024). AI Impact on Hotel Guest Satisfaction via Tailor-Made Services: A Case Study of Serbia and Hungary. *Information (Basel)*, *15*(11), 700. https://doi.org/10.3390/info15110700DOI: 10.3390/info15110700

Mandić, D., Panić, A., & Čičin-Šain, M. (2023). Automated service delivery in hotels: balancing efficiency and human interaction for optimal guest satisfaction. *International Scientific Conference on Economy, Management and Information Technologies*, Vol. 1: 167-174. DOI - DOI: 10.46793/ICEMIT23.137M

Manoharan, G., & Ashtikar, S. P. (2025). AI Era in Hospitality: The New Frontier in Guest Satisfaction and Operations. In Digital Disruption in Hospitality, AI, and Emerging Technologies: A Roadmap to Personalized Experiences, Enhanced Operations, and Revenue Growth (pp. 59-77). Emerald Publishing Limited.

Marriott International Annual Report 2024.pdf. (n.d.).

Marriott-2024-Annual-Report. (n.d.). Marriott Investor Relations. https://marriott.gcs-web.com/financial-information/annual-reports

Ma, S. (2024). Enhancing Tourists' Satisfaction: Leveraging Artificial Intelligence in the Tourism Sector. *Pacific International Journal, 7*(3), 89–98. DOI: 10.55014/pij.v7i3.624

Mathieson, K. (1991). Predicting User Intentions: Comparing the Technology Acceptance Model with the Theory of Planned Behavior. *Information Systems Research, 2*(3), 173–191. DOI: 10.1287/isre.2.3.173

Mbarushimana, S. (2024). The Impact of Smart Room Technology on Guest Satisfaction and Operational Efficiency. *Journal of Modern Hospitality., 3*(2), 1–13. DOI: 10.47941/jmh.1956

McColl-Kennedy, J. R., Vargo, S. L., Dagger, T. S., Sweeney, J. C., & Kasteren, Y. V. (2012). Health Care Customer Value Cocreation Practice Styles. *Journal of Service Research, 15*(4), 370–389. DOI: 10.1177/1094670512442806

McLean, G., & Barhorst, J. B. (2022). Living the experience before you go... but did it meet expectations? The role of virtual reality during hotel bookings. *Journal of Travel Research, 61*(6), 1233–1251. DOI: 10.1177/00472875211028313

Melián-González, S., & Bulchand-Gidumal, J. (2016). The role of information technology in increasing job satisfaction in the hotel sector. *Tourism Management, 55*, 220–229. DOI: 10.1016/j.tourman.2016.02.008

Millennium. (2024). *M Social Hotel Auckland.* Millennium Hotels and Resorts. https://www.millenniumhotels.com/en/auckland/m-social-auckland/

Milton, T.Dr.T. Milton. (2024). Artificial Intelligence Transforming Hotel Gastronomy: An In-depth Review of AI-driven Innovations in Menu Design, Food Preparation, and Customer Interaction, with a Focus on Sustainability and Future Trends in the Hospitality Industry. *International Journal for Multidimensional Research Perspectives, 2*(3), 47–61. DOI: 10.61877/ijmrp.v2i3.126

Mishra, D., Das, S., & Patnaik, R. (2024). Application of AI Technology for the Development of Destination Tourism towards an Intelligent Information System. *Economic Affairs*, *69*(02), 1083–1095.

Mithas, S., Ramasubbu, N., & Sambamurthy, V. (2011). How Information Management Capability Influences Firm Performance1. *Management Information Systems Quarterly*, *35*(1), 237–256. DOI: 10.2307/23043496

Mohanty, P., Singh, K., & Lakshmi, R. B. (2025). Can Artificial Intelligence Revolutionize Crisis Management in Tourism?: Exploring the Implications and Challenges. In *Tracking Tourism Patterns and Improving Travel Experiences With Innovative Technologies* (pp. 83–104). IGI Global.

Morosan, C. (2016). An empirical examination of U.S. travelers' intentions to use biometric e-gates in airports. *Journal of Air Transport Management*, *55*, 120–128. DOI: 10.1016/j.jairtraman.2016.05.005

Morosan, C., & DeFranco, A. (2016). Modeling guests' intentions to use mobile apps in hotels: The roles of personalization, privacy, and involvement. *International Journal of Contemporary Hospitality Management*, *28*(9), 1968–1991. DOI: 10.1108/IJCHM-07-2015-0349

Moser, J. (2022). Look but don't touch: Contactless solutions to enhance the customer journey in hotels (Master's thesis, Universidade Catolica Portuguesa, Portugal).

Mujtaba, B. G. (2025). Human-AI Intersection: Understanding the Ethical Challenges, Opportunities, and Governance Protocols for a Changing Data-Driven Digital World. *Business Ethics and Leadership*, *9*(1), 109–126. DOI: 10.61093/bel.9(1).109-126.2025

Mukherjee, S., Baral, M. M., Venkataiah, C., Pal, S. K., & Nagariya, R. (2021). Service robots are an option for contactless services due to the COVID-19 pandemic in hotels. *Decision (Washington, D.C.)*, *48*(4), 445–460. DOI: 10.1007/s40622-021-00300-x

Mukherjee, S., Baral, M. M., Venkataiah, C., Pal, S. K., & Nagariya, R. (2021). Service robots are an option for contactless services due to the COVID-19 pandemic in the hotels. *Decision (Washington, D.C.)*, *48*(4), 445–460.

Müller, C., & Kutzbach, N. (2022). World Robotics 2022 – Service Robots. In B. Graf & K. Pfeiffer (Eds.), *IFR Statistical Department, VDMA Services GmbH*. VDMA Services GmbH. https://ifr.org/img/worldrobotics/Executive_Summary_WR_Service_Robots_2022.pdf

Murphy, J., Hofacker, C., & Gretzel, U. (2017). The robots are coming: Implications for the tourism industry. *Tourism Management Perspectives*, *21*, 5–7. DOI: 10.1016/j.tmp.2016.11.006

Mustikasari, A., Hurriyati, R., Dirgantari, D. P., Sultan, A. M., & Sugiana, S. N. (2025). The Role of Artificial Intelligence in Brand Experience: Shaping Consumer Behavior and Driving Repurchase Decisions. *International Journal of Advanced Computer Science and Applications*, *16*(4), •••. DOI: 10.14569/IJACSA.2025.0160432

Nagadeepa, C., Jaheer Mukthar, K. P., Ramirez-Asís, E., Huaranga-Toledo, H., Espinoza-Maguiña, M., & Mory-Guarnizo, S. (2023). A reinforcer to become a foodpreneur through cloud kitchen: Sustainable technology-driven business model during pandemic. In Al Mubarak, M., & Hamdan, A. (Eds.), *Technological Sustainability and Business Competitive Advantage* (pp. 401–413). Springer., DOI: 10.1007/978-3-031-35525-7_25

Nair, A. J., Manohar, S., Mittal, A., & Khanna, V. (2023). Revolutionizing Tourism and Hospitality Services: Integrating AI in the Metaverse. *2023 6th International Conference on Contemporary Computing and Informatics (IC3I), 6*, 1206–1212.

Naumov, N. (2019). The impact of robots, artificial intelligence, and service automation on service quality and service experience in hospitality. In *Emerald Publishing Limited eBooks* (pp. 123–133). DOI: 10.1108/978-1-78756-687-320191007

Neuhofer, B., Buhalis, D., & Ladkin, A. (2015). Smart tourism ecosystem: Frameworks for smart destination development. *International Journal of Tourism Research*, *17*(5), 458–469. DOI: 10.1002/jtr.2001

Neuhofer, B., Buhalis, D., & Ladkin, A. (2015). Technology as a Catalyst of Change: Enablers and Barriers of the Tourist Experience and Their Consequences. In Tussyadiah, I., & Inversini, A. (Eds.), *Information and Communication Technologies in Tourism 2015* (pp. 789–802). Springer International Publishing., DOI: 10.1007/978-3-319-14343-9_57

Nica, I., Tazl, O. A., & Wotawa, F. (2018). Chatbot-based Tourist Recommendations Using Model-based Reasoning. *ConfWS*, 25–30.

Nira, R. A. (2025). AI-Driven Hyper-Personalization in Hospitality: Application, Present and Future Opportunities, Challenges, and Guest Trust Issues. *International Journal of Research and Innovation in Social Science*, *9*(4), 5562–5573. DOI: 10.47772/IJRISS.2025.90400397

Nomura, T., Kanda, T., Suzuki, T., & Kato, K. (2008). Prediction of human behaviour in human-robot interaction using psychological scales for anxiety and negative attitudes toward robots. *IEEE Transactions on Robotics*, *24*(2), 442–451. DOI: 10.1109/TRO.2007.914004

Ocean. (2024a). *"horigin" metaverse hotel platform*. Ocean Sky Hotel Management Limited. https://www.oceansky.com.hk/en/horigin-metaverse-hotel-platform-en/

Ocean. (2024b). *Popway hotel hong kong*. Ocean Sky Hotel Management Limited. https://www.oceansky.com.hk/en/popway-hotel-hong-kong-en/

Omar, S. I., Aliah, N., & Kwong, T. A. (2024). The Ethical Implications of AI-Powered Personalization in Hospitality and Tourism. Impact of AI and Tech- Driven Solutions in Hospitality and Tourism. IGI Global, 103-122. DOI:10.4018/979-8-3693-6755-1.ch006

Oracle's top hospitality industry predictions for 2023. (n.d.). Retrieved October 7, 2025, from https://www.duettocloud.com/library/oracles-top-hospitality-industry-predictions-for-2023

Osawa, H., Minato, T., Shimada, M., & MacDorman, K. F. (2017). The uncanny valley: A pervasive phenomenon. *Frontiers in Psychology*, *8*, 1943. DOI: 10.3389/fpsyg.2017.01943

Ottenbacher, M., & Harrington, R. J. (2007). The innovation development process of Michelin-starred chefs. *International Journal of Contemporary Hospitality Management*, *19*(6), 444–460. DOI: 10.1108/09596110710775110

Palrão, T., Rodrigues, R. I., Madeira, A., Mendes, A. S., & Lopes, S. (2023). Robots in tourism and Hospitality: The perception of future professionals. *Human Behavior and Emerging Technologies*, *2023*, 1–12. DOI: 10.1155/2023/7172152

Pappas, I., Mikalef, P., Dwivedi, K. Y., Jaccheri, L., & Krogstie, J. (2023). Responsible Digital Transformation for a Sustainable Society. *Information Systems Frontiers*, *25*(3), 945–953. https://doi.org/10.1007/s10796-023-10406-5DOI: 10.1007/s10796-023-10406-5 PMID: 37287710

Parasuraman, A. (2000). Technology Readiness Index (Tri): A Multiple-Item Scale to Measure Readiness to Embrace New Technologies. *Journal of Service Research*, 2(4), 307–320. DOI: 10.1177/109467050024001

Parasuraman, A., & Colby, C. L. (2015). An Updated and Streamlined Technology Readiness Index: TRI 2.0. *Journal of Service Research*, *18*(1), 59–74. DOI: 10.1177/1094670514539730

Park, S. (2020). Multifaceted trust in tourism service robots. *Annals of Tourism Research, 81*, 102888. DOI: 10.1016/j.annals.2020.102888

Park, S.-M., & Kim, Y.-G. (2022). A metaverse: Taxonomy, components, applications, and open challenges. *IEEE Access : Practical Innovations, Open Solutions, 10*, 4209–4251. DOI: 10.1109/ACCESS.2021.3140175

Parthipan, S., Guruprasath, R., & Maha Sriram, S. (2025, June). Optimizing Hotel Performance through AI-Powered Revenue Forecasting and IoT Data Analysis. In *2025 International Conference on Emerging Technologies in Engineering Applications (ICETEA)* (pp. 1-5). IEEE.

Peter, R., Roshith, V., Lawrence, S., Mona, A. E., Narayanan, K. B., & Yusaira, F. (2025). Gen AI – Gen Z: Understanding Gen Z's emotional responses and brand experiences with Gen AI-driven, hyper-personalized advertising. *Frontiers in Communication, 10*, 1554551. DOI: 10.3389/fcomm.2025.1554551

Post, K. (2019). The neuroscience of trust. *Science, 363*(6432), 1187.16-1189. DOI: 10.1126/science.363.6432.1187-p

Praharaj, S., Mishra, B., Mishra, U., Panigrahi, R. R., & Mishra, P. C. (2023). Role of service automation on guest experience of hotel industry. *Tourism and Hospitality Management, 29*(2), 265–278. DOI: 10.20867/thm.29.2.11

Pratisto, E. H., Thompson, N., & Potdar, V. (2022). Immersive technologies for tourism: A systematic review. *Information Technology & Tourism, 24*(2), 181–219. DOI: 10.1007/s40558-022-00228-7

Qi, H., Han, Z., & Feng, X. (2024). Research on Hotel Service Robot Management System Based on Artificial Intelligence. *2024 3rd International Conference on Artificial Intelligence and Autonomous Robot Systems (AIARS)*, 265–269.

Qiu, H., Li, M., Shu, B., & Bai, B. (2019). Enhancing hospitality experience with service robots: The mediating role of rapport building. *Journal of Hospitality Marketing & Management, 29*(3), 247–268. DOI: 10.1080/19368623.2019.1645073

Raghuram, V. (2022). *A study on the functions and applications of robots within the housekeeping department of a hotel. International Journal of Advanced Research in Science, Communication and Technology (IJARSCT), 2*(4). Anjuman-I-Islam's Institute of Hospitality Management, Mumbai, India. https://ijarsct.co.in/A13483.pdf

Rahi, S., Ghani, M. A., Al-Okaily, M., Rashid, A., Alghizzawi, M., & Shiyyab, F. S. (2024). Breaking into the black box of customer perception towards robot service: Empirical evidence from service sector. *Heliyon, 10*(19), e38117. DOI: 10.1016/j.heliyon.2024.e38117 PMID: 39386864

Ramachandran, K. K., Vinjamuri, L. P., Ramesh, R., Al-Taee, M., & Alazzam, M. B. (2023). Using AI for risk management and improved business resilience. *2023 3rd International Conference on Advance Computing and Innovative Technologies in Engineering (ICACITE)*, 978–982.

Rane, N., Choudhary, S., & Rane, J. (2024). Artificial Intelligence (AI), Internet of Things (IoT), and blockchain-powered chatbots for improved customer satisfaction, experience, and loyalty. *Internet of Things (IoT), and Blockchain-Powered Chatbots for Improved Customer Satisfaction, Experience, and Loyalty (May 29, 2024).*

Rane, N., Choudhary, S., & Rane, J. (2023, October 31). Sustainable tourism development using leading-edge Artificial Intelligence (AI), Blockchain, Internet of Things (IoT), Augmented Reality (AR) and Virtual Reality (VR) technologies. *Blockchain, Internet of Things (IoT), Augmented Reality (AR) and Virtual Reality (VR). Technologies.*

Ravichandra, T., Murugeswari, P., Revathi, M., Senthilkumar, S., Rathore, D. S., & Sudhakar, M. (2025). The Future of Machine Learning and Robotics in Digital Technology for Hospitality. In *Cutting-Edge Technologies for Business Sectors* (pp. 401–428). IGI Global.

Ray, N., Dey, P., Chakraborty, M., & Bandyopadhyay, M. A. (2024). A Study on AI Application in Travel and Tourism Industry. *Redefining Commerce and Management: New Paradigms for the Digital Age (Volume 1)*, 222.

Ray, A., Dhir, A., Bala, P. K., & Kaur, P. (2019). Why do people use food delivery apps (FDA)? A uses and gratification theory perspective. *Journal of Retailing and Consumer Services*, *51*, 221–230. DOI: 10.1016/j.jretconser.2019.05.025

Reis, J., Melão, N., Salvadorinho, J., Soares, B., & Rosete, A. (2020). Service robots in the hospitality industry: The case of Henn-na hotel, Japan. *Technology in Society*, *63*, 101423. Advance online publication. DOI: 10.1016/j.techsoc.2020.101423

Rekiek, S., Jebari, H., & Reklaoui, K. (2024). Prediction of Booking Trends and Customer Demand in the Tourism and Hospitality Sector Using AIBased Models: A Case Study of Major Hotel Chain. *International Journal of Advanced Computer Science and Applications*, *15*(10). Advance online publication. DOI: 10.14569/IJACSA.2024.0151043

Rodríguez-Lizundia, E., Marcos, S., Zalama, E., Gómez-García-Bermejo, J., & Gordaliza, A. (2015). A bellboy robot: Study of the effects of robot behaviour on user engagement and comfort. *International Journal of Human-Computer Studies*, *82*, 83–95. DOI: 10.1016/j.ijhcs.2015.06.001

Rogers, E. M. (2003). *Diffusion of innovations* (5th ed.). Free Press.

Sabharwal, D., & Jain, V. (2025). Transforming Hospitality through AI, Big Data, and Automation: The New Era of Service Innovation. In Addressing Contemporary Challenges in the B2B Hospitality Sector (pp. 309-336). IGI Global Scientific Publishing.

Samala, N., Katkam, B. S., Bellamkonda, R. S., & Rodriguez, R. V. (2020). Impact of AI and robotics in the tourism sector: A critical insight. *Journal of Tourism Futures*, *8*(1), 73–87. DOI: 10.1108/JTF-07-2019-0065

Sam, S. J. I., & Jasim, K. M. (2025). Diving into the technology: A systematic literature review on strategic use of chatbots in hospitality service encounters. *Management Review Quarterly*, *75*(1), 527–555. DOI: 10.1007/s11301-023-00388-8

Saravanakumar, S., & Narayanan, M. B. (2018). The service automation and robotics in hospitality industry: A study on business implications. *International Journal of Mechanical and Production Engineering Research and Development*, *8*(6), 91–100. DOI: 10.24247/ijmperddec201810

Saurabh, D. (2024). *Hopper Travel app: Business Model, Development Cost, & features*. https://www.addwebsolution.com/blog/hopper-travel-app-business-model-app-development-cost-features

Seyitoğlu, F., & Ivanov, S. (2023). Humans and/or robots? Tourists' preferences toward the humans-robots mix in the service delivery systems of tourism and hospitality companies. *Frontiers in Psychology*, *14*, 9786514. DOI: 10.3389/fpsyg.2023.9786514

Sharma, R. (2024). Genetic Algorithm Based Personalized Travel Recommendation System. *2024 2nd International Conference on Intelligent Data Communication Technologies and Internet of Things (IDCIoT)*, 867–874.

Sharma, M., & Singh, A. (2024a). Enhancing Competitive Advantages Through Virtual Reality Technology in the Hotels of India. In Kumar, S., Talukder, M., & Pego, A. (Eds.), *Utilizing Smart Technology and AI in Hybrid Tourism and Hospitality* (pp. 243–256). IGI Global., DOI: 10.4018/979-8-3693-1978-9.ch011

Sharma, M., & Singh, A. (2024b). Embracing Technological Innovation: A Review of Hi-Tech Services in Hospitality Industry. *Evergreen*, *11*(4), 2818–2830. DOI: 10.5109/7326926

Sharma, M., Tyagi, P. K., Semwal, R., Tyagi, P., Singh, V. J., & Kumari, N. (2024, June). Robotic concierges and smart spaces: A vision for the future of hospitality services. In *International Conference on Innovation and Regenerative Trends in Tourism and Hospitality Industry (IRTTHI 2024)* (pp. 511–525). Atlantis Press. DOI: 10.2991/978-94-6463-437-2_34

Sharma, R., & Singh, A. (2024c). Blockchain Technologies and Call for an Open Financial System: Decentralised Finance. In Vardari, L., & Qabrati, I. (Eds.), *Decentralized Finance and Tokenization in FinTech* (pp. 21–32). IGI Global., DOI: 10.4018/979-8-3693-3346-4.ch002

Shelke, P. M., Dedgaonkar, S., & Bhimanpallewar, R. N. (2024). Powering User Interface Design of Tourism Recommendation System with AI and ML. In *Artificial Intelligence, Machine Learning and User Interface Design* (pp. 108–135). Bentham Science Publishers. DOI: 10.2174/9789815179606124010008

Shin, H., & Kang, J. (2020). The effect of service robots on customer emotional experience and behavioral intention in the hotel industry during the COVID-19 pandemic. *Journal of Hospitality and Tourism Management*, *44*, 290–298. DOI: 10.1016/j.jhtm.2020.07.009

Shirzad, A., & Rahmani, A. (2024). Smart Treasury: Leveraging Artificial Intelligence and Robotic Process Automation for Financial Excellence. *Knowledge Economy Studies*, *1*(1), 65–86.

Shi, Y., Zhan, L., & Lin, J. (2025). Adaptive learning for service robots in unstructured hospitality environments. *Artificial Intelligence in Tourism and Hospitality*, *2*(1), 44–63. DOI: 10.1016/j.aihth.2025.01.004

Singh, A., & Bathla, G. (2023). Fostering Creativity and Innovation: Tourism and Hospitality Perspective. In P. Tyagi, V. Nadda, V. Bharti, & E. Kemer (Eds.), Embracing Business Sustainability through Innovation and Creativity in the Service Sector (pp. 70-83). IGI Global.

Singh, A., & Kumar, V. (2025a). Organizational Innovation and Adaptation in Hospitality and Tourism: Strategies for Resilience and Leadership. In C. Ramos, T. Costa, R. Madeira, & J. Ferreira do Rosário (Eds.), Open Innovation and Technology in Tourism and Hospitality (pp. 355-372). IGI Global Scientific Publishing. DOI: 10.4018/979-8-3693-8633-0.ch018

Singh, A., & Kumar, V. (2025b). User-Centered Hospitality: UI/UX as the New Frontier in Hotel and Restaurant Innovation. In C. Ramos, T. Costa, R. Madeira, & J. Ferreira do Rosário (Eds.), Open Innovation and Technology in Tourism and Hospitality (pp. 395-410). IGI Global Scientific Publishing. DOI: 10.4018/979-8-3693-8633-0.ch020

Singh, A. (2024). Virtual Research Collaboration and Technology Application: Drivers, Motivations, and Constraints. In Chakraborty, S. (Ed.), *Challenges of Globalization and Inclusivity in Academic Research* (pp. 250–258). IGI Global., DOI: 10.4018/979-8-3693-1371-8.ch016

Singh, A., & Hassan, S. C. (2024). Service Innovation Through Blockchain Technology in the Tourism and Hospitality Industry: Applications, Trends, and Benefits. In Singh, S. (Ed.), *Service Innovations in Tourism: Metaverse, Immersive Technologies, and Digital Twin* (pp. 205–214). IGI Global., DOI: 10.4018/979-8-3693-1103-5.ch010

Söderberg Granström, D. E. A. D. E. A., Pronk, A. M., & Criscione-Naylor, N.. (2023). Robotic services in the hotel industry: An examination of Henn-Na Hotels. In *International Journal of Gaming Hospitality and Tourism* (Vol. 1). https://stockton.edu/light/documents/ijght_vol.3no.1/robotic_services_henn_na_hotels-6.7.23.pdf

Sofi, M. R., Bashir, I., Alshiha, A., Alnasser, E., & Alkhozaim, S. (2025). Creating exceptional guest experiences: The role of engagement and relationship building in hospitality. *Journal of Hospitality and Tourism Insights*, *8*(3), 891–914. DOI: 10.1108/JHTI-04-2024-0318

Solakis, K., Katsoni, V., Mahmoud, A. B., & Grigoriou, N. (2024). Factors affecting value co-creation through artificial intelligence in tourism: A general literature review. *Journal of Tourism Futures*, *10*(1), 116–130. DOI: 10.1108/JTF-06-2021-0157

Soliman, M., Ivanov, S., Elgammal, I., Webster, C., & Al-Alawi, A. (2024). Pursuing the drivers of consumer behaviour toward service robots in the hotel industry. *International Journal of Hospitality Management*, *120*, 103752. DOI: 10.1016/j.ijhm.2024.103752

Song, B., Cai, Z., & Xu, H. (2025). How to shape consumer preference for service robots? A meta-analysis of use intention. *Journal of Service Theory and Practice*, •••, 1–22. DOI: 10.1108/JSTP-12-2024-0425

Soofastaei, A. (2025). Intelligent scheduling: how AI and advanced analytics are revolutionizing time optimization. In *Mastering Time-Innovative Solutions to Complex Scheduling Problems*. IntechOpen. DOI: 10.5772/intechopen.1008531

Soori, M., Arezoo, B., & Dastres, R. (2023). Artificial intelligence, machine learning and deep learning in advanced robotics, a review. *Cognitive Robotics*, *3*, 54–70. DOI: 10.1016/j.cogr.2023.04.001

Sousa, A. E., Cardoso, P., & Dias, F. (2024). The Use of Artificial Intelligence Systems in Tourism and Hospitality: The Tourists' Perspective. *Administrative Sciences*, *14*(8), 165. DOI: 10.3390/admsci14080165

Srivastava, A., Singh, B. M., Singh, A., & Singh, A. (2025). Customer Immersiveness and Customer Retention: A Metaverse-Driven Study for Brands. In M. Al Aqad, A. SorayyaeiAzar, A. Albattat, & A. Singh (Eds.), Navigating AI and the Metaverse in Scientific Research (pp. 475-500). IGI Global Scientific Publishing. DOI: 10.4018/979-8-3373-0340-6.ch023

Srivastava, P. & Tyagi, K. P. (2024). Implications of AI in Hotel Housekeeping Operations. Hotel and Travel Management in the AI Era, IGI Global. DOI - DOI: 10.4018/979-8-3693-7898-4.ch016

Steinhoff, L., & Martin, K. D. (2023). Putting data privacy regulation into action: The differential capabilities of service frontline interfaces. *Journal of Service Research*, *26*(3), 330–350. DOI: 10.1177/10946705221141925

Štilić, A., Nicić, M., & Puška, A. (2023). Check-in to the future: Exploring the impact of contemporary information technologies and artificial intelligence on the hotel industry. *Turističko poslovanje, (31).*

Stringam, B. B., & Gerdes, J. H. (2021). *Hotel and guest room technology* (Vol. 17). University of South Florida M3 Publishing.

Sultana, F., Nikhat, R., & Ahmed, B. (2025). Feeding the future: Trust, technology, and sustainability in cloud kitchens. *International Journal of Management*, *16*(3), 154–171. DOI: 10.34218/IJM_16_03_011

Sun, J., & Zhang, R. (2023). Ethical AI in hospitality: Data privacy and customer trust. *Journal of Hospitality Technology*, *19*(3), 145–163.

Sun, Y., Konar, R., & Wei, C. K. (2025). Unlocking the power of employee advocacy: The role of internal communication and engagement in five-star hotels. *International Journal of Hospitality Management*, *126*, 104086. DOI: 10.1016/j.ijhm.2025.104086

Švancár, S., Chrpa, L., Dvořák, F., & Balyo, T. (2024). Cloud kitchen: Using planning-based composite AI to optimize food delivery processes. *arXiv preprint* arXiv:2402.10725. https://doi.org//arXiv.2402.10725DOI: 10.48550

Syarul, M. (2023). Potensi Dan Hambatan Metaverse Di Era Terkini. *Jurnal Mahasiswa Teknik Informatika, 7*(3).

Tai, Y.-H., Nwachukwu, P. T. T., LePage, B. A., & Fang, W.-T. (2024). Examining customer intentions to purchase intelligent robotic products and services in Taiwan using the theory of planned behaviour. *BMC Psychology, 12*(1), 351. DOI: 10.1186/s40359-024-01683-z PMID: 38879510

Talukder, M. B., Rahman, M. Z., & Khan, M. R. (2025). Enhancing Guest Experience through Smart Hotel Technologies. In *Adapting to Evolving Consumer Experiences in Hospitality and Tourism* (pp. 301–324). IGI Global.

Tamuliene, V., Diaz-Meneses, G., & Vilkaite-Vaitone, N. (2024). The Cultural Roots of Green Stays: Understanding Touristic Accommodation Choices Through the Lens of the Theory of Planned Behavior. *Sustainability (Basel), 16*(22), 9947. DOI: 10.3390/su16229947

Tang, L., & Jiang, J. (2024). Enhancing the Combined-TAM-TPB model with trust in the sharing economy context: A meta-analytic structural equation modeling approach. *Journal of Cleaner Production, 442*, 141168. DOI: 10.1016/j.jclepro.2024.141168

Taniguchi, A., Isobe, S., El Hafi, L., Hagiwara, Y., & Taniguchi, T. (2020). Autonomous planning based on spatial concepts to tidy up home environments with service robots. *arXiv.* https://arxiv.org/abs/2002.03671

Tan, K. L., Hofman, P. S., Noor, N., Tan, S. R., Hii, I. S. H., & Cham, T. H. (2024). Does artificial intelligence improve hospitality employees' individual competitive productivity? A time-lagged moderated-mediation model involving job crafting and meaningful work. *Current Issues in Tourism, 28*(20), 3284–3301. DOI: 10.1080/13683500.2024.2391114

Taufiqur Rahman, B. S., & Waked, N. H. (2025). The impact of artificial intelligence on enhancing guest experience in luxury hotels in Kuala Lumpur, Malaysia. [DOI: 10.22271/27069583.2025.v7.i1c.141]. *International Journal of Tourism Hotel Management, 7*(1), 208–216.

Thakur, A., Kaipa, K., Banerjee, A. G., Cappelleri, D. J., Krovi, V. N., & Gupta, S. K. (2025). Physical AI for Powering the Next Revolution in Robotics. *Journal of Computing and Information Science in Engineering*, •••, 1–21.

Thea, S. (2018). *Hopper's AI-driven prediction models can help you save big on airfare.* https://outsideinsight.com/insights/hoppers-ai-driven-prediction-models-can-help-you-save-big-on-airfare/

Thenahandi, H. (2023). *The effects of self-service automation on customer experience during the check-in stage at Helsinki Vantaa airport*. https://urn.fi/URN:NBN:fi:amk-202302082195

Tian, S., & Song, Q. (2023). Study on the Technical Model Design of a Tourist Attraction from the Metaverse Perspective. *Business. Economics and Management BEMS*, *6*, 371–380.

Tiwari, S. (2021). Space tourism: An initiative pushing limits. *Journal of Tourism Leisure and Hospitality*, *3*(1), 38–46. DOI: 10.48119/toleho.862636

Toichoa Eyam, A., Mohammed, W. M., & Martinez Lastra, J. L. (2021). Emotion-driven analysis and control of human-robot interactions in collaborative applications. *Sensors (Basel)*, *21*(14), 4626. DOI: 10.3390/s21144626 PMID: 34300366

Tornatzky, L. G., & Fleischer, M. (1990). *The processes of technological innovation*. Lexington Books.

Torres-Ruiz, M., Mata, F., Zagal, R., Guzmán, G., Quintero, R., & Moreno-Ibarra, M. (2020). A recommender system to generate museum itineraries applying augmented reality and social-sensor mining techniques. *Virtual Reality (Waltham Cross)*, *24*(1), 175–189. DOI: 10.1007/s10055-018-0366-z

Torrey, C., Fussell, S. R., & Kiesler, S. (2013). How a robot should give advice. *2016 11th ACM/IEEE International Conference on Human-Robot Interaction (HRI)*, 275–282. DOI: 10.1109/HRI.2013.6483599

Trihas, N., Zozonakis, G., & Kourgiantakis, M. (2022, May). Contactless Tourism in the Post-COVID-19 Era: How Ready Are Greek Tourists? In International Conference on Modern Trends in Business Hospitality and Tourism (pp. 103-118). Cham: Springer Nature Switzerland.

Troise, C., O'Driscoll, A., Tani, M., & Prisco, A. (2020). Online food delivery services and behavioural intention – a test of an integrated TAM and TPB framework. *British Food Journal*, *123*(2), 664–683. DOI: 10.1108/BFJ-05-2020-0418

Tul-Krzyszczuk, A. (2023). The metaverse in hospitality management in the age of digital transformation. *Zeszyty Naukowe Szkoły Głównej Gospodarstwa Wiejskiego w Warszawie. Ekonomika i Organizacja Logistyki*, *8*(4), 105–120. DOI: 10.22630/EIOL.2022.7.4.31

Tung, V. W. S., & Au, N. (2018). Exploring customer experiences with robotics in hospitality. *International Journal of Contemporary Hospitality Management*, *30*(7), 2680–2697. DOI: 10.1108/IJCHM-06-2017-0322

Tuo, Y., Wu, J., Zhao, J., & Si, X. (2024). Artificial intelligence in tourism: insights and future research agenda. *Tourism Review, ahead-of-print*(ahead-of-print). DOI: 10.1108/TR-03-2024-0180

Tuomi, A., Tussyadiah, I. P., & Stienmetz, J. (2020). Applications and implications of service robots in hospitality. *Cornell Hospitality Quarterly, 62*(2), 232–247. DOI: 10.1177/1938965520923961

Tuomi, A., Tussyadiah, I., & Stienmetz, J. (2020). Service robots and the changing roles of employees in restaurants: A cross-cultural study. *Ereview of Tourism Research, 17*(5), 836–841.

Tussyadiah, I. (2020a). A review of research into automation in tourism: Launching the Annals of Tourism Research Curated Collection on Artificial Intelligence and Robotics in Tourism. *Annals of Tourism Research, 81*, 102883. DOI: 10.1016/j.annals.2020.102883

Tussyadiah, I. P., & Park, S. (2018). Consumer Evaluation of Hotel Service Robots. In Stangl, B., & Pesonen, J. (Eds.), *Information and Communication Technologies in Tourism 2018* (pp. 308–320). Springer International Publishing., DOI: 10.1007/978-3-319-72923-7_24

Tussyadiah, I. P., & Park, S. (2018). Consumer evaluation of robotic service: The role of anthropomorphism. *International Journal of Tourism Research, 20*(7), 804–815. DOI: 10.1002/jtr.2198

Tussyadiah, I. P., Zach, F. J., & Wang, J. (2020). Do travelers trust intelligent service robots? *Annals of Tourism Research, 81*, 102886. DOI: 10.1016/j.annals.2020.102886

Tussyadiah, I. P., Zach, F., & Wang, J. (2020). Attitudes toward service robots in the hotel industry: Insights from an online review analysis. *Journal of Hospitality and Tourism Technology, 11*(1), 173–193. DOI: 10.1108/JHTT-05-2018-0050

U.S. Food and Drug Administration. (2017). *Food Code 2017: Recommendations of the United States Public Health Service.* https://www.fda.gov/media/110822/download

United Nations World Tourism Organization (UNWTO). (2023). The Glasgow Declaration on Climate Action in Tourism.

United Nations World Tourism Organization (UNWTO). (2024). Tourism and Climate Action: Global Commitments and Progress.

Uniyal, R. S. D. M., & Sharma, N. (2020). Emerging use of artificial intelligence, robots & service automation in hotel industry. *International Journal of Future Generation Communication and Networking*, *13*(3), 2785–2790.

Valtonen, A. & Holopainen, M.,(2025). Mitigating employee resistance and achieving well-being in digital transformation. Information Technology and People, Vol. 38 (8): 42-72. DOI-DOI: 10.1108/ITP-05-2024-0701

Vargo, S. L., & Lusch, R. F. (2004). Evolving to a New Dominant Logic for Marketing. *Journal of Marketing*, *68*(1), 1–17. DOI: 10.1509/jmkg.68.1.1.24036

Vargo, S. L., & Lusch, R. F. (2008). Service-dominant logic: Continuing the evolution. *Journal of the Academy of Marketing Science*, *36*(1), 1–10. DOI: 10.1007/s11747-007-0069-6

Vashishth, T. K., Sharma, V., Sharma, M. K., & Sharma, R. (2025). Enhancing Hotel Customer Service with AI-Powered Chatbots. In Marketing Technology-Infused Hospitality: Upskilling Frontline Employees for Competitiveness (pp. 83-114). IGI Global Scientific Publishing.

Vashishth, T. K., Sharma, V., Sharma, K. K., Chaudhary, S., Kumar, B., & Kumar, P. (2025). Graphs, language models, and NLP: The future of search engines. In *Applied Graph Data Science* (pp. 91–104). Morgan Kaufmann. DOI: 10.1016/B978-0-443-29654-3.00011-9

Velikova, N., Bouzdine-Chameeva, T., Ben Nasr, I., Sanz-Ibáñez, C., Anton Clavé, S., & Rabassa-Figueras, N. (2025). Drivers of green wine tourism among young adults: Intentions, attitudes, and expansion to sustainable behaviour. *Current Issues in Tourism*, •••, 1–19. DOI: 10.1080/13683500.2025.2559313

Venkatesh, V., & Davis, F. D. (2000). A Theoretical Extension of the Technology Acceptance Model: Four Longitudinal Field Studies. *Management Science*, *46*(2), 186–204. DOI: 10.1287/mnsc.46.2.186.11926

Venkatesh, V., Morris, M. G., Davis, G. B., & Davis, F. D. (2003). User Acceptance of Information Technology: Toward A Unified View1. *Management Information Systems Quarterly*, *27*(3), 425–478. DOI: 10.2307/30036540

Venkatesh, V., Thong, J. Y. L., & Xu, X. (2012). Consumer Acceptance and Use of Information Technology: Extending the Unified Theory of Acceptance and Use of Technology1. *Management Information Systems Quarterly*, *36*(1), 157–178. DOI: 10.2307/41410412

Venkateswaran, P. S., Sriramkumar, M., Vaddy, R. K., Kotagiri, A., Rajest, S. S., & Regin, R. (2024). *Applications of artificial intelligence and robots in service industries. In advancing intelligent networks through distributed optimization*. IGI Global.

Vo, N. T. (2019). Development of restaurant serving robot using line following approach. *Journal of Science and Technology Issue on Information and Communications Technology, 17*(12.1), 1. DOI: 10.31130/JST-UD2019-087

Wade, M., & Hulland, J. (2004). *Review:* The Resource-Based View and Information Systems Research: Review, Extension, and Suggestions For Future Research1. *Management Information Systems Quarterly, 28*(1), 107–142. DOI: 10.2307/25148626

Walters, M. L., Syrdal, D. S., Dautenhahn, K., Boekhorst, R. T., & Koay, K. L. (2007). Avoiding the uncanny valley: Robot appearance, personality and consistency of behaviour in an attention-seeking home scenario for a robot companion. *Autonomous Robots, 24*(2), 159–178. DOI: 10.1007/s10514-007-9058-3

Wang, H., Ning, H., Lin, Y., Wang, W., Dhelim, S., Farha, F., Ding, J., & Daneshmand, M. (2023). A survey on the metaverse: The state-of-the-art, technologies, applications, and challenges. *IEEE Internet of Things Journal, 10*(16), 14671–14688. DOI: 10.1109/JIOT.2023.3278329

Wang, X. V., & Wang, L. (2021). A literature survey of the robotic technologies during the COVID-19 pandemic. *Journal of Manufacturing Systems, 60*, 823–836. Advance online publication. DOI: 10.1016/j.jmsy.2021.02.005 PMID: 33612914

Wang, Y.-C., & Uysal, M. (2024). Artificial intelligence-assisted mindfulness in tourism, hospitality, and events. *International Journal of Contemporary Hospitality Management, 36*(4), 1262–1278. DOI: 10.1108/IJCHM-11-2022-1444

Wan, Y., Chan, E., & Luo, J. (2020). *The rise of service robots in the hospitality industry: Some actionable insights*. Boston University Hospitality Review.

Watts, S., & Munir, T. (2025). Bridging the gap: Exploring innovation enablers, challenges and AI adoption for enhanced workforce productivity. *International Journal of Productivity and Performance Management*, •••, 1–17. DOI: 10.1108/IJPPM-01-2025-0001

Wirtz, J., Patterson, P. G., Kunz, W. H., Gruber, T., Lu, V. N., Paluch, S., & Martins, A. (2018). Brave new world: Service robots in the frontline. *Journal of Service Management, 29*(5), 907–931. DOI: 10.1108/JOSM-04-2018-0119

World Economic Forum. (2023). *Travel & tourism development index 2023: Rebuilding for a sustainable and resilient future*. https://www.weforum.org/reports/travel-tourism-development-index-2023

World Travel & Tourism Council (WTTC). (2024). [A new era for Travel & Tourism.]. *Sustainability Report*, •••, 2024.

Xiaoxin, L., Konar, R., & Kim, M. J. (2026). The impact of human-robot collaboration on hotel employees: A meta-analytic exploration of behavioral intentions and influencing factors from the UTAUT perspective. *International Journal of Hospitality Management*, *133*, 104492. DOI: 10.1016/j.ijhm.2025.104492

Xiaoxin, L., Liqun, C., Konar, R., Sthapit, E., Garrod, B., & Weimei, C. (2025). Towards a better understanding of enhancing task performance in resort hotels. *Tourism and Hospitality Research*, •••, 14673584251314282. DOI: 10.1177/14673584251314282

Xu, T., Zheng, Y., Zhang, J., & Wang, Z. (2025). "They" threaten my work: How AI service robots negatively impact frontline hotel employees through organisational dehumanisation. *International Journal of Hospitality Management*, *128*, 104162. DOI: 10.1016/j.ijhm.2025.104162

Yan, Z., Konar, R., & Balasubramanian, K. (2024). The future of hotel robotics, AI, and service automation in practice. In *Advances in e-business research series* (pp. 249–278). DOI: 10.4018/979-8-3693-9699-5.ch011

Yan, Z., Konar, R., & Balasubramanian, K. (2025). The Future of Hotels Robotics, AI, and Service Automation in Practice. In Sustainability, Innovation, and Consumer Preference (pp. 249-278). IGI Global Scientific Publishing.

Yan, Z., Konar, R., Sthapit, E., Balasubramanian, K., Chen, L. and Prentice, C. (2025a), Opting for smart hotels: do digital immigrants perceive differently than digital natives do?, *Journal of Hospitality and Tourism Insights*, Vol. ahead-of-print No. ahead-of-print. https://doi.org/DOI: 10.1108/JHTI-08-2024-0785

Yang, J. (2024). The investigation of bionic-companionship framework for Humanoid Service Robotics (HSR) with deep learning image captioning (Doctoral dissertation, Cardiff Metropolitan University).

Yang, X., Zhang, L., & Feng, Z. (2024). Personalized tourism recommendations and the E-tourism user experience. *Journal of Travel Research*, *63*(5), 1183–1200. DOI: 10.1177/00472875231187332

Yan, Z., Balasubramanian, K., Konar, R., Chen, L., & Wei, Y. (2024). Smart hotel in the eyes of the silver: Developing and testing the silver tourists' behavioural intention scale. *Current Issues in Tourism*, •••, 1–20. DOI: 10.1080/13683500.2024.2381249

Yan, Z., Konar, R., & Balasubramanian, K. (2025). The future of hotels: Robotics, AI, and service automation in practice. In *Sustainability, innovation, and consumer preference* (pp. 249–278). IGI Global Scientific Publishing.

Ye, H., Sun, S., & Law, R. (2022). A review of Robotic applications in Hospitality and tourism research. *Sustainability (Basel), 14*(17), 10827. DOI: 10.3390/su141710827

Yoganathan, V., Osburg, V.-S., Fronzetti Colladon, A., Charles, V., & Toporowski, W. (2025). Societal Attitudes Toward Service Robots: Adore, Abhor, Ignore, or Unsure? *Journal of Service Research, 28*(1), 93–111. DOI: 10.1177/10946705241295841

Yu, C. (2019). Humanlike robots as employees in the hotel industry: Thematic content analysis of online reviews. *Journal of Hospitality Marketing & Management, 29*(1), 22–38. DOI: 10.1080/19368623.2019.1592733

Zahidi, F., Kaluvilla, B. B., & Mulla, T. (2024). Embracing the new era: Artificial intelligence and its multifaceted impact on the hospitality industry. Journal of Open Innovation: Technology, Market, and Complexity. DOI. Vol. 10(4 https://doi.org/10.1016/j.joitmc.2024.100390

Zaman, A., Majib, M. S., Tanjim, S. A., Siddique, S. M. A., Islam, S., Aadeeb, M. S., & Islam, M. N. (2022). UVC-PURGE: A novel cost-effective disinfection robot for combating COVID-19 pandemic. *IEEE Access : Practical Innovations, Open Solutions, 10*, 37613–37634. DOI: 10.1109/ACCESS.2022.3163243 PMID: 35582495

Zaman, M., Hasan, P. R., Vo-Thanh, T., Shams, R., Rahman, M., & Jasim, K. M. (2025). Adopting the metaverse in the luxury hotel business: A cost–benefit perspective. *International Journal of Contemporary Hospitality Management, 37*(4), 1309–1331. DOI: 10.1108/IJCHM-08-2023-1265

Zeng, Z., Chen, P., & Lew, A. A. (2020). From high-touch to high-tech: COVID-19 drives robotics adoption. *Tourism Geographies, 22*(3), 724–734. DOI: 10.1080/14616688.2020.1762118

Zhang, J. (2018). Big data and tourism geographies – an emerging paradigm for future study? *Tourism Geographies, 20*(5), 899–904. DOI: 10.1080/14616688.2018.1519719

Zhang, W. (2022). Research on the Application of Computer Artificial Intelligence Technology in the Response Mechanism of Emergency Logistics System. *2022 IEEE International Conference on Advances in Electrical Engineering and Computer Applications (AEECA)*, 676–680. DOI: 10.1109/AEECA55500.2022.9918994

Zhao, Y.. (2024). Emotional intelligence in service robots: Adaptive human–robot interaction models. *IEEE Transactions on Affective Computing*.

Zhong, L., Coca-Stefaniak, J. A., Morrison, A. M., Yang, L., & Deng, B. (2022). Technology acceptance before and after COVID-19: No-touch service from hotel robots. *Tourism Review, 77*(4), 1062–1080. DOI: 10.1108/TR-06-2021-0276

Zhong, L., Xu, Z., Morrison, A. M., Li, Y., & Zhu, M. (2024). Metaverse customer journeys in tourism: Building viable virtual worlds. *Tourism Review*, *79*(8), 1409–1426. DOI: 10.1108/TR-07-2023-0492

Zhou, M., & Yan, Y. (2023). The Research on the Application of Metaverse in the Cultural and Tourism Industry. *Advances in Economics. Management and Political Sciences*, *47*(1), 109–113.

About the Contributors

Ankit Shukla is an accomplished educator and researcher with a robust background in the hospitality and tourism industry. His career spans various roles, from frontline food and beverage service to academic leadership and extensive research contributions. Ankit's journey exemplifies a blend of practical experience, academic acumen, and a passion for advancing knowledge in his field. Early Career and Industry Experience Ankit's career in the hospitality industry began at the Hyatt Regency Chandigarh in September 2015, where he worked in the Food & Beverage Service department. His responsibilities included taking room service orders, managing point-of-sale systems, and setting up banquets for events. This role provided Ankit with hands-on experience in the operational aspects of the hospitality industry. Following his time at the Hyatt Regency, Ankit transitioned into academia, joining Kumaun University Nainital as a teaching personnel in September 2016. He took on multiple responsibilities, including lecturing MBA Tourism students, overseeing practical training, and designing course curricula. This marked the beginning of his journey in education, where he contributed to shaping the next generation of hospitality professionals. Academic Leadership and Teaching Ankit's academic career continued to flourish as he assumed the role of Assistant Professor at SIHMCT & AN Tehri Garhwal in July 2017. He became the head of the Food & Beverage Service department, where he coordinated events, designed course curricula, and managed the department's overall functioning. From February 2018 to April 2022, Ankit served as an Assistant Professor at Shoolini University Solan. Here, he expanded his responsibilities to include training and placement coordination, NAAC coordination for the Hospitality Management department, and handling operations of a hotel with eight rooms. His leadership skills and ability to manage complex tasks were instrumental in the smooth functioning of the university's hospitality department. Contributions to Research and Academia Ankit's passion for research is evident through his significant contributions to academic literature. He has presented research papers at various international conferences, focusing on topics such as ecology, local and regional food in destination marketing, and the influence of technological amenities on hotel guest satisfaction. His research has been published in reputable journals, with manuscripts addressing safety and security measures in hotels, tobacco consumption in India, and the role of Indian spices in Indian history. In addition to his research publications, Ankit has filed 40 patents, copyrights, and utility applications, with several already granted. His commitment to innovation and intellectual property rights underscores his dedication to advancing knowledge in his field. Current Role and Future Endeavors As of July 2023, Ankit is a lecturer at Raja Mangala University of Technology Thanyaburi in Thailand. In this role, he continues to inspire and educate students in the field of hospitality while pursuing his research interests. Ankit's extensive experience in both industry and academia, coupled with his passion for teaching and research, positions him as a key figure in the ongoing development of the hospitality and tourism sector. Ankit Shukla's journey from industry practitioner to esteemed academic is a testament to his dedication, expertise, and unwavering commitment to excellence in hospitality and education.

Nagendra Yadav is a highly regarded expert in hospitality, bringing over 12 years of extensive experience in both academia and the hotel industry. He holds a Ph.D. in Hospitality and a Master's degree in Hotel Management from Kurukshetra University. Currently, he serves as a distinguished faculty member at the Welcomgroup Graduate School of Hotel Administration (WGSHA) at Manipal Academy of Higher Education (MAHE) in Manipal, India. With a robust professional background, Dr. Yadav has successfully collaborated with prestigious hotel brands, including Taj Hotels, The Leela Group, Lemon Tree Hotels, and Piccadilly Hotel, in addition to his academic tenure at Shoolini University. His significant contributions to research and academia include over 45 patents and copyrights, alongside more than 15 Scopus-indexed papers and books published in leading journals. Dr. Yadav's research interests are pivotal in the field, focusing on gastronomic tourism, culinary heritage, hospitality innovation, and experiential learning. His ability to merge industry insights with academic rigor positions him as an influential leader in shaping the future of hospitality professionals.

Rohit Chauhan is an Assistant Professor at the Mittal School of Business, Lovely Professional University, India. His core expertise spans sustainable tourism development, business analytics, digital marketing, and computer applications. He has extensive teaching and research experience, with publications in journals and book chapters indexed in Scopus and Web of Science. Dr. Chauhan is also skilled in quantitative data analysis, curriculum development, and academic administration, and is committed to fostering research-driven innovation in the fields of tourism and business analytics.

Abhishek Ghai is an accomplished academic and hospitality professional with over 15 years of combined experience in teaching and the hospitality industry. He currently serves as an Associate Professor at Maharishi Markandeshwar (Deemed to be University), Mullana, Ambala. He has previously held esteemed academic positions at Panjab University, Chandigarh, and Kurukshetra University. Dr. Ghai holds a Ph.D. in Tourism and Hotel Management from Kurukshetra University, specializing in bar service operations within the Indian hotel industry. An alumnus of both Bangalore University and Kurukshetra University, he is a recognized expert in food and beverage service, alcoholic beverages, and molecular mixology. His hands-on training at international bars in the UAE and professional experience with luxury hotel brands such as One & Only Royal Mirage (Dubai), ITC Hotels, Taj Hotels, Radisson bring practical insights to his academic work. A prolific researcher and published author, Dr. Ghai has contributed numerous peer-reviewed articles, book chapters, and a co-edited book. His research interests span bar operations, food and beverage management and event management.

Preeti Agrawat is an accomplished academic and professional with over 15 years of combined experience in industry and academia. She is currently serving as an Associate Professor at Welcomgroup Graduate School of Hotel Administration (WGSHA), Manipal, where she contributes her expertise in Human Resource Management, Organizational Behavior, HR Analytics, Learning & Development, Public Policy, and Development Administration. Throughout her career, Dr. Agrawat has successfully bridged theory and practice, bringing industry insights into academic settings while fostering research-driven approaches to human resource development. Her areas of specialization include human capital management, organizational behavior analysis, and the application of HR analytics in decision-making, enabling her to mentor students and professionals in navigating the evolving business landscape. Dr. Agrawat's professional journey reflects her passion for nurturing talent, advancing organizational effectiveness, and contributing to policy-level discussions in development administration. With her dedication to both teaching and research, she continues to make a significant impact in academia and beyond.

Shahanawaj Ahamad is member of International Association of Engineers, The Society of Digital Information and Wireless Communications, Global Professors Welfare Association, International Association of Academic plus Corporate (IAAC), Computer Science Teachers Association and Internet Society .

Anamika Ahirwar, Ph.D., is a Professor in the Department of Computer Science & Engineering at Compucom Institute of Technology & Management, Jaipur, Rajasthan, India. With over 21+ years of academic and research experience, she earned her Ph.D. in Computer Applications from Rajiv Gandhi Technical University, Bhopal, where her doctoral research focused on data mining schemes for medical imaging. Her areas of expertise include medical imaging, data science, machine learning, sentiment analysis, and celestial sound. Dr. Ahirwar is a prolific researcher with more than 80 publications in SCI-indexed journals, Scopus, as well as in international and national journals and conference proceedings. She has authored and edited several books with reputed international publishers such as Wiley, CRC Press, IGI Global, Cambridge Scholars, and Nova Science. She is the recipient of several awards, including the I2OR National Eminent Researcher Award, the Academic Influencer Award, and the Faculty Excellence Award. Dr. Ahirwar holds five published patents and has supervised numerous postgraduate and doctoral research scholars in computer science. As a reviewer and editorial board member for various reputed journals and conferences, she plays an active role in promoting research in emerging technologies. She is a life member of esteemed technical bodies such as IAENG and I2OR and has contributed significantly as a session chair, expert speaker, and convener at numerous international academic forums.

Sandhya Anilkumar Professional commitment, constant focus on creative ability and learning through research are some of the best practices she has followed in her academic career of 32 years. She is currently working as Assistant Professor and HOD Housekeeping at the Faculty of Hospitality Management and Catering Technology, M. S. Ramaiah University of Applied Sciences. She began her career in the hotel industry with the Taj group of Hotels, before moving into academia. She has diverse exposure in teaching and research. She holds a degree in Hotel Management from IHM Trivandrum and has also completed Masters Degree in Tourism Management. She alongside her student has been granted a patent this year for their finding- An intelligent dustbin. Was recognised as the best teacher in the recent Teachers Day celebration. She has been a proud member of the Professional Housekeepers Association for about 10 years now. She is currently the Chief Mentor of PHA YUVA and is the Communications Lead of PHA. She was awarded by The World of Hospitality for 'Excellence in Housekeeping' in June 2023. She received the decade of dedication award- 2025 for serving PHA and PHA YUVA. She ensures holistic growth in her personal sphere, she supports initiatives that drive local community, societal development and animal welfare.

Rachna Chandan, a distinguished academician and esteemed expert in the hospitality management. As Associate Professor and HOD-Housekeeping at the prestigious Banarsidas Chandiwala Institute of Hotel Management and Catering Technology, New Delhi, Dr. Chandan boasts an impressive 21-year tenure in both academia and industry. Holding a Ph.D. in Management from the revered Banasthali Vidyapeeth, she shines as a luminary in research-based experiential learning, propelling academic excellence to new heights. Dr. Chandan's esteemed editorial roles include steering Voice of Housekeepers, a bi-annual magazine by PHA, and co-leading the "Indian Journal of Applied Hospitality and Tourism Research." Her research accomplishments include publications and presentations at esteemed international conferences, culminating in the prestigious "Best Paper Award" at a prominent online International Conference.

Madhu Chandok, with 40+ years in hospitality, has excelled in planning, design, and managing projects for brands like Hilton and Crowne Plaza. She contributed to Hilton's India-specific prototypes and leads conversions as a founding partner at PIP 2020. As State President of WICCI Hospitality & Tourism Council, she mentors students and artisans while promoting sustainability and skill development in the hospitality sector.

Sachin Chaudhary With a strong academic foundation and a passion for teaching, he plays a vital role in shaping the technical skills of his students. His primary area of interest lies in Software Engineering, where he focuses on modern development methodologies, software design principles, and emerging technologies.

Weimei Chen is currently pursuing her PhD in Hospitality and Tourism Management at Taylor's University in Malaysia, with research interests in hospitality and tourism. As an active PhD student, she has published four articles in international journals and has been invited to participate in several international seminars. Her research not only delves into cutting-edge topics in the hospitality and tourism industry, but also provides valuable insights into the industry.

Swati Haridas Deshani is an Assistant Professor at WGSHA (MAHE) & Research Scholar at Vivekananda Global University, Jaipur, India, in the field of Hospitality and Tourism Management. With a strong academic background in hospitality studies & tourism management, she has been actively engaged in research on emerging technologies in hospitality, including robotics, artificial intelligence, and service automation and Generation Z's perception towards hospitality as a consumer and as an employee. Deshani's research interests focus on technology adoption in hotel services, human–robot interaction, service quality, and the role of generational dynamics, particularly Generation Z, in shaping the future of hospitality. She has presented research at national and international conferences and contributed to scholarly publications in hospitality and tourism. Passionate about bridging theory and practice, Deshani aims to advance innovative and sustainable service models that balance technological efficiency with the human essence of hospitality.

Aruditya Jasrotia is currently working as an Assistant Professor in Amity Institute of Travel and Tourism at Amity University, Noida, India. He completed his Ph.D. from the Department of Tourism and Travel Management,Central University of Jammu, Jammu and Kashmir, India. His areas of expertise are Smart Tourism, Smart Tourism Destinations, Smart Cities, Tourism Technologies, Urban Planning, Community Based Tourism and Sustainable Tourism. He has 30 Scopus Indexed Publications, 2 Patent and more than 40 research publications in total including research papers and book chapters.He is appointed as Associate Editor of ESTEEM Journal of Social Science and Humanities and International Journal of Hospitality & Tourism Management.

Vikesh Kashyap is an experienced academic and hospitality professional currently serving as Associate Professor at Bahra University, Shimla Hills, and formerly Head of Department at Chandigarh College of Hospitality, CGC Landran. With over 13 years of academic experience and a strong foundation in the hospitality industry, he holds a Ph.D. in Tourism Studies and an MBA in International Hotel & Tourism from Vatel International School, France. Dr. Kashyap has published extensively in reputed journals and has authored multiple books in the field of hospitality and tourism. He has served as a co-convener and session chair at major national and international conferences, and holds a patent in tourism recommendation systems. He has also successfully led an AICTE-funded MOBROB project.

Rupam Konar is the Programme Director for the Doctor of Philosophy (Hospitality & Tourism) and Master of Science (Tourism) programmes. Dr. Konar is also a Senior Lecturer in the School of Hospitality, Tourism & Events at Taylor's University, Malaysia. He also serves as a managing editor for the SCOPUS-indexed Asia Pacific Journal of Innovation in Hospitality and Tourism and as an Associate Director at the Centre for Research and Innovation in Tourism (CRiT). Dr. Rupam is a Certified Microsoft Innovative Educator. His specialization area includes service innovation, service delivery, and service design within the hospitality industry. Within his research interest, he has published in many international refereed journals, and books and has successfully complemented many national and privately funded research projects. With his gained years of experience in the hospitality sector and academia, he is now really intrigued by the possibilities and innovative changes he can contribute to these industries.

Vikas Rana is an Associate Professor at Vivekananda Global University (VGU), Jaipur, with over 13 years of academic and industry experience in hospitality and culinary management. He began his teaching career in 2010 as a Food Production Instructor at S.D.S.J.I.T.I., Solan, and later served as Assistant Professor at Bells Institute of Management and Technology, Shimla, followed by NIMS University, Jaipur. He is currently pursuing a Ph.D. in Hospitality and Tourism Management from MAU, Baddi, and holds an M.Sc. in HMCTT from PTU Jalandhar along with a Diploma in Food Production from IHM Kufri, Shimla. Dr. Rana has published five research papers in UGC-approved journals, contributed a book chapter, and presented at five national conferences. His research interests span culinary innovation, hospitality education, and technology integration in hotel services.

Shivansu Sachan Pursuing Ph.D. in Management, a Master's holder from Integral University, Lucknow, Bachelor's from IHM, Lucknow. Academic and professional accomplishments of 6 years with renowned organizations and institutes. My Teaching Career has been very satisfying and have continuously upgraded the skills and knowledge by attending many courses and training programs, seminars, conferences organized by different organizations and also delivered many lectures as a Hospitality sector.

Md. Mehedi Hasan Sawon is a dedicated academic and researcher currently serving as a Senior Lecturer in the Department of Tourism and Hospitality Management at Daffodil International University, Bangladesh. With a strong passion for advancing knowledge in tourism and hospitality, he has made significant contributions through his research and publications. His research interests are diverse and focused on contemporary topics within the tourism and hospitality sectors. Sawon is particularly interested in the intersection of Artificial Intelligence and its applications in tourism, exploring innovative ways to enhance customer experiences and optimize operational efficiency focus on tourism education, seeking to develop new teaching methods and curricula that better align with industry trends. Furthermore, Sawon is interested in social media marketing in the hospitality industry, analyzing how digital platforms shape consumer behavior and business strategies. His work also extends to sustainable tourism, promoting practices that contribute to the long-term preservation of tourism destinations and the environment.

Swati Sharma , Ph.D. is currently working as an Associate Professor with Amity University, India. She holds 18 years of teaching and industry experience. She is Vice- President of the Women's Indian Chamber of Commerce and Industry . She is the reviewer of Consumer Behavior in Tourism & Hospitality & IGI Global . She has authored & Edited 15 books & published research papers in National & International journals of repute. (IGI Global, Taylor & Francis, Springer, Apple

Academy Press, CABI, Edward & Elgar). Her area of interest is Sustainable Tourism, MICE, Community based Tourism, Destination branding.

Vikas Sharma is an active academician and researcher in the field of computer science with 8 years of experience. He earned M. Tech degree specialized in Computer Science. He is involved in academic development and scholarly activities. He is member of International Association of Engineers, The Society of Digital Information and Wireless Communications, Global Professors Welfare Association, International Association of Academic plus Corporate (IAAC), Computer Science Teachers Association and Internet Society. His research interest includes Cloud Computing, Artificial Intelligence, Machine Learning; published more than 65 research articles with 4 books and 90 book chapters in edited books. He is contributing as member of editorial and reviewers' boards in conferences and various computer journals and edited books published by Elsevier, CRC Press, Taylor and Francis, Springer, IGI global, Wiley Publication, IET Publication, De Gruyter, River Publication etc.

Ankit Shukla is an accomplished educator and researcher with a robust background in the hospitality and tourism industry. His career spans various roles, from frontline food and beverage service to academic leadership and extensive research contributions. Ankit's journey exemplifies a blend of practical experience, academic acumen, and a passion for advancing knowledge in his field. Early Career and Industry Experience Ankit's career in the hospitality industry began at the Hyatt Regency Chandigarh in September 2015, where he worked in the Food & Beverage Service department. His responsibilities included taking room service orders, managing point-of-sale systems, and setting up banquets for events. This role provided Ankit with hands-on experience in the operational aspects of the hospitality industry. Following his time at the Hyatt Regency, Ankit transitioned into academia, joining Kumaun University Nainital as a teaching personnel in September 2016. He took on multiple responsibilities, including lecturing MBA Tourism students, overseeing practical training, and designing course curricula. This marked the beginning of his journey in education, where he contributed to shaping the next generation of hospitality professionals. Academic Leadership and Teaching Ankit's academic career continued to flourish as he assumed the role of Assistant Professor at SIHMCT & AN Tehri Garhwal in July 2017. He became the head of the Food & Beverage Service department, where he coordinated events, designed course curricula, and managed the department's overall functioning. From February 2018 to April 2022, Ankit served as an Assistant Professor at Shoolini University Solan. Here, he expanded his responsibilities to include training and placement coordination, NAAC coordination for the Hospitality Management department, and handling operations of a hotel with eight rooms. His leadership skills and ability to manage complex tasks were instrumental in the smooth functioning of the university's hospitality department. Contributions to Research and Academia Ankit's passion for research is evident through his significant contributions to academic literature. He has presented research papers at various international conferences, focusing on topics such as ecology, local and regional food in destination marketing, and the influence of technological amenities on hotel guest satisfaction. His research has been published in reputable journals, with manuscripts addressing safety and security measures in hotels, tobacco consumption in India, and the role of Indian spices in Indian history. In addition to his research publications, Ankit has filed 40 patents, copyrights, and utility applications, with several already granted. His commitment to innovation and intellectual property rights underscores his dedication to advancing knowledge in his field. Current Role and Future Endeavors As of July 2023, Ankit is a lecturer at Raja Mangala University of Technology Thanyaburi in Thailand. In this role, he continues to inspire and educate students in the field of hospitality while pursuing his research interests. Ankit's extensive experience in both industry and academia, coupled with his passion for teaching and research, positions him as a key figure in the ongoing development of the hospitality and tourism sector. Ankit Shukla's journey from industry practitioner to esteemed academic is a testament to his dedication, expertise, and unwavering commitment to excellence in hospitality and education

Dharna Shukla is a hotel management educationist and an experienced hotelier with over two decades of experience in academia hospitality operations. Dr. Shukla has worked in India and abroad in the hospitality sector with chains of The Park, The Taj and Coral Hotels. She has also taught hospitality and tourism management at IHM Lucknow, Christ University Bangalore and was the founding faculty and program leader, hospitality and tourism, at Amity University Dubai. Dr.Shukla has reviewed hospitality and tourism content of Aptech and has been a consultant inspector for NEBOSH certifications in India for a UK-based organization. Dr. Shukla has authored and co-authored many research papers and book chapters. She holds a PhD in Management from Banasthali University, Rajasthan and currently is the Program Leader of BBA (Tourism Management) and an Assistant Professor at Amity School of Hospitality, Amity University Uttar Pradesh, Lucknow Campus.

Amrik Singh is working as Professor and HOD in the School of Hotel Management and Tourism at Lovely Professional University, Punjab, India. He obtained his Ph.D. degree in Hotel Management from Kurukshetra University, Haryana, India. He has published 115 Scopus-indexed publications. He has published 12 patents, and 01 patent has been granted in the inter-disciplinary domain. Dr.

Amrik Singh participated and acted as a resource person in various national and international conferences, seminars, research workshops, and industry talks. He has published 20 books with IGI Global Publication, Emerald Publication, Bentham Publication, Nova Publication etc., on trending areas such as Food waste, Artificial Intelligence, Mountain Tourism, Virtual tourism, Healthcare Tourism, Guest behavior in tourism and hospitality. His area of research interest are Accommodation management, sustainability, Green practices, Human Resource Management in Hospitality, Artificial Intelligence, AR VR in hospitality, Hotel pricing, Employee Disability, Work life balance, women empowerment etc. He is currently guiding 8 Ph.D. scholars, and 5 Ph.D. scholars have been awarded Ph.D.

Gurucharan Singh is an accomplished academic and hospitality professional with over a decade of experience spanning both the education and industry sectors. Currently serving as an Assistant Professor at MMICT&BM (Hotel Management), MMDU, Mullana, Ambala, he specializes in hospitality operations. His professional journey includes managerial roles in training and development within the hospitality industry and operational stints at renowned establishments like Marriott International and Country Inn & Suites by Carlson. An active researcher and contributor to academic literature, he has published numerous articles in reputable journals, focusing on topics such as Artificial Intelligence, guest experience, sustainability, technology, and operational management in the hospitality industry.

Kuldeep Singh currently serves as an Assistant Professor in Amity School of Hospitality, Amity University, Haryana, India. He completed his Ph.D. in tourism from Maharshi Dayanand University (Rohtak) India in 2021. He is also a UGC (Net- JRF qualified). Dr. Singh has also served the tourism industry for a couple of years and more than three years in academics. Dr. Singh has so far published more than 40 research articles in both international and national referred journals as well as in edited books in the field of Tourism. Currently, He is serving as an editor of book series in various reputed publications (Emerald, IIP series). Dr. Singh is passionate about the academic areas of Service Quality Management, Rural tourism, Ecotourism, and Sustainable tourism. He also won an aspiring researchers welcome award from Indian Hospitality Congress. His credentials may be verified on various research platforms like Google Scholar, SSRN, LinkedIn, Academia, and Research Gate.

Sanskriti Singh is a dedicated hospitality professional, currently working as an Assistant Professor in SRM University, Sonepat. She began her hospitality education with B.Sc. in Hospitality & Hotel Administration from IHM Jaipur and M.Sc. in Hotel Administration from NCHMCT, Noida with specialization in Sales &

Marketing. To enhance her knowledge and skills, she did her PhD in Hospitality Management from CT University Ludhiana, Punjab.

Devika Grover is having 20 + industry & Teaching Experience and presently working as Assistant Professor with Amity University, Noida

Mohammad Badruddoza Talukder is an Associate Professor at the College of Tourism and Hospitality Management at the International University of Business Agriculture and Technology (IUBAT) in Dhaka, Bangladesh. Holding a PhD from Lovely Professional University, India, he is renowned for his innovative integration of cutting-edge technologies such as Artificial Intelligence (AI), the Internet of Things (IoT), and Blockchain into tourism and hospitality management. His research explores the transformative impact of these technologies on smart tourism, cloud kitchens, and the broader industry landscape. As a prolific author, Dr. Talukder has published numerous scholarly articles and books, with forthcoming titles that address AI innovations in hospitality, the evolution of cloud kitchens, and the enhancement of digital literacy within hospitality education. His work shapes contemporary tourism and hospitality practices, skillfully marrying technological advancements with practical industry applications. Since 2008, Dr. Talukder has imparted his knowledge through a diverse array of courses in tourism and hospitality across various universities in Bangladesh.

Gautam Tanwar is a seasoned hospitality professional, culinary expert, and academician with over 15 years of experience in the hospitality industry and education. Currently serving as an Assistant Professor and Culinary Expert at the School of Hospitality & Tourism, Bahra University, he is also pursuing a Ph.D. in Hotel Management. His career includes key roles with The Oberoi Group, and his professional foundation was laid through the Systematic Training & Education Programme (STEP) under the esteemed Oberoi Centre of Learning & Development. With an eye for detail and a heart for hospitality, Chef Gautam combines academic insight and hands-on expertise, skillfully blending tradition with modernity and a fervent passion for traditional Indian cuisine." making him a sought-after mentor and an influential voice in the Indian hospitality landscape.

Akshita Tiwari is a dedicated hospitality professional with a passion for learning and expanding her knowledge in the field. She began her hospitality education in 2014 with a Bachelor's in Hospitality and Hotel Administration from IHM, Faridabad, and went on to complete her Master's in Hospitality Administration with a specialization in Human Resource Management from IHM, Pusa, New Delhi in 2019.In pursuit of her interest in hospitality and a desire to deepen her knowledge, she joined the Ph.D. in Hospitality at Amity School of Hospitality, Amity University,

Noida, in January 2020.. Dr. Tiwari is currently a faculty member at Amity School of Hospitality, Amity University, Noida, Uttar Pradesh, where she teaches Front Office Operations and Management to Bachelor's in Hotel Management students and Master's in Hospitality students.

Tarun Kumar Vashishth is an active academician and researcher in the field of Computer Science with 22 years of experience. He earned Ph.D. Mathematics degree specialized in Operations Research; served several academic positions such as Dy. Director, HoD, NAAC Coordinator, Academic Coordinator, Member Secretary of Department Research Committee , Assistant Center Superintendent and Head Examiner in University Examinations. He is involved in academic development and scholarly activities. He is member of International Association of Engineers, The Society of Digital Information and Wireless Communications, Global Professors Welfare Association, International Association of Academic plus Corporate (IAAC), Computer Science Teachers Association and Internet Society . His research interest includes Cloud Computing, Artificial Intelligence, Machine Learning and Operations Research; published more than 50 research articles & conference papers with 2 book and 90 book chapters. He is contributing as member of editorial and reviewer's boards in conferences and various computer journals and edited books published by Elsevier, CRC Press, Taylor and Francis, Springer, IGI global, Wiley Publication, IET Publication, De Gruyter, River Publication etc.

Pushpendra Verma is member of International Association of Engineers, The Society of Digital Information and Wireless Communications, Global Professors Welfare Association, International Association of Academic plus Corporate (IAAC), Computer Science Teachers Association and Internet Society . His research interest includes Cloud Computing, Artificial Intelligence, Machine Learning and Operations Research; published more than 10 research articles & conference papers with 2 book and 5 book chapters. He is contributing as member of editorial and reviewer's boards in conferences and various computer journals and edited books published by Elsevier, CRC Press, Taylor and Francis, Springer, IGI global, Wiley Publication, IET Publication, De Gruyter, River Publication etc.

Index

G

H

I

L

M

T

V

www.ingramcontent.com/pod-product-compliance
Lightning Source LLC
LaVergne TN
LVHW080906300926
849162LV00003B/576
* 9 7 9 8 3 3 7 3 6 5 4 1 1 *